Clinical Chemistry

Clinical Chemistry

Principles and Procedures

Fourth Edition

Joseph S. Annino

Director, Clin-Chem Laboratories, Boston

Roger W. Giese, Ph.D.

Assistant Professor of Clinical Chemistry,
Northeastern University, Boston

Little, Brown and Company
Boston

This book is dedicated to the memory of John P. Peters, M.D.
a founding father and guiding genius of the profession
of clinical chemistry

Preface

This introductory book is designed to help students and analysts to understand, use, and control analytical technics in clinical chemistry.

The preparation of this fourth edition involved extensive revision of the previous edition. However, the basic format, style, and organization have remained unchanged.

The first eight chapters (Part I) deal with fundamental information and basic technics that apply to clinical chemistry. Since the book is not designed to be a textbook of quantitative analysis, many aspects of this topic are omitted or presented only briefly. The reader who desires further information on any particular aspect of this topic should consult one of the cited references.

The remainder of the book (Part II) is devoted to descriptions and explanations of methods of analysis. Most of the chapters include a brief survey of current methods for determining each substance, a description of a recommended method, directions for standardization, discussion of the underlying principles, capabilities, and limitations of the method, and a note on the clinical significance of the results. Where appropriate, references are made to Part I for fundamental explanations.

The selection of a suitable method of analysis for a substance from the many methods proposed in the literature is a difficult problem. In the clinical chemistry laboratory, simplicity, efficiency, speed, and accuracy all must be considered. The methods recommended in this text are not necessarily the best ones available by all of these criteria; however, most have been tried and proved in many laboratories, including the laboratories of the authors, so that their value is thus well established. The discussion that follows each method should help the analyst to work with greater understanding.

The large numbers of tests requested from clinical laboratories have required increasing use of automated equipment. Although essentially an entire chapter (Chapter 6) is devoted to automation, individual automated methods are not presented in this text, because the basic principles involved generally are the same as those in the manual versions, and the details depend on the particular automated instrument available.

The rapid development of clinical chemistry requires that new material be added frequently, but seldom can anything be deleted. It was therefore a real challenge to keep the size of this text at a minimum without omitting important information. The resultant omissions, revisions, and compromises reflect the considered judgment of the authors.

In Part I, chapters have been added dealing with automation, separation technics, and competitive binding radioassay. The subjects of quality control, spectrophotometry, and emission spectrophotometry have been expanded. A section on atomic absorption has been added to the flame

vii

emission chapter, and the review of quantitative analysis has been sub-stantially revised.

In Part II, chapters on thyroid function tests and vitamins have been added. Additional methods in other chapters include blood ammonia, immunoglobulins, triglycerides, vanilmandelic acid, digoxin, and car-bon monoxide. The chapter on proteins, and the one on drugs and poisons have been expanded greatly. A discussion of the fundamentals of enzyme methodology was added to Chapter 20. All of the other methods chapters have been revised to at least some extent. In some cases, methods appearing in the third edition have been replaced with more modern ones.

References are provided both at the ends of the chapters and in an annotated bibliography at the end of the book. The annotated bibliog-raphy contains references to general books and journals in the field of clinical chemistry.

A deliberate attempt has been made to avoid extensive bibliographies at the ends of the chapters. The references generally are limited to those that refer to a specific method, to recent significant articles, and to books and reviews. Comprehensive bibliographies may be found by consulting these sources as well as the books and journals cited in the annotated bibliography.

The number of illustrations has been substantially increased. Only eight have been carried over from the third edition. The new illustrations were skillfully drawn by George Robinson.

The generous support of Louis A. Williams, the professional colleague of Joseph S. Annino, is acknowledged by this author. Both authors are grateful to Joanne Puopolo, who has been immensely helpful and pro-ficient with the preparation of the manuscript. They also are particularly indebted to their wives, Louise A. and Mary-Ann G., for patience and support during a difficult period.

It is the hope and expectation of the authors that this edition will prove useful to teachers, students, and practitioners of clinical chemistry and that the book will be a positive influence toward the advancement of clinical chemistry as a science and a profession.

J.S.A.
R.W.G.

Contents

Illustrations

Tables

I. Basic Technics and Fundamental Information

1. A Review of Quantitative Analysis

It is important that technicians know and understand the rudiments of quantitative analysis. The theory and technic are best learned in appropriate college courses and it is assumed that clinical chemistry technicians have had such exposure. A few aspects of quantitative analysis that are of fundamental importance in clinical chemistry are reviewed in this chapter. The reader is referred to appropriate textbooks [3, 4] for more thorough presentations of these and other related topics.

Units and Abbreviations
The units and abbreviations used in clinical chemistry and in this book are given in Table 1-1. It is important that these expressions be understood and used exactly as listed.

Concentrations
Some concentration units used in clinical chemistry are given in Table 1-2. The units are defined both descriptively and by means of directions for preparing appropriate solutions.

Table 1-1. Units and Abbreviations

g	gram
kg	kilogram (1000 g)
mg	milligram (10^{-3} g)
μg	microgram (10^{-6} g)
ng	nanogram (10^{-9} g)
pg	picogram (10^{-12} g)
l	liter
ml	milliliter (10^{-3} liter)
dl	deciliter (10^{-1} liter or 100 ml)
mol	mole (g molecular weight)
mmol	millimole (10^{-3} mol)
μmol	micromole (10^{-6} mol)
eq	equivalent or equivalent weight (fraction of formula weight reacting with or providing 1 mol of H^+ or OH^-)
meq	milliequivalent (10^{-3} eq)
M	molarity (mol of solute per liter of solution or mmol of solute per ml of solution)
N	normality (eq of solute per liter of solution or meq of solute per ml of solution)
m	molality (mol of solute per kg of solvent)
h	hour
d	day (24 h)
ppm	parts per million (ng of solute per g of solution or mg of solute per kilogram of solution; essentially μg per liter for dilute aqueous solutions because the density of water is 1.0)

Table 1-2. Definitions of Concentrations

Solution	Description	Preparation
1% (w/w)	1 g solute/100 g solution	Add 1 g of solute to 99 g of solvent.
1% (w/v)	1 g solute/100 ml solution	Dilute 1 g of dissolved solute to 100 ml with the same solvent.
1% (v/v)	1 ml solute/100 ml solution	Dilute 1 ml of solute to 100 ml with the same solvent.
1M	1 mol solute/liter solution	Dilute 1 mol of dissolved solute to 1 liter with the same solvent.
1N	1 eq solute/liter solution	Dilute 1 eq of dissolved solute to 1 liter with the same solvent.
1m	1 mol solute/kg solvent	Add 1 mol of solute to 1 kg of solvent.
1 ppm	1 ng solute/g solution or 1 mg solute/kg solution	Dilute 1 mg of dissolved solute to 1 kg with the same solvent. (When water is the solvent dilution to 1 liter usually is acceptable.)

Equivalents

The concept of *equivalent* (equivalent weight) requires special comment. The equivalent weight of a substance refers to its behavior in a specific chemical reaction. Some compounds will have different equivalent weights for different chemical reactions. For most purposes in clinical chemistry, the equivalent weight of a substance may be considered to be the fraction of its gram formula weight which could react with or provide 1 mole of hydrogen ions (H^+) or 1 mole of hydroxyl ions (OH^-). Some examples are given in Table 1-3.

Table 1-3. Typical Equivalent Weights

Substance	Equivalent Weight (g)
NaCl	gwf/1 = 58.5/1 = 58.5[a]
NaOH	gfw/1 = 40/1 = 40
Na_2SO_4	gfw/2 = 142/2 = 71
H_2SO_4	gfw/2 = 98.08/2 = 49.04
H_3PO_4	gfw/3 = 98/3 = 32.7
$BaSO_4$	gfw/2 = 233.4/2 = 116.7
$BaCl_2$	gfw/2 = 208.24/2 = 104.12
$BaCl_2 \cdot 2H_2O$	gfw/2 = 244.28/2 = 122.14

[a] The term gfw is an abbreviation for gram formula weight.

Table 1-4. Atomic Weights of Some Common Elements[a]

Element	Symbol	Atomic Weight	Element	Symbol	Atomic Weight
Aluminum	Al	26.98	Lithium	Li	6.94
Arsenic	As	74.92	Magnesium	Mg	24.31
Barium	Ba	137.34	Manganese	Mn	54.94
Bromine	Br	79.91	Mercury	Hg	200.59
Cadmium	Cd	112.40	Molybdenum	Mo	95.94
Calcium	Ca	40.08	Nickel	Ni	58.71
Carbon	C	12.01	Nitrogen	N	14.01
Cerium	Ce	140.12	Oxygen	O	16.00
Chlorine	Cl	35.45	Phosphorus	P	30.97
Chromium	Cr	52.00	Platinum	Pt	195.09
Cobalt	Co	58.93	Potassium	K	39.10
Copper	Cu	63.54	Silver	Ag	107.87
Fluorine	F	19.00	Sodium	Na	22.99
Gold	Au	196.97	Sulfur	S	32.06
Hydrogen	H	1.01	Tin	Sn	118.69
Iodine	I	126.90	Tungsten	W	183.85
Iron	Fe	55.85	Uranium	U	238.03
Lead	Pb	207.19	Zinc	Zn	65.37

[a] Weights given here are rounded off to two significant decimal places.

Atomic Weights

For convenience in solving the sample problems in this chapter as well as in preparing reagent solutions in the laboratory, a list of atomic weights of some of the elements commonly encountered in clinical chemistry is given in Table 1-4.

SAMPLE PROBLEMS

A few sample problems demonstrating the use of the concepts reviewed above are presented here.

1. How many g of NaCl are required to make 1 liter of $0.05M$ NaCl?
   0.05 mol of NaCl/liter is 0.05×58.5 g/liter $= 2.925$ g NaCl/liter of solution

2. How many g of NaCl are required for 1500 ml of $2M$ NaCl?

$$1500 \text{ ml} = 1.5 \text{ liters}$$

$$2 \text{ mol/liter of NaCl} = 2 \times 58.5 \text{ g/liter} = 117 \text{ g/liter}$$

$$1.5 \text{ liter} \times 117 \text{ g/liter} = 175.5 \text{ g NaCl}$$

3. Convert a chloride concentration of 300 mg/dl into meq/liter.

$$300 \text{ mg Cl} \times \frac{1 \text{ meq Cl}}{35.5 \text{ mg Cl}} = 8.45 \text{ meq Cl}$$

$$1 \text{ dl} = 100 \text{ ml} = 0.1 \text{ liter}$$

$$8.45 \text{ meq Cl}/0.1 \text{ liter} = 84.5 \text{ meq Cl/liter}$$

4. Convert a calcium concentration of 5 meq/liter to mg/dl.
   The equivalent of Ca = ½ its molecular weight (40). Hence

$$5 \text{ meq Ca} \times \frac{20 \text{ mg Ca}}{1 \text{ meq Ca}} = 100 \text{ mg Ca}$$

$$1 \text{ liter} = 10 \text{ dl}$$

$$100 \text{ mg Ca}/10 \text{ dl} = 10 \text{ mg/dl}$$

5. What is the molarity of a 5% (w/v) solution of NaCl?
   A 5% (w/v) solution of NaCl is 5 g NaCl/100 ml solution

$$5 \text{ g NaCl} \times \frac{1 \text{ mol NaCl}}{58.5 \text{ g NaCl}} = 0.0855 \text{ mol NaCl}$$

$$100 \text{ ml} = 0.1 \text{ liter}$$

$$0.0855 \text{ mol NaCl}/0.1 \text{ liter} = 0.855 \text{ mol/liter} = 0.855M$$

6. What is the normality of a 22% (w/v) solution of Na_2SO_4?
   The equivalent weight of Na_2SO_4 is ½ its molecular weight (142).
   Normality is eq/liter.
   A 22% solution of Na_2SO_4 is 22 g Na_2SO_4 per 100 ml of solution.

$$1 \text{ eq } Na_2SO_4 = \text{mol wt}/2 = 142/2 = 71$$

$$22 \text{ g } Na_2SO_4 \times \frac{1 \text{ eq } Na_2SO_4}{71 \text{ g } Na_2SO_4} = 0.31 \text{ eq } Na_2SO_4$$

$$100 \text{ ml} = 0.1 \text{ liter}$$

$$0.31 \text{ eq } Na_2SO_4/0.1 \text{ liter} = 3.1 \text{ eq/liter} = 3.1N \text{ } Na_2SO_4$$

7. What is the normality of $0.05M$ H_2SO_4?
   Normality is eq/liter

$$1 \text{ mol } H_2SO_4 = 2 \text{ eq } H_2SO_4$$

$$0.05 \text{ mol } H_2SO_4 = 0.10 \text{ eq } H_2SO_4$$

$$0.1 \text{ eq } H_2SO_4/\text{liter} = 0.1N \text{ } H_2SO_4$$

8. What is the molarity of $2N$ H_3PO_4?

$$3 \text{ eq } H_3PO_4 = 1 \text{ mol } H_3PO_4$$

$$2N \text{ } H_3PO_4 = 2 \text{ eq } H_3PO_4/\text{liter}$$

$$2 \text{ eq} \times \frac{1 \text{ mol}}{3 \text{ eq}} = \frac{2}{3} \text{ mol} = 0.667 \text{ mol}$$

$$0.667 \text{ mol/liter} = 0.667M$$

Dilutions

Dilutions involve adding more solvent to a given solution. For instance: A 1:10 dilution results when 1 volume of a given solution is diluted to a total volume of 10 with solvent. Generally it would be acceptable to add 9 volumes of solvent to obtain the total volume of 10.[1] The concentration of solute in the final solution then is $\frac{1}{10}$ of its concentration in the original solution. If two successive 1:10 dilutions are carried out, the concentration of the final solution is $\frac{1}{10} \times \frac{1}{10}$ or $\frac{1}{100}$ times the concentration of the original solution. It is useful to carry out successive dilutions not only to avoid diluting very small volumes (which are difficult to measure accurately), but also to avoid diluting into very large volumes (which require large containers and an excessive amount of solvent). For instance, a 1:5000 dilution can be carried out by successive 1:50 and 1:100 dilutions (1 ml of original solution diluted to 50 ml, and then 1 ml of this solution diluted to 100 ml) rather than a straight 1:5000 dilution, which would require a 5-liter container and nearly 5 liters of solvent if 1 ml of the original solution were taken for this dilution.

SAMPLE PROBLEMS

1. What is the final concentration when a $5N$ solution is diluted successively first 1:5, then 2:15, and finally 3:25?

$$\frac{1}{5} \times \frac{2}{15} \times \frac{3}{25} \times 5 = \frac{2}{125} = 0.016$$

The final solution is $0.016N$

2. How many ml of water must be added to 5 ml of a $2M$ aqueous NaCl solution in order to obtain a final NaCl concentration of $0.1M$?

   a. First calculate the fractional dilution.

$$\frac{0.1}{2} = \frac{1}{20}$$

---

[1] This sometimes is not the case when two different solvents are involved; for example, 1 volume of water plus 1 volume of alcohol gives less than 2 volumes because the water and the alcohol molecules fit together very closely.

The original solution must be diluted $1:20$
b. Next calculate the total final volume.

$$\frac{5}{x} = \frac{1}{20} \qquad x = 100 \text{ ml} = \text{total final volume}$$

Add 95 ml of water to 5 ml of the original solution to obtain a final volume of 100 ml.

A DILUTION FORMULA

Consider that a solution designated A is diluted to give a solution B. If volume V_A of solution A is diluted to give volume V_B of solution B and the concentrations of solutions A and B are C_A and C_B, respectively, then the following very useful relationship is true whenever the same volume and concentration units are used for both solutions:

$$V_A \times C_A = V_B \times C_B$$

For instance, Solution A Solution B

$$\text{ml} \times M = \text{ml} \times M$$

$$\text{ml} \times \% = \text{ml} \times \%$$

$$\text{liters} \times N = \text{liters} \times N$$

This formula is derived from the fact that the amount of solute in volume V_A is the same as the amount of solute in volume V_B. This amount is equal to the concentration times the volume in each case, that is,

Amount of solute in $V_A = V_A \times C_A$

Amount of solute in $V_B = V_B \times C_B$

For instance, the total amount of NaCl in 10 ml of a solution with a NaCl concentration of 2 mg per ml is 10 ml $\times$ 2 mg/ml $=$ 20 mg of NaCl. If 10 ml of water were added to the original solution, the total volume would then be 20 ml. The concentration of this second solution can be determined with the above formula.

$$V_A \times C_A = V_B \times C_B$$

$$10 \text{ ml} \times 2 \text{ mg/ml} = 20 \text{ ml} \times C_B$$

$$C_B = 1 \text{ mg/ml}$$

SAMPLE PROBLEMS

1. How much of a 0.2M solution of HCl is required to prepare 600 ml of a 0.01M solution of HCl?

$$V_A \times C_A = V_B \times C_B$$

$$V_A \times 0.2 = 600 \times 0.01$$

$$V_A = \frac{600 \times 0.01}{0.2} = 30$$

Diluting 30 ml of 0.2M HCl to 600 ml will give a 0.01M solution of HCl.

2. Prepare 200 ml of 70% (v/v) ethanol from 95% (v/v) ethanol.

$$V_A \times C_A = V_B \times C_B$$

$$V_A \times 95 = 200 \times 70$$

$$V_A = 147.4$$

Diluting 147.4 ml of 95% ethanol to 200 ml will give a 70% ethanol solution.

3. What volume of 0.5M solution can be prepared from 85 ml of a 1.5M solution?

$$V_A \times C_A = V_B \times C_B$$

$$85 \times 1.5 = V_B \times 0.5$$

$$V_B = 255 \text{ ml}$$

4. Prepare 1 liter of a 0.02N solution from a 0.5N solution.

$$V_A \times C_A = V_B \times C_B$$

$$V_A \times 0.5 = 1 \times 0.02$$

$$V_A = 0.04 \text{ liter}$$

Diluting 0.04 liter, or 40 ml, of the 0.5N solution to 1 liter will give a solution which is 0.02N.

pH and Buffers

The pH scale is used to express the concentration of H^+ in a solution. Mathematically stated, pH is the logarithm of the reciprocal of the hydrogen-ion concentration in moles per liter.

$$pH = \log \frac{1}{[H^+]} = -\log [H^+]$$

A liter of pure water contains 10^{-7} moles of H^+. Therefore: pH = $-\log 10^{-7} = 7$ (neutral). Because of this reciprocal relationship, the pH *decreases* as the H^+ concentration *increases*. Therefore the portion of the pH scale between zero and 7 is the *acid* range because the H^+ are in excess of the OH^-. Between 7 and 14 the opposite situation prevails, and the pH is *alkaline*. At pH 7 the ions are present in equal concentration, and the solution is neutral.

Buffers are substances dissolved in solutions which cause the solutions to resist changes in pH upon addition of hydrogen (H^+) or hydroxyl (OH^-) ions. Weak acids and bases form buffer systems because they react with added H^+ or OH^-. For instance, addition of HCl to a bicarbonate solution results in a fewer number of H^+ ions than addition of HCl to a corresponding bicarbonate-free solution. The bicarbonate therefore functions as a buffer. Some of the added H^+ react with the bicarbonate (HCO_3^-) to form carbonic acid (H_2CO_3).

$$H^+ + HCO_3^- \rightleftharpoons H_2CO_3$$

To the extent that H_2CO_3 molecules are formed when HCl is added, the change in pH is less because only H^+ that are free will be measured as pH. Those H^+ which end up in a bound state, as in H_2CO_3, are still acidic, but they are not measured as pH, which refers only to free H^+.

A similar explanation can be presented for the reduced change in pH which occurs when H_2CO_3 is present in a solution to which base is added. The added base (OH^-) partly reacts with the H_2CO_3 to form HCO_3^- and H_2O. Consequently the change in pH is minimized.

When both a weak acid and its corresponding base, for example H_2CO_3 and HCO_3^-, are present in significant quantities, the solution is said to be *optimally buffered* because only then is it maximally resistant to pH changes in either direction, that is, from the addition of either acid or base.

This effect occurs in a different region of the pH scale for each buffer system. Consequently, different buffers are used to buffer different regions of the pH scale.

Since buffer solutions may be prepared to produce a specific pH and since they tend to maintain that pH, they are used to calibrate pH meters. Suitable solutions are available commercially as are preweighed powders for the preparation of such solutions. It is important that the temperature of buffer solutions be controlled or noted when they are used since the pH may vary with temperature.

Standard Acids and Bases

Various concentrations of standard HCl and NaOH solutions suitable for standardizing other acid or base solutions in the laboratory are commercially available. When additional assurance is desired, a primary reference solution can be prepared in the laboratory. Working solutions of acids may be standardized by titration with a solution of base immediately after the base solution itself has been standardized by titration with a standard acid solution. The concentration values of standard acid solutions tend to be more accurate than those of standard base solutions because the latter pick up carbon dioxide (CO_2) from the air every time the cap is removed from the bottle. The CO_2 forms H_2CO_3 in water, and the H_2CO_3 neutralizes some of the base. For this reason, standard base solutions should be capped whenever they are not being poured. The concentration of a base can be checked by titration with a standard acid solution of HCl, or with potassium hydrogen phthalate, as described below.

POTASSIUM HYDROGEN PHTHALATE ($KHC_8H_4O_4$)
PRIMARY STANDARD,[2] $0.100N$

1. Dry approximately 22 g of primary standard grade phthalate at 110 C for 1 h, then cool to room temperature in a desiccator.
2. Accurately weigh 20.423 g of the salt, dissolve it in approximately 500 ml of purified water, and dilute to 1 liter in a volumetric flask.

This primary standard may be used to standardize a base as follows:

Place the base to be standardized in a 25-ml buret. Measure 10 ml of standard $0.1N$ acid into a 125-ml Erlenmeyer flask. Add 1 drop of 0.1% (w/v) phenolphthalein (in 95% ethanol) to the acid and carefully titrate with base until a pink endpoint persists. By noting the volume of base required to accomplish the neutralization and by applying the known data, the exact normality of the base can be calculated (see page 12).

Once this is accomplished, the base may be used to standardize another acid, and these solutions may be used routinely in the laboratory, the primary-standard acid being used only occasionally as a reference solution. More details regarding the mathematics of acid-base calculations are presented later (see page 12).

CARBONATE-FREE SODIUM HYDROXIDE

Solid sodium hydroxide (NaOH) as purchased commercially contains some sodium carbonate which will precipitate from a concentrated solution of sodium hydroxide. The carbonate may interfere with some tests and, for this reason, should be removed from the NaOH. A carbonate-free solution may be prepared as follows:

Place equal weights of solid reagent-grade sodium hydroxide and purified water in an Erlenmeyer flask. Swirl the flask in a sink, cooling it intermittently until the NaOH is dissolved. Cover the solution, and allow it to stand overnight during which time some sodium carbonate will settle out. Pour off the

[2] This material, as well as other standard reference materials, is available from the United States National Bureau of Standards.

supernatant, and store it in a tightly capped plastic bottle. The strength of the concentrated sodium hydroxide solution will be approximately $19N$. Although additional Na_2CO_3 will precipitate from this 50% solution because of the air exposure that occurs every time it is used, this precipitate will settle out, and the supernatant will be carbonate free. The lip should be wiped clean to avoid build-up of a Na_2CO_3 precipitate. This stock solution may be used to prepare carbonate-free NaOH solutions of any concentration. Each of these solutions must be titrated with standard acid if it is necessary to know the exact NaOH concentration, because it is difficult to measure the viscous 50% NaOH solution accurately and because the normality of the concentrated solution may change because of CO_2 absorption. These 50% NaOH solutions also are available commercially.

PRECAUTIONS IN USING STANDARD SOLUTIONS

Never introduce a pipet into a primary-standard solution because this risks contamination. Pour out a little more than is necessary into a second container, carefully wipe off the pouring edge, pipet out of the second container, and discard whatever is left in this container. If any moisture condenses on the inside of the container of the primary-standard solution, swirl this back into solution before removing any of the standard solution.

SAMPLE PROBLEMS

A useful formula for acid-base calculations is:

volume of acid (V_A) × normality of acid (N_A)

$$= \text{volume of base } (V_B) \times \text{normality of base } (N_B)$$

A more general form of this equation was presented for dilution calculations on page 8.

1. The volume of base required to neutralize 10.0 ml of $0.100N$ acid was 9.80 ml. What is the normality of this base?

$$V_A \times N_A = V_B \times N_B$$
$$10.0 \times 0.100 = 9.80 \times N_B$$
$$N_B = 0.102N$$

2. How much $0.1N$ acid is required to neutralize 15 ml of $0.3N$ base?

$$V_A \times N_A = V_B \times N_B$$
$$V_A \times 0.1 = 15 \times 0.3$$
$$V_A = 45 \text{ ml}$$

The Analytical Balance

The balance most commonly used for accurate weighing in clinical chemistry laboratories is the analytical projection balance. The basic features of this instrument are illustrated in Figure 1-1. Only a single pan and a set of dials are used by the operator. The pan is suspended on one side of an arm (beam) along with a set of weights. These weights cannot be seen unless the top cover on the balance is removed. The pan and weights are kept in suspension by a counterweight on the other end of the beam. The beam is supported in the middle by a knife-edge resting on a hard surface to give a relatively frictionless fulcrum. Placement of an object on the pan results in an imbalance. The degree of this imbalance, if it is small, is read directly from a calibrated optical scale attached to the beam. This scale is illuminated and its image *projected* onto a small screen, hence the term *projection balance*. If the imbalance is larger, then reattainment of balance or near-balance requires that some of the weights on the pan side of the beam be removed by turning appropriate dials. The settings on the dials correspond to weights that are removed. In other words, the imbalance is corrected by operator-directed mechanical removal of weights from the beam. When the balance point is nearly restored, the mass of weights removed plus the small degree of residual imbalance projected onto the screen give the mass of the object on the pan.

The weights and other components are enclosed in the balance to protect them from exposure to the atmosphere and to chemicals which are weighed on the balance. These balances are sensitive, delicate instruments and require service (cleaning, sensitivity adjustment, standardization) at least annually by qualified personnel even when they receive the

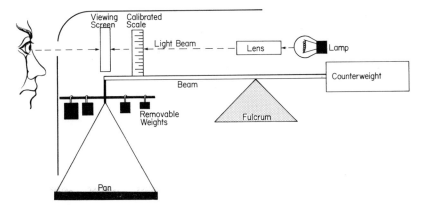

Figure 1-1
Components of a Single-Pan Projection Balance

best of care during daily use. Some important rules for the use of these balances are as follows:

1. Never weigh chemicals directly on the pan; use a container or weighing paper.
2. Do not weigh hot objects. Convection air currents around the object will cause faulty and unstable readings.
3. Do not try to weigh objects which may be too heavy for the balance. Check the weight of a heavy object on a less exact balance first.
4. Take care not to spill chemicals on the pan or floor of the balance.
5. If a level indicator is present, check that the balance is level before weighing.
6. Do not handle any of the internal components of the balance since fingerprints may add weight and cause corrosion.
7. Keep the balance clean and covered when it is not in use.

Pipets

There are many types of volumetric pipets available. The recommendations presented here are not meant to preclude the use of other types. However, technicians are strongly urged to use only pipets bearing the inscriptions of reputable manufacturers who publish the tolerances of their pipets and abide by these tolerances. Care and discretion should be used in selecting pipets, since the accuracy of analytical methods depends upon the use of good volumetric glassware. Further discussion of these and other points has been published [1].

Since *precision* refers only to repeatability while *accuracy* relates to some absolute known value, it is possible that a pipet may be quite precise but very inaccurate—that is, it can deliver the same wrong volume repeatedly with good precision. Almost all pipets are capable of good precision, but the accuracy is of primary importance.

Before the types of pipets to be used in a laboratory are selected, a basic decision must be made regarding the degree of accuracy to be demanded from the volumetric apparatus. The tolerances on volumetric pipets set by manufacturers sometimes are in terms of absolute amounts rather than percentage. Therefore, the accuracy of pipets varies with the volume delivered so that, while an accuracy of $\pm 0.2\%$ might be expected with a 10 ml pipet, a 0.5 ml pipet might have a tolerance of $\pm 1.0\%$. In order to calculate and control limits of error in laboratory methods, it is essential to have a standard (percentage) and minimal tolerance for the accuracy of all volumetric pipets. Since, by and large, the smaller volume pipets are used more frequently and for the more significant measurements, they should certainly be no less accurate than pipets of larger volume.

A tolerance of $\pm 0.5\%$ is considered adequate without placing unreasonable demands on the equipment. This degree of tolerance should be the criterion in selecting the types of pipet to be used. The following are recommended for routine laboratory work.

VOLUMETRIC TRANSFER PIPET

For accurate volumetric measurements of 0.5 ml or greater, the volumetric transfer or the Ostwald type of pipet is recommended (see A and B in Figure 1-2). Although both pipets are calibrated *to deliver* the appropriate volume, the transfer pipet is allowed only to drain whereas the Ostwald pipet requires that the last drop be blown out. Since the *blowing-out* technic affords a certain degree of standardization and eliminates the possibility that the busy technician may not allow a slow-draining pipet to drain completely, it is the pipet of choice here. Pipets calibrated to be blown out are identified as such by a pair of etched rings near the top of the pipet.

Some general instructions for using pipets are given below. However, the technician should always use the pipetting technic approved by his supervisor. The volumetric transfer pipet should be used as follows:

1. Draw the solution above the calibration mark and hold the column in place.
2. Wipe the outside tip of the pipet with a tissue.
3. With the tip of the pipet against a surface (glass or solution) and the pipet in a vertical position bring the bottom of the solution meniscus to the mark.

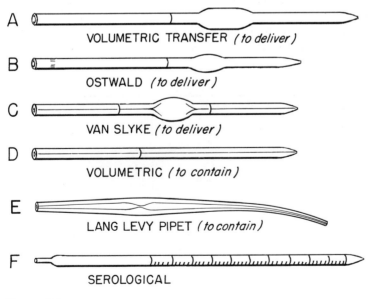

A   VOLUMETRIC TRANSFER *(to deliver)*

B   OSTWALD *(to deliver)*

C   VAN SLYKE *(to deliver)*

D   VOLUMETRIC *(to contain)*

E   LANG LEVY PIPET *(to contain)*

F   SEROLOGICAL

Figure 1-2
Types of Pipets

4. Place the tip of the pipet in the vessel to receive the sample and allow to *drain freely,* touching the tip to a surface to complete the measurement. There will always be a small amount of solution remaining in the tip, but this is constant and has been compensated for in the calibration of the pipet. Blowing on this pipet to hasten delivery changes the drainage time and may cause significant errors.

The Ostwald pipet is used in the same manner as the volumetric transfer type except that the last drop is blown out.

It is good policy to check the delivery calibration of a few pipets from each new batch, especially those of 1 ml size or smaller.

The following is a method for checking volumetric pipets:

1. Weigh a small weighing bottle (with a stopper) on the analytical balance.
2. With the pipet in question, measure into the weighing bottle a sample of water, employing good pipetting technic, and restopper the bottle.
3. Weigh the stoppered bottle plus the water.
4. Subtract the weight of the stoppered bottle from the weight of the stoppered bottle plus the water to obtain the weight of water delivered by the pipet.
5. Consult Table 1-5 to locate the weight of exactly 1 ml of water at the temperature being used. Multiply this weight by the volume of the pipet being checked to obtain the weight of water that the pipet *should* deliver at that particular temperature.

The difference between the weight of water delivered and the weight that should have been delivered is the error of the pipet. This should be divided by the volume of the pipet and multiplied by 100 to find the error as a percentage.

Table 1-5. Weights[a] of Water and Mercury in Air

Temperature (°C)	Weight (g) of 1 ml Water	Weight (g) of 1 ml Mercury
20	0.9972	13.547
21	0.9970	13.545
22	0.9968	13.543
23	0.9966	13.541
24	0.9964	13.539
25	0.9961	13.537
26	0.9959	13.534
27	0.9956	13.532
28	0.9954	13.530
29	0.9951	13.528
30	0.9948	13.526

[a] These values include effects of buoyancy and dissolved oxygen and differ from those given in most handbooks by approximately 0.1%.

EXAMPLE

A 0.5-ml pipet is tested as described above, and the following data are obtained:

Weighing bottle: 15.3650 g
Bottle + pipetted volume: 15.8528 g
Weight of water delivered: 0.4878 g

The ambient temperature is 25 C, so 0.5 ml of water should weigh

$$\frac{0.9961}{2} = 0.4980 \text{ g}$$

Since the pipet delivered a smaller amount than this, it has a *negative* error of 0.0102 g, which is

$$\frac{0.0102}{0.4980} \times 100 = 2.0\%$$

VAN SLYKE PIPET

The Van Slyke pipet (see C in Figure 1-2) is constructed of thick-walled capillary tubing with a bulb in the center and one mark above the bulb and one below. The specified volume is delivered between these two marks. Pipets of this type are more precise than the Ostwald or transfer pipet, but they are more tedious to use, especially if many measurements are to be made.

TO-CONTAIN PIPET

For volumes smaller than 0.5 ml, the "to-deliver" types of pipets (Figure 1-2) described above usually are not sufficiently precise. Apparently small variations in the amount of liquid adhering to the walls of the pipet during delivery prevent consistent accuracy. For this reason, a *to-contain* type of pipet (see D in Figure 1-2) is recommended. The difference between a to-contain and a to-deliver type of pipet is that the former must be *rinsed out* (twice generally is satisfactory) after complete delivery in order to wash all of the fluid contained in the pipet into the diluent.

Calibrations of to-contain pipets should be checked by using mercury. Mercury is a nonwetting liquid that drains completely, thereby simulating the washing-out technic employed with aqueous solutions. Furthermore, mercury is approximately 13.5 times as heavy as water, making possible the accurate weighing of small volumes. The calculations are as described in the example given above except that the weight of mercury from Table 1-5 is used.

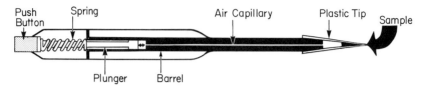

Figure 1-3
Semiautomatic Pipet

If mercury is used, extreme caution must be exercised to avoid contact of the mercury with the balance pans or weights.

Lang-Levy Pipet

For volumes in the microliter range the Lang-Levy type of pipet (E in Figure 1-2) is highly recommended. This pipet has a small constriction in the upper part of the bore, which provides a semiautomatic zero adjustment. The tip is drawn to a fine bore and curved to allow smooth and complete delivery of the sample. For pipetting, suction must be applied using an aspirator tube or similar device.

Serological Pipet

The serological type of pipet (F in Figure 1-2) is not suitable for accurate measurements of less than 2 ml, and it should be used with discretion in the chemistry laboratory. If only a portion of the total capacity of a serological pipet is to be delivered, it is advisable to deliver between two of the marks rather than to use the lower portion of the pipet. Many serological pipets have large calibration errors near the tip. However, these errors usually are small relative to the total volume of the pipet, so that, if proper technic is employed and the entire volume of the pipet is delivered, the serological pipet is reasonably accurate. It is useful also in making smaller multiple measurements which do not require maximum accuracy.

Semiautomatic Pipets

As a result of increased need for multiple measurements, numerous types of semiautomatic pipets are in use. The most popular of these devices operates on a syringe or displacement principle (Figure 1-3). When the button is depressed, the plunger displaces a calibrated volume of air from the barrel. The disposable plastic tip is immersed in the sample and, when the button is released, the calibrated volume of sample is drawn into the tip. The sample is then delivered by pushing the button again. Generally these pipets are satisfactory but there are several potential problems:

1. The calibration may be off significantly, and there is no way to adjust it. The accuracy may be tested in the same manner as described for manual pipets.
2. The precision (repeatability) may not be good, although good precision is one of the stronger features of these pipets.
3. Accuracy or precision or both may change with use. Regular cleaning and lubrication may help avoid this problem.
4. Some of these pipets will draw up significantly different amounts of aqueous solution and sample (serum or blood). This is a problem when aqueous standards are used.
5. Even when the pipet is functionally sound, defective plastic tips can cause poor performance.
6. Because the operation of the pipet is simple and semiautomatic, technicians sometimes overlook fine points in technic. This results in inaccurate and erratic measurements.

As with any other piece of analytical equipment, if these pipets are of good quality and are used and maintained properly, they can be an important asset in the chemistry laboratory.

Concentrations of Commercial Solutions

Since many solutions are prepared from concentrated commercial reagents, approximate concentrations of the most common of these reagents are given in Table 1-6 along with the amounts of each required to prepare 1 liter of approximately $1M$ solution.

Table 1-6. Concentrations of Some Common Commercial Solutions

Reagent (concentrated)	Approximate % (w/w)	Approximate Molarity	ml Required for 1 l of (Approx) $1M$ Solution
Hydrochloric acid (HCl)	37	12	83
Nitric acid (HNO_3)	70	16	63
Perchloric acid ($HClO_4$)	70	12	86
Acetic acid (CH_3CO_2H)	100	17	58
Sulfuric acid (H_2SO_4)	96	18	56
Phosphoric acid (H_3PO_4)	85	15	68
Ammonium Hydroxide (NH_4OH)[a]	58	15	69
Sodium Hydroxide (NaOH, 50%)	50	19	52
Potassium Hydroxide (KOH, 45%)	45	12	86

[a] 28% Ammonia (NH_3)

Hydrates

Some salts are available not only in anhydrous (water-free) form but also as one or more hydrates, that is, with a certain number of water molecules attached to each molecule of salt. Since the water of hydration must be counted in determining the molecular weight of the salt, it is very important that the proper molecular weight be used in calculating the amount of salt needed to prepare a given solution. Directions should always specify the hydrate if the salt comes in more than one form. In the absence of this information, the most common form must be assumed.

Copper sulfate comes in an anhydrous form ($CuSO_4$), as a monohydrate ($CuSO_4 \cdot H_2O$), or as a pentahydrate ($CuSO_4 \cdot 5H_2O$). One liter of a $1M$ solution of copper sulfate would require 160 g of anhydrous salt, 178 g of the monohydrate, or 250 g of the pentahydrate. If directions called for a 5% (w/v) solution of copper sulfate, the hydrate to be used should be specified since 5 g of each of the above-mentioned salts would contain a different quantity of copper sulfate.

Purified Water

Water usually is not considered a reagent because of its abundance and generally constant appearance. However, water is recognized as a reagent when (not infrequently) problems develop with chemical tests which eventually are traced to contaminated water. The production, storage, and distribution of water therefore must be carefully controlled. Whenever reagents are prepared or a method is performed, purified water should be used unless the direction specifically state that the use of tap water is permissible. A discussion of some of the more important aspects of purified water is presented below. More comprehensive discussions of the subject are available [2, 5].

Three important questions should be asked about water: (1) How can pure water be obtained? (2) How can the purity of water be tested? (3) How pure must water be in the clinical chemistry laboratory?

Before answering these questions, it is important to recognize the types of impurities sometimes found in water. As might be expected, water contains at least a trace amount of almost everything, including metals (Al, Fe, Mn, Na, K, Zn, Ca, Mg), anions (nitrate, phosphate, silicate, sulfate), organic matter (organisms, vegetation, animal decay products, detergents, industrial-waste chemicals), dissolved gases (NH_3, CO_2, O_2), and suspended solids (dust and aggregates of the above materials). The problem of water purification would not be so severe if the types and quantities of impurities present were constant. However, they vary with geographic location, with seasons, and sometimes even from one faucet to another in a given building because of variations in the pipes. Complete removal of all of these contaminants is impossible. It is no small problem just to keep most of them regularly at an acceptable level. A suitable procedure for one water source may not be suitable for another because of differences in contaminants.

1. How Can Pure Water Be Obtained?

The two general procedures for purifying water are distillation and deionization. Either can provide water which is acceptable for use in the clinical chemistry laboratory. Certain advantages and disadvantages exist for each and for the use of both together.

Both metal and glass stills are used for distillation. Removal of all contaminants is incomplete because small quantities of the contaminated water are blown into the air by the turbulent action of the boiling and are carried into the distillate along with the water which is being distilled. Even when this spattering is minimized, the distillate will still contain, along with dissolved gases, sodium silicate from the glass condenser or some of the metal ions from the still when a metal still is used. The main advantage of distillation is that much of the organic matter is removed.

Deionization columns are more effective than stills in removing inorganic salts from water but are generally less effective in removing organic matter. Special filters or activated charcoal can be incorporated to lower the amount of organic matter to acceptable levels. Usually an anion and a cation exchange resin are mixed in one tank to produce a "mixed-bed resin column." These resins are highly effective in removing dissolved ionized particles from water.

Periodic maintenance of both systems is necessary. Stills must be cleaned regularly unless they are preceded by a deionization column. The deionization column must be regenerated regularly in any case because it eventually becomes saturated with the removed ions.

2. How Can Purity Of Water Be Tested?

Each testing procedure measures only certain types of contaminants. The most common test is a *conductivity* measurement using a conductivity meter. (Conductivity is the reciprocal of electrical resistance.) Although this measures only the charged substances which are present (like inorganic salts) and does not measure most of the organic matter, which is mostly uncharged, the measurement is easy to perform and serves as an overall check on the quality of the water. Some commercial units continually monitor the conductivity of the water output. The other tests which can be carried out are specific chemical tests, each of which quantitates or detects a particular contaminant. The usual clinical chemistry methods may be used (with reagents prepared from another water source of known purity), for example, phenolhypochlorite reagent for ammonia, flame emission spectrophotometry for sodium, atomic absorption spectrophotometry for the metals (Ca, Pb, Zn, Mg, etc.), molybdenum blue for phosphate, etc. Details of these tests as well as further information about purified water have been published [5].

3. How Pure Must Water Be in the Clinical Chemistry Laboratory?

The following specifications have been recommended for purified water:

Distilled Water
    Carbon dioxide, less than 5 ppm (mg/liter)
    Ammonia, less than 0.1 ppm (mg/liter)
    pH, 5.5 to 7.0
    Metals, less than 0.01 ppm (mg/liter)
    Resistance, no less than 100,000 ohms (that is, conductivity no greater than 10 micromhos)

Deionized Water
    Silica, negative
    Ammonia, less than 0.2 ppm (mg/liter)
    Sodium, negative
    pH, 6.0–7.0
    Resistance, greater than 1,000,000 ohms (1 megohm) (that is, conductivity less than 1 micromho)

The specifications adopted by each laboratory and the particular purification equipment selected will depend on the water source available, the most sensitive tests carried out by the laboratory, and the amount of purified water required.

Cleaning Glassware

The importance of clean glassware in the clinical chemistry laboratory cannot be overemphasized. Most detergents advertised for cleaning laboratory glassware are satisfactory. Uneven wetting or the presence of droplets of water that cling to the glass indicate that the glass is not sufficiently clean. A long soak in a concentrated hot detergent solution is a very effective cleaning procedure. Petroleum ether (a mixture of hydrocarbons such as pentane, hexane, etc.) is good for removing fat or grease residues.

When all else fails, chromic-sulfuric cleaning solution should be used. The solution is commercially available or can be prepared as follows:

1. Prepare a saturated, aqueous solution of potassium dichromate. (Approximately equal weights of this salt and water are combined; it is convenient to keep this as a stock solution.)

2. To 25 ml of this stock solution in a large beaker or Erlenmeyer flask in a sink, slowly add 1 liter of concentrated sulfuric acid with stirring.

3. After the solution cools, transfer it to a glass bottle and stopper. (An empty sulfuric acid bottle is satisfactory.) This solution is highly

corrosive and should be handled near a sink with the water running. In addition, rubber gloves and an apron should be worn.

When a little water is added to a vessel before the addition of some of the chromic-sulfuric cleaning solution, considerable heat results, which often facilitates rapid cleaning. If this is not effective, the vessel should be soaked in undiluted cleaning solution overnight. The solution may be returned to the stock supply unless it has turned green, which indicates a loss of effectiveness. A green solution should be washed down the sink with large quantities of water. One disadvantage of this cleaning agent is that it leaves traces of chromium ions on glassware, which may contaminate certain procedures, especially enzyme reactions. Four washes with tap water followed by three with purified water are recommended as a minimum washing procedure.

An alternate, strong cleaning solution is a 1:1 mixture of sulfuric and nitric acids.

The Centrifuge

The centrifuge is an important piece of equipment in the chemistry laboratory. It is used primarily for separating particles from solutions by centrifugal force. Directions for centrifuging (spinning) blood, protein precipitates, and mixtures of nonmiscible liquids frequently are given in terms of speed, that is, revolutions per minute (rpm). It is fortunate that accurate reproduction of this particular step in a method seldom is important because the centrifugal force is a function not only of speed but also of the diameter and angle of the head used in the centrifuge. Data are available from manufacturers of centrifuges which permit one to determine the centrifugal force under various conditions. A few basic points on the proper use of centrifuges are presented here:

1. A given setting of the rheostat does not always produce the same speed. Variation in weight load and aging of the centrifuge can result in changes in the relationship of power input to rpm. A tachometer (speed indicator) rather than a rheostat setting should be used as a guide if reproducible centrifugation is desired.
2. Loads should be balanced before being spun. This is best done by using blank tubes containing water; water should not be put directly into the shields (cups) for balancing.
3. Before placing tubes in shields, the latter should be checked to see that rubber cushions are in place. When out of the centrifuge, shields should be stood in an upright position to avoid losing the cushions.
4. Centrifuges should be equipped with automatic time switches that will turn off the centrifuge after the preset time has elapsed.
5. If a centrifuge is equipped with a brake, it should be used with caution. Braking may cause resuspension of some of the sediment.

6. Centrifuges should be checked, cleaned, and lubricated regularly. If sparking is noted, the brushes should be changed as soon as possible to avoid damage to the armature.
7. The lid should always be closed while a centrifuge is in operation to avoid accidents.

Sample Problems in Quantitative Analysis

The problems presented here are designed to test the reader's understanding of the various expressions of concentration and dilution and their relationships to each other. The answers are on page 25 but students are urged to work the problems without reference to the answers.

Na = 23	S = 32	Ca = 40	Cl = 35.5	K = 39
O = 16	Cu = 64	H = 1	Ag = 108	N = 14

1. How many grams of a salt would be required to make each of the following solutions?
   a. 100 ml of a 10% (w/v) solution
   b. 500 ml of a 5% (w/v) solution
   c. 50 ml of a 1% (w/v) solution
2. How many grams of each of the following substances would be required to make 1 liter of a $1M$ solution?
   a. NaOH
   b. H_2SO_4
   c. $CaCl_2$
3. How many grams of each of the following substances would be required to make 1 liter of a $1N$ solution?
   a. HCl
   b. Na_2SO_4
   c. $CuSO_4$
4. Express the following in terms of molarity:
   a. $1N$ NaOH   c. $0.5N$ $AgNO_3$
   b. $1N$ H_2SO_4 d. $2N$ $CaCl_2$
5. Express the following in terms of normality:
   a. $2M$ HCl   c. $1M$ $CaCl_2$
   b. $0.1M$ $CuSO_4$ d. $0.5M$ H_2SO_4
6. What is the concentration, expressed in percent (w/v), of each of the following solutions?
   a. 5 g of $AgNO_3$ in 500 ml
   b. 250 g of KCl in 1000 ml
   c. 1 g of $CuSO_4$ in 25 ml
   d. 2 g of $CaCl_2$ in 10 ml
7. If 40 g of NaOH are diluted to 1 liter, what is the concentration in terms of:
   a. molarity   b. normality   c. percent (w/v)

8. If 55.5 g of $CaCl_2$ are made to 1 liter, what is the concentration of the solution in terms of:
   a. molarity   b. normality   c. percent (w/v)

9. a. If a solution of concentration 10 mg per 100 ml is diluted 1:10, what is the final concentration?
   b. If the resulting solution is rediluted 1:5, what is the final concentration?

10. What is the final concentration of an x normal solution that is diluted 5:50, rediluted 10:100, then rediluted 5:25?

11. Given a series of ten tubes, each of which contains 4 ml of diluent. One milliliter of fluid is added to the first tube, and a serial dilution, using 1 ml is carried out in the remaining tubes. What is the dilution of the fluid in the fifth tube? Tenth tube?

12. Twenty milliliters of $1N$ NaOH will neutralize how many milliliters of a $0.1N$ HCl solution?

13. How much 70% (v/v) alcohol can be made from 200 ml of 95% alcohol?

14. A $1N$ solution of NaOH is diluted 5:25, then rediluted 3:100. What is the final normality? How many grams of NaOH are present in 100 ml of the final dilution?

15. Eighty grams of $CuSO_4$ are dissolved and diluted to 1000 ml. This solution is diluted 1:10, then rediluted 5:25. How many grams of $CuSO_4$ would be present in 1 liter of the final solution? What is the concentration of the final solution in terms of normality, molarity, and percent (w/v)?

16. What is the concentration of 250 ml of 0.85% saline in terms of molarity and normality?

17. Convert the following concentrations from mg per 100 ml to meq per liter.
    a. 10 mg per 100 ml NaCl
    b. 10 mg per 100 ml Ca
    c. 20 mg per 100 ml K

18. Convert the following concentrations from meq per liter to mg per 100 ml.
    a. 150 meq/liter Na
    b. 4 meq/liter Ca
    c. 100 meq/liter NaCl

ANSWERS TO SAMPLE PROBLEMS
1. a. 10   b. 25   c. 0.5
2. a. 40   b. 98   c. 111
3. a. 36.5   b. 71   c. 80
4. a. 1   b. 0.5   c. 0.5   d. 1
5. a. 2   b. 0.2   c. 2   d. 1
6. a. 1   b. 25   c. 4   d. 20
7. 1M, 1N, 4% (w/v)

8. $0.5M$, $1N$, 5.55% (w/v)
9.   *a.* 1 mg per 100 ml   *b.* 0.2 mg per 100 ml
10. $(x/500)N$ or $0.002xN$
11. *a.* $1:3125 (1:5^5)$ *b.* $1:9,765,625 (1:5^{10})$
12. 200 ml
13. 271.4 ml
14. *a.* $0.006N(\frac{3}{500})$ *b.* 0.024 g
15. *a.* 1.6 g/liter   *b.* $0.02N$, $0.01M$, 0.16% (w/v)
16. $0.145M$, $0.145N$ (the 250 ml is irrelevant)
17. *a.* 1.7   *b.* 5   *c.* 5.1
18. *a.* 345   *b.* 8   *c.* 585

References

1. Annino, J. S.   Capabilities and limitations of volumetric pipets. *Am. J. Clin. Pathol.* 29:479, 1958.
2. Hamilton, P. B., and Myoda, T. T.   Contamination of distilled water, HCl, and NH_4OH with amino acids, proteins, and bacteria. *Clin. Chem.* 20:687, 1974.
3. Peters, D. G., Hayes, J. M., and Hieftje, G. M.   *Chemical Separations and Measurements: Theory and Practice of Analytical Chemistry.* Philadelphia: Saunders, 1974.
4. Skoog, D. A., and West, D. M.   *Fundamentals of Analytical Chemistry* (3rd ed.). New York: Holt, Rinehart, & Winston, 1976.
5. Winstead, M.   *Reagent Grade Water.* Austin: Steck Company, 1967.

2. Quality Control

The need for systems to control the quality of products was acknowledged by industry long before clinical chemistry evolved into a profession. The product of the clinical chemistry laboratory is test results which are used for diagnosis and treatment of disease. Erroneous laboratory data can result in unnecessary medical treatment, prolonged hospital stays, or missed diagnoses. Because of the effect of test results on health and health care costs, the quality of such data must be controlled [3, 6].

A complete quality control system includes three interrelated activities: (1) assessment, (2) maintenance, and (3) improvement.

Assessment

Assuming that a laboratory is already in operation, the quality of data being produced must be *assessed* or *evaluated*. Two terms that are important in this regard are *accuracy* and *precision*. *Accuracy* refers to how close a measurement (or the average of many measurements) is to the true value. This true value is also referred to as the actual, correct, most accepted, or right value. The closer the measured value is to the true value, the greater is the accuracy. *Precision* refers to how close together are multiple measurements of the *same* sample. The closer together a pair, group, or series of measurements are on a given sample, the higher the precision. It follows from the definitions that demonstrated precision is not a guarantee of accuracy.

The concepts of accuracy and precision are illustrated in Figure 2-1. In the figure, curves A, B, C, and D are plotted from groups of data obtained from analyzing a single lot of control serum. Group A shows data which are tightly grouped with the mean lying on the correct value; therefore these results are both precise and accurate. In group B, the grouping (precision) is also good, but since the mean is not close to the correct value, the accuracy is poor. In group C, the data are spread over a wide range of values indicating poor precision. However, since the mean value coincides with the correct value, the overall accuracy is good. In group D the data are widely scattered and the mean is not near the correct value. This indicates a lack of both precision and accuracy.

Accuracy and precision are assessed with quality control materials obtained either from commercial sources or from pooled materials in the chemistry laboratory. Commercial materials usually are lyophilyzed serum or urine from humans or animals. A suitable control serum may be prepared by pooling serum samples after all requested analyses have been completed. Pour the serum through a funnel, containing a plug of glass wool, into a plastic bottle. This removes any large solid particles that might be present. Store the bottle in a freezer, and remove it only when more serum is to be added. When a sufficient quantity is collected (for example, 2 liters), thaw and mix the serum pool thoroughly. Since

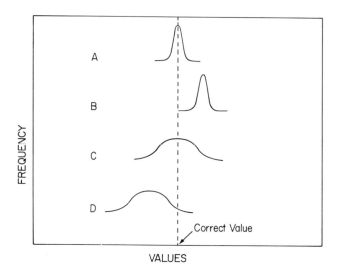

Figure 2-1
Accuracy and Precision

the bilirubin concentration probably will be low, it is advisable to add sufficient bilirubin to bring the concentration in the serum to approximately 1.5 mg per dl by dissolving 20 mg of bilirubin in 5 ml of 2% (w/v) sodium carbonate and adding this to the serum pool.

Adjust the pH of this serum to 7.4 by adding concentrated sulfuric acid dropwise while stirring vigorously and check the pH intermittently with a pH meter. Then measure the serum into plastic vials in amounts sufficient for each day's use (10 to 20 ml). Cap the vials tightly and store in a freezer. When serum is prepared in this manner, the following constituents are reasonably stable for at least 4 months: bicarbonate, chloride, sodium, potassium, urea, glucose, proteins, cholesterol, triglycerides, uric acid, creatinine, bilirubin, calcium, phosphorus, thyroxine, iron, transaminase, lactic dehydrogenase, amylase, lipase, alkaline phosphatase, and most drugs.

Determinations should be performed for each substance at least 10 times (preferably in duplicate). The mean (average) is calculated for each group of determinations as described below.

It would be unreasonable to expect to obtain the mean value every time a test is performed. Therefore some plus-or-minus deviation from the mean must be allowed. To arrive at a realistic allowance, data should be accumulated using the control serum for each method over a period of 4 to 6 months. During this time the control serum should receive no preferential treatment so that the subsequent calculations will reflect all variations in methods and technic. These data then are evaluated by the use of statistics.

STATISTICS

Statistical methods are required to quantitate some of the aspects of quality assessment [2, 4]. The fundamental basis for any statistical analysis is variation, and there is plenty of that in laboratory values. The variation may be due to decomposition of standards or reagents, batch-to-batch differences in reagents, instrument failure, technical errors, or errors in calculation or transcription of results. Sometimes the control material itself may change. Fortunately, a limited number of straightforward statistical concepts substantially deal with most of the problems of variation in clinical chemistry. These concepts are mean, standard deviation, and coefficient of variation.

MEAN

The mean or arithmetic average ($\bar{x}$) of a set of individual measurements (x) is given by

$$\bar{x} = \frac{\Sigma x}{n}$$

where Σ means *sum of* and n is the number of measurements. If a glucose level is measured 3 times ($n = 3$) on a serum sample, resulting in values of 86, 90, and 92 mg per dl, the mean is

$$\bar{x} = \frac{86 + 90 + 92}{3} = \frac{268}{3} = 89.3 \text{ mg/dl}$$

STANDARD DEVIATION

The standard deviation (SD) measures precision. It is essentially an expression of average error or average variation. A small SD indicates that the measurements are grouped closely about the mean, whereas a large SD indicates that there is a wide variation among the individual measurements. It is calculated as follows:

1. Calculate the mean ($\bar{x}$).
2. Subtract each of the individual values (x) from the mean to obtain the deviations ($\bar{x} - x$).
3. Square each of these deviations.
4. Add these squares.
5. Divide the sum by the number of individual values minus one.
6. The square root of this figure is the SD.

The mathematical expression for standard deviation is as follows:

$$SD = \sqrt{\frac{\Sigma (\bar{x} - x)^2}{n - 1}}$$

where *SD* is the standard deviation, Σ is the sum, $\bar{x}$ is the mean, x is each individual value, and n is the number of individual values.

If the individual values in a normal statistical distribution are plotted against their respective frequencies of occurrence, a curve similar to that in Figure 2-2 will be obtained.

Once the SD for a method has been calculated using the control, then, on a statistical basis it may be expected that about 67% of the subsequent determinations will fall within ±1 SD from the mean, approximately 95% will fall within ±2 SD, and approximately 99% will fall within ±3 SD. This variation should be distributed symmetrically about the mean, as illustrated in Figure 2-2. The ±1 SD and ±2 SD intervals about the mean value are shown in this figure.

The SD for a method should not change appreciably unless there is some significant change in methodology, instrumentation, or technic. Therefore, for each new pool of control serum, only the mean values need be determined and the existing SD values may be applied. Knowing the SD value for each substance helps the analyst determine how much variation might normally be expected if everything is working properly.

COEFFICIENT OF VARIATION

The ratio of the SD to the mean value is called the *coefficient of variation*. Usually it is expressed as a percentage by multiplying by 100.

$$CV = \frac{SD}{\bar{x}} \times 100$$

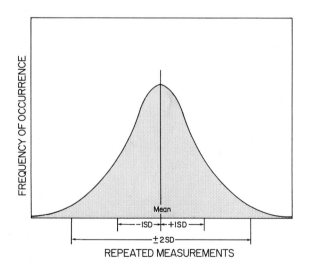

Figure 2-2
Normal Statistical Distribution Curve

The CV is a useful working guide to the precision of a test. For instance, consider a test with an SD of 3. If the test is for sodium, where $\bar{x} = 140$ meq per liter, the CV is 2.1% ($\frac{3}{140} \times 100 = 2.1\%$), which is good precision. However, if the test is for potassium, where $\bar{x}$ for normal serum is 4.0 meq per liter, the CV is 75%, which, of course, is quite unacceptable. Although it would be most desirable to have CVs for all methods in the 2% range, most methods have much larger variations than this at the present time.

QUALITY CONTROL CHARTS

Data from sequential analysis of the same sample may be plotted on a chart to produce a graphic display of performance. This is known as a quality control chart [6]. One may be set up for each test. Whether the tolerance limits should be 2 SD, 2.5 SD, or 3 SD (or some other arbitrary number) is the concern of the laboratory director. Usually the mean and tolerance limits are shown as horizontal lines on the chart.

Figure 2-3 shows a composite quality control chart illustrating a number of situations in which the chart is useful. The usual interpretations of these situations are given in Table 2-1. For this substance, the mean value is 20 and the control limits (± 2 SD) are 16–24. Of course, in many of these situations the control serum itself can be at fault.

Quality control charts frequently are hung on the walls in the laboratory or corridors. Although this policy may be of some advantage to the director, it also may have undesirable effects on the technical staff by causing undue embarrassment and encouraging bias or unhealthy competition.

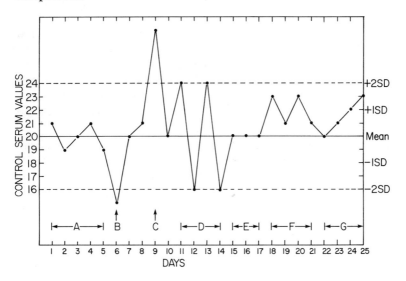

Figure 2-3
Quality Control Chart

Table 2-1. Interpretations of Situations Illustrated in Figure 2-3

Day	Interpretation
A (days 1–5)	Normal variation about the mean; method in control.
B (day 6)	Value slightly outside control limit. Although this could signify a problem, this situation may be expected to occur statistically once in 20 times (± 2 SD includes only 95% of the normal variation).
C (day 9)	Value is far outside the limit; quite probable that a problem exists.
D (days 11–14)	Although the values are not outside the limits, the variation is great, signifying poor precision (faulty technic?).
E (days 15–17)	No variation from the mean over a period of time suggests possible bias by the technician toward the mean value. (See text under Quality Control and later under Human Bias and Error.)
F (days 18–21)	When a series of values are all on one side of the mean something has changed to produce a new mean. Usually a deteriorated standard is at fault.
G (days 22–25)	These points show a trend in one direction; something is changing. Usually it is a reagent or a standard.

REFERENCE MATERIALS AND METHODS

It should be emphasized that the quality control scheme described above is useful for monitoring precision. Accuracy is determined in analytical chemistry by the use of reference materials and reference methods. Reference materials may be primary or secondary standards. A *primary standard* must be a stable substance of well-defined composition which is not hygroscopic (water absorbing). Such material is weighed and dissolved to prepare a primary standard solution. Substances or solutions which are standardized by comparison with a primary standard are *secondary standards*. Sometimes commercial control materials with assigned values are used as standards. In most cases the values on these materials are assay values and are subject to inaccuracies. Wherever possible, methods should be standardized with pure primary or secondary standards.

The US National Bureau of Standards (NBS) produces Standard Reference Materials for clinical chemistry laboratories. These materials are recommended for use in primary standards. A current listing of such materials is available from the US Department of Commerce, National Bureau of Standards, Washington, DC 20234.

Maintenance

Data from the quality assessment system are used to guide the *maintenance* or *assurance* of quality. Results may be in error for a number of reasons. To maintain high quality results the following considerations are essential:

1. Detailed instructions for methods should be available at each bench.
2. Technicians should be trained and checked out on each method.
3. Preparation details, storage specifications, and expiration dates for all reagents should be closely observed.
4. Primary standards should be used whenever feasible.
5. Standards and controls should accompany each group of tests.
6. Instrument performances should be checked frequently.
7. All glassware should be washed and rinsed thoroughly. In many cases disposable glassware is recommended.
8. Volumetric apparatus should be of the highest quality.
9. Distilled or deionized water should be checked frequently for contamination.
10. Calculations should be stated as simply as possible and performed with a calculator if possible.
11. Experienced supervisors should be involved in the laboratory activities at all times.

Improvement

It is a mistake to become so involved with the mechanics of assessing and maintaining existing quality that no effort is made toward improvement. There is always room for improvement of the quality of performance in a clinical laboratory. Usually this is accomplished through better methods, instruments, training, and quality control programs, all of which are the concern of the director. However, technicians can contribute significantly by assuming a professional attitude toward their work. Not only should they do the best job they can but they should also communicate observations and suggestions to their supervisor for appropriate action [5].

Laboratory personnel at all levels should strive to improve their own status and capabilities by joining appropriate professional societies, reading technical journals and books, attending scientific meetings, and taking continuing education courses.

Reports and Records

No less important than all the foregoing considerations are the systems for issuing reports and keeping records. The technician's workbook should contain all pertinent data such as readings, dilutions, calculations, and special observations (for example, "Sample is turbid") and should be dated and initialed daily. Results should be transcribed with maximum legibility and reported to the appropriate decimal place. Variations in rounding off figures for a given test may lead to misinterpretation of results or may mislead the clinicians regarding the accuracy of a method.

Careful records should be maintained in the laboratory for those instances where reports might go astray or a physician might wish confirmation of a result he received.

Human Bias and Error

When a test is repeated or a control sample with a known value is analyzed, there is a natural tendency, called "human bias," to take readings or round off calculations to favor the expected result. However, a conscious effort must be made to avoid this type of bias.

Technicians should, of course, make every effort to avoid mistakes, but when mistakes do occur, they should be reported to a supervisor for appropriate action. A good supervisor will recognize honest technicians as an essential part of the quality control program and will encourage their cooperation in ensuring the accuracy of all reported data. To err is human; this is no less true in clinical chemistry than in any other profession. Complete honesty by the entire laboratory staff is very important, since *questionable results are potentially more harmful to the patient than no results at all.*

General Considerations

In addition to the topics already discussed, a few other points relevant to quality control are worth considering.

1. A laboratory should be of sufficient size and should be clean, orderly, and efficient. There should be adequate lighting and utilities. Although the basic environment must be provided by management, proper maintenance of it is the responsibility of the technicians.
2. Technicians must have an active interest in the quality control program and should be encouraged to assist the director of the laboratory in his efforts to report accurate data.
3. Detailed written instructions should exist for operating and repairing equipment and for performing the tests for all other procedures in the laboratory.
4. A preventive maintenance schedule should be prepared. This includes checking such things as oven and bath temperatures, calibration of balances, spectrophotometers, and other instrumentation, and checks on wearable parts of instruments [1].
5. Written records should be kept of performance checks of maintenance and repairs, of calculations and results, and of quality control data.
6. All laboratory personnel, and particularly the director, should be active in professional societies which offer educational meetings, lectures, and journals and afford opportunities for the personnel of various laboratories to discuss common problems with one another.
7. Laboratory personnel should keep abreast of the current scientific literature in order to be informed of new methods and developments. This may be done by reading appropriate technical journals, abstracts, reviews, or new books (see the annotated bibliography at the end of the text).

The importance of a sound philosophy among laboratory personnel cannot be overemphasized. The growing importance of the significance of laboratory data in clinical medicine and research demands that, in the interest of good patient care and medical progress, a proper relationship exists between laboratory and medical staffs. Experience reveals that if laboratory personnel conduct themselves in a professional manner and exhibit a sincere desire to do the best they can under the prevailing conditions, the medical staff, by and large, is cooperative and appreciative.

References

1. Ackermann, P. G.   *Electronic Instrumentation in the Clinical Laboratory.* Boston: Little, Brown, 1972.
2. Barnett, R. N.   *Clinical Laboratory Statistics.* Boston: Little, Brown, 1971.
3. Gabrieli, E. R. (Ed.).   The use of data mechanization and computers in clinical medicine. *Ann. N.Y. Acad. Sci.* 161(2), 1969. Entire issue.
4. Henry, R. J., and Dryer, R. L.   Some Applications of Statistics to Clinical Chemistry. In Seligson, D. (Ed.), *Standard Methods of Clinical Chemistry.* New York: Academic, 1963.
5. Newell, J. E.   *Laboratory Management.* Boston: Little, Brown, 1972.
6. Whitby, L. G., Mitchell, F. L., and Moss, D. W.   Quality control in routine clinical chemistry. In Sobotka, H., and Stewart, C. D. (Eds.), *Advances in Clinical Chemistry.* New York: Academic, 1967. Vol. 10.

3. Biochemical Specimens and Normal Values

Accurate analysis of samples of biological fluids depends upon the proper collection and preparation of the samples as well as upon proper technic and methods of analysis. Some general considerations are discussed here and more specific ones are given in the chapters on methods.

Whole Blood, Serum, and Plasma

COMPOSITION

The great majority of analyses performed in a clinical chemistry laboratory require whole blood, serum, or plasma. The liquid portion of the circulating or uncoagulated blood is plasma; serum is plasma with the fibrinogen removed by means of the clotting process.

COLLECTION

The technic of collecting blood samples (phlebotomy) is outside the scope of this book. However four general considerations relevant to the accuracy of subsequent chemical analysis are:

1. Stasis (stoppage of blood flow, as by the use of a tourniquet) should be used for a minimum period of time since prolonged stasis may result in alteration of some chemical values.
2. Blood should not be taken while intravenous solutions are being administered since these solutions may influence the chemical assay.
3. Syringes or evacuated tubes (for example, Vacutainers[1]) used to obtain blood should be clean and dry to avoid contamination or hemolysis.
4. Some tests require blood collection into tubes which contain anticoagulants. Blood should be collected in a container appropriate to the requested tests. If an anticoagulant is used, thorough (but gentle) mixing is necessary to avoid clotting.

The chemistry technician should be certain that the people collecting blood are aware of these general considerations as well as the specific considerations that are cited in the chapters on methods.

EFFECTS OF EATING

Although the concentrations of some chemical constituents of blood may not be affected measurably by the intake of food [1], others are affected markedly. Therefore, as a matter of general policy, it is advisable that patients fast for at least 12 hours before blood collection for chemical

[1] Becton, Dickinson of Rutherford, N.J.

36

analysis. Obviously, the best time to obtain such a sample is before breakfast.

CONTAINERS AND ANTICOAGULANTS

A blood sample may be obtained with a syringe and transferred to a clean test tube. However, in most instances the blood is drawn directly into an evacuated tube (Vacutainer).

When whole blood or plasma is required, an anticoagulant must be used. The three anticoagulants most commonly used are heparin, oxalate, and ethylenediaminetetraacetic acid (EDTA). Vacuum tubes are color coded to indicate which anticoagulant is present.

Heparin is a normal blood constituent, but its physiological concentration is not high enough to prevent clotting in freshly drawn blood. It is a polysaccharide derivative and apparently functions by inhibiting some of the enzymes involved in clotting. Although it is the anticoagulant that interferes the least with clinical chemistry tests, it is more expensive than the others and its effect is temporary. Heparin is suitable when either whole blood or plasma is required.

Oxalates act by precipitating calcium from the blood. This prevents clotting because removal of calcium inactivates some of the enzymes involved in the clotting mechanism. Potassium oxalate salts are used most commonly, although lithium, sodium, and ammonium salts also are used. The oxalates shrink the red cells by causing the water in these cells to diffuse into the plasma. For this reason, oxalate should be used only when whole blood is required for the test and the integrity of the cells is not important.

EDTA acts similarly to oxalates, except that it renders the calcium unavailable for the clotting enzymes by chelating it (binding it tightly) rather than precipitating it.

Centrifugation and removal of serum or plasma from the cells helps preserve the integrity of many constituents. Although the exchange of intracellular and extracellular material is much slower in clotted samples, it still can result in significant changes.

Refrigeration is perhaps the simplest and most reliable method for the preservation of specimens. Glycolysis and other enzymatic and bacteriological processes are slowed down considerably at lower temperatures. Refrigeration of specimens until such time as they can be processed is helpful, but even though the rate is slower, changes still take place and eventually become significant. Refrigerated specimens must be brought to room temperature before they can be measured accurately.

The same preservation technics apply to both plasma and serum. Refrigeration is an effective method for preserving many of the chemical constituents of plasma or serum. If a sample is to be kept, it should be placed in a stoppered tube and stored in a refrigerator. The concentrations of most serum constituents will remain stable for several days if treated in this manner.

Freezing whole blood results in rupture of the red cells, but it does not injure plasma or serum. Therefore storage of samples of plasma or serum in a freezer is effective and is particularly helpful in preserving most enzymatic activities. Practically all of the constituents of serum are stable for longer periods of time in the frozen state. When serum is frozen, it separates into layers of various composition. Therefore it is very important, after thawing a frozen sample, to invert it many times to restore the homogeneity of the solution.

There is, of course, no perfect substitute for a fresh sample of blood or serum. Preservation should therefore be considered the exception rather than the rule.

It is always advisable to save all samples of biological fluids for at least 24 hours after the analyses are completed. This affords an opportunity to repeat the test when necessary. Should the results of a test seem questionable in relation to other analytical data or to the clinical observations, the analysis may be repeated with the same sample. Refrigeration provides adequate preservation for most samples over a period of 1 or 2 days.

Other Body Fluids

It is sometimes necessary to perform chemical analyses with body fluids other than blood or urine. These fluids include cerebrospinal fluid, gastric juice, and transudates or exudates that may collect in the abdominal cavity (ascitic fluid), in the chest cavity (pleural fluid), in the joints (synovial fluid), or from draining wounds. The collection of these fluids is the concern of the physician or nurse, but the technician should be prepared to analyze them. The analysis of cerebrospinal fluid and gastric juice is discussed in Chapter 27. The analyses most commonly requested with other fluids are protein, chloride, sodium, potassium, glucose, and specific gravity. Most of the analyses requested may be performed with the methods used for serum or blood determinations. However, these fluids may contain certain substances in concentrations very different from those found in the blood. For this reason, some adjustment may be required. Often it is easier for the technician to decide on the amount of fluid to be used if the physician can give a rough estimate of the concentration of the substance he desires to have quantitated.

If a specific gravity measurement on a fluid is requested, several methods may be used. If there is enough fluid, the specific gravity may be determined with a calibrated float (hydrometer) such as is used for determining the specific gravity of urine. If the amount of fluid is insufficient or if a highly accurate specific gravity is desired, a carefully measured amount of fluid may be weighed on the analytical balance and its weight compared with the weight of the same volume of water at the ambient temperature.

EXAMPLE

Exactly 1.0 ml of fluid is weighed on the balance, and its weight is found to be 1.0110 g. The temperature is 25 C and, according to Table 1-5, 1 ml of water at 25 C weighs 0.9961 g. Therefore, the specific gravity of the fluid is

$$\frac{1.0110}{0.9961} = 1.015$$

Specific gravity may be estimated also by using an instrument to measure the refractive index, as discussed in Chapter 28.

Normal Values[2]

Although the normal ranges for most of the commonly measured chemical constituents of the blood are reasonably well established, some discrepancy still exists among the reported data. There are several factors which bear on this problem:

1. Some of the data reported are based on invalid or insufficient sampling.
2. Factors such as age, sex, and diet may influence the concentration of certain substances in blood or urine.
3. In normal individuals, concentrations of some substances vary greatly from day to day or even during a given day.
4. A stated normal range usually is statistically calculated to include approximately 95% of the values found in healthy people. Therefore approximately 5% of the healthy population will show values outside the stated range. (It is important to realize that an average of one out of every twenty results from *normal* subjects therefore will "seem to be" abnormal.) By the same reasoning, some sick people will show values within the normal range [3–5].
5. Many methods exist for the determination of each chemical constituent of the blood. In some cases different methods do not give the same values, and some disagreement exists regarding which are the more correct values. However, much progress has been made in recent years toward the standardization of methodology so that this problem is becoming less intense.
6. Normal ranges usually are established with subjects who are in good health. Diseased, malnourished, or nonambulatory (bed-confined)

---

[2] Summaries of normal values for the chemical constituents of most biological materials have been published [2, 6–8].

persons may exhibit changes in the concentrations of certain blood or urine constituents which are not related to a pathological condition. For example, most hospitalized patients, regardless of their clinical condition, have lower serum albumin concentrations than do ambulatory persons in good health.

7. With the advent of automation, tests on a large number of normal subjects can be, and have been, performed with very little effort. However the new (different) normal ranges defined by such studies sometimes are more the result of lack of specificity of the automated method than of the larger number of persons included.

Determinations of substances such as enzymes, which are expressed in terms of empirical units, are not well standardized among various laboratories. Therefore, the normal values for these tests should always be made known to the physicians who are utilizing the services of the chemistry laboratory.

A table of normal values such as Table 3-1 should be available from each laboratory to the physicians who are using the laboratory's services.

Table 3-1. Normal Values

Constituent[a]	Normal Range
Acid phosphatase, prostatic	up to 0.15 units
Acid phosphatase, total	up to 0.63 units
Albumin	3.6–5.0 g/dl
Alkaline phosphatase	up to 2.3 units
Amino acid nitrogen, urine	200–700 mg/d
Ammonia nitrogen, blood	50–150 μg/dl
Ammonia nitrogen, urine	200–700 mg/d
Amylase	up to 150 units
Ascorbic acid (vitamin C)	0.5–2.0 mg/dl
Bilirubin, direct	up to 0.4 mg/dl
Bilirubin, total	up to 1.0 mg/dl
Butanol-extractable iodine (BEI)	3.2–6.4 μg/dl
Calcium	8.6–11.0 mg/dl
Calcium, urine	50–400 mg/d
Carbon dioxide content	25–32 meq/liter
Carotene	25–250 μg/dl
Catecholamines, urine	up to 150 μg/d
Cephalin flocculation	up to 2+ in 48 h
Chloride, csf	120–130 meq/liter
Chloride	100–108 meq/liter
Chloride, urine	75–200 meq/d

[a] Material is serum unless otherwise indicated.

Table 3-1 (Continued)

Constituent[a]	Normal Range
Cholesterol[b]	100–250 mg/dl
Cholesterol esters	70–78% of total
Colloidal gold, csf	0000000000
Creatine	0.2–0.7 mg/dl
Creatine, urine	up to 200 mg/d
Creatinine	0.3–1.1 mg/dl
Creatinine, urine	Male: 20–26 mg/kg/d
	Female: 14–22 mg/kg/d
Creatine phosphokinase (CPK)	up to 12 units
Fatty acids, total	200–425 mg/dl
Folic acid	5–16 ng/ml
Free thyroxine (Free T4)	1.0–2.2 ng/dl
Gamma globulin (IgG), csf	up to 15% of total
Globulins, total	2.5–3.5 g/dl
Alpha-1	0.1–0.4 g/dl
Alpha-2	0.5–1.0 g/dl
Beta	0.6–1.2 g/dl
Gamma	0.6–1.6 g/dl
Glucose, csf	50–70 mg/dl
Glucose	65 to 100 mg/dl
Glucose, urine (quantitative)	up to 150 mg/d
17-Hydroxycorticosteroids, urine	Male: 5–15 mg/d
	Female: 2–10 mg/d
5-Hydroxyindoleacetic acid, urine	up to 10 mg/d
Immunoglobulins	see page 198
Iron	60–180 μg/dl
Iron-binding capacity, total	250–400 μg/dl
17-Ketosteroids, urine	Male: 8–25 mg/d
	Female: 5–15 mg/d
Lactic dehydrogenase (LDH)	up to 140 units
Lead, blood	up to 60 μg/dl
Lipase	up to 1.0 unit
Lipids, total	400–1000 mg/dl
Magnesium	1.7–2.8 mg/dl
Nitrogen, total, urine	10–18 g/d
Nonprotein nitrogen (NPN)	25–40 mg/dl
Osmolality	275–295 mosmol/kg
Osmolality, urine	300–1000 mosmol/kg
Phospholipids	100–300 mg/dl
Phosphorus, inorganic	2.6–4.5 mg/dl
Phosphorus, inorganic, urine	500–1200 mg/d

[a] Material is serum unless otherwise indicated.
[b] See Chapter 21.

Table 3-1 (Continued)

Constituent[a]	Normal Range
Porphobilinogen, urine	none detectable
Porphyrins, urine	Copro: up to 150 μg/d
	Uro: up to 25 μg/d
Protein, total, csf	20–45 mg/dl
Protein, urine	up to 150 mg/d
Protein, total	6.5–8.0 g/dl
Protein-bound iodine (PBI)	3.5–8.0 μg/dl
Sodium	138–146 meq/liter
Sodium, urine	75–200 meq/d
Potassium	3.8–5.0 meq/liter
Potassium, urine	40–80 meq/d
Thymol turbidity	up to 2 units
Thyroxine (T4)	4.0–11.0 μg/dl
Thyroxine, free (Free T4)	1.0–2.2 ng/dl
Transaminase, GO	up to 40 units
Transaminase, GP	up to 35 units
Triglycerides[b]	50–190 mg/dl
Triiodothyronine (T3) uptake	25–35%
Urea nitrogen	10–20 mg/dl
Urea nitrogen, urine	7–16 g/d
Uric acid	3.0–7.0 mg/dl
Uric acid, urine	up to 800 mg/d
Urobilinogen, urine	0.3–2.1 mg/2 h
Vanilmandelic acid (VMA)	up to 10 mg/d
Vitamin A	15–60 μg/dl
Vitamin B_{12}	150–1000 pg/ml
Vitamin C (ascorbic acid)	0.5–2.0 mg/dl

[a] Material is serum unless otherwise indicated.
[b] See Chapter 21.

For the various reasons given above, stated normal ranges should be considered only a *guide* to interpreting laboratory data. Clinical experience with each disease and test is necessary before the physician can use a stated normal range as an effective guide.

References

1. Annino, J. S., and Relman, A. S.   The effect of eating on some of the clinically important chemical constituents of the blood. *Am. J. Clin. Pathol.* 31:155, 1959.
2. Castleman, B. (Ed.).   Normal values. *N. Engl. J. Med.* 290:39, 1974.
3. Gabrieli, E. R. (Ed.).   The use of data mechanization and computers in clinical medicine. *Ann. N.Y. Acad. Sci.* 161(2):371, 1969.
4. Mainland, D.   Remarks on clinical norms. *Clin. Chem.* 17:267, 1971.

5. Reed, A. H., Henry, R. J., and Mason, W. B.   Influence of statistical method used on the resulting estimate of normal range. *Clin. Chem.* 17:275, 1971.
6. Siest, G. (Ed.).   *Reference Values in Human Chemistry.* Basel: Karger, 1973.
7. Wallach, J.   *Interpretation of Diagnostic Tests: A Handbook Synopsis of Laboratory Medicine* (2nd ed.). Boston: Little, Brown, 1974.
8. Young, D. S.   Normal laboratory values (case records of the Massachusetts General Hospital) in SI units. *N. Engl. J. Med.* 292:795, 1975.

4. Spectrophotometry

Most methods in clinical chemistry are based on quantitative measurement of a colored compound (*chromophore*) produced when a sample containing the substance to be measured is mixed with appropriate reagents and subjected to certain reaction conditions. Such methods usually are referred to as *colorimetric* methods and measurements are made with instruments called *photoelectric colorimeters* or *spectrophotometers*. This chapter deals with the basic principles of this analytic technic and some practical aspects of its application in the laboratory. Spectrophotometry is treated in greater detail in other textbooks [1, 2, 8, 10].

Color and Wavelength

Daylight, or white light, is a combination of all colors. If white light is dispersed or partially absorbed, certain of its component colors become visible. The colors seen depend upon the wavelength of the emergent light. For example, droplets of moisture in the air disperse the sun's rays to create a spectrum of colors known as the rainbow. Many solutions have the property of absorbing certain wavelengths of light and transmitting others. These solutions appear to have certain characteristic colors, which are determined by the wavelength of the light that is permitted to pass through (transmitted). For example, a solution of hemoglobin appears to be red because it absorbs blue-green light and transmits the complementary color of red.

Every color belongs to a particular wavelength region of the visible spectrum. Light travels in a wavelike manner. The wavelength of a band of light is defined as the distance between the wave peaks. The *visible* spectrum extends from about 400 nanometers (nm), violet light, to about 700 nm, red light. Below 400 nm is *ultraviolet* and above 700 nm is *infrared,* both of which are invisible to the human eye. Various wavelengths of light possess different energies, and various substances are capable of absorbing only certain energies of light. These facts form the basis for a certain amount of spectrophotometric specificity.

Light Intensity and Beer's Law

Besides wavelength, the other property of light that is of principle interest in spectrophotometry is intensity. When light is absorbed, its intensity is reduced. The amount of light which is absorbed by a given solution depends on the concentration of the absorbing substance in the solution and on the thickness or depth of solution through which the light travels. This has been formalized into Beer's law:

$$A = alc$$

where A is the absorbance, which is a measure of the amount of light

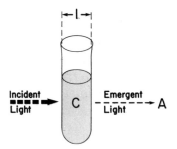

Figure 4-1
Absorbance and Concentration of Light. Since the path length, l (I.D. of the cuvet), is constant, the absorbance, A, of a solution is linearly related to concentration, c, of the absorbing substance, according to Beer's law.

absorbed by the solution; a is a proportionality constant called the *molar absorptivity* or *extinction coefficient, l* is the length (thickness) of the solution in cm, and c is the molar concentration of the absorbing substance. This concept is illustrated in Figure 4-1.

It is reasonable that a more concentrated solution or longer light path should absorb more light, since in either case there are more light-absorbing molecules placed in the path of the light. In other words, if the concentration of an absorbing substance is doubled, or twice the amount of solution is placed in the light path, the absorbance is twice as great. The quantitative application of Beer's law is illustrated in Figure 4-2. Further discussion of calculations is given in this chapter.

Another measure of change in light intensity is transmittance, T. Whereas absorbance, A, measures the amount of light absorbed by a

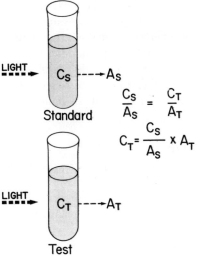

$$\frac{C_S}{A_S} = \frac{C_T}{A_T}$$

$$C_T = \frac{C_S}{A_S} \times A_T$$

Figure 4-2
Application of Beer's Law

solution, T measures the amount of light which passes through the solution. A simple reciprocal or subtractive relationship between A and T does not exist, however, since A relates to the *rate* of absorbance of light passing through the sample, while T gives the *total* fraction of transmitted light. The actual relationship between A and T is a logarithmic one, $A = -\log T$. Since some instruments are calibrated in percent transmittance $(\%T)$ and $\%T = 100 \times T$, then

$$A = -\log \frac{\%T}{100}$$

Applying the laws of logarithms,

$$A = -(\log \%T - \log 100) = -\log \%T + \log 100$$

$$A = -\log \%T + 2$$

$$A = 2 - \log \%T$$

If readings are made in $\%T$, calculation of concentration requires conversion of these readings to A or that the $\%T$ values be plotted against concentration on semi-log paper.

DEVIATIONS FROM BEER'S LAW

Not all colorimetric reactions follow Beer's law. Theoretically the law applies to substances whose chemical environment is constant as their concentration changes and which are being measured by a narrow band of light. Deviations tend to occur when these conditions are not met, for instance, (1) when more than one absorbing substance is present, (2) when the chemical nature of the medium changes (for example, at higher concentrations, the molecules of the absorbing substance may crowd each other enough to change each other's chemical environment), and (3) when more than one wavelength of light (for example, a band of light or significant amount of stray or uncontrolled light) is incident on the sample. Higher quality spectrophotometers give fewer deviations from Beer's law because they provide very narrow bands of light and a minimum amount of stray light.

Visual and Nonvisual Spectrophotometry

Most spectrophotometric measurements in clinical chemistry are made in the visible region (400–700 nm), where the normally visible colors occur, and these measurements, therefore, are called colorimetric analyses. In order to be measured by colorimetric analysis, a substance must either have a characteristic color or participate in a reaction which produces such a color. Furthermore, the amount of color produced should be proportional to the amount of substance being measured.

Measurements also are made in the ultraviolet (UV) and, much less so, in the infrared (IR) regions of the spectrum. The principles already discussed, including Beer's law, apply equally to these regions, although the convenience of knowing which solutions have more absorbance just by looking at them is not present, because colors exist for human vision only in the visible region. Solvents, buffers, and most substances absorb light in the UV and IR regions rendering measurements more susceptible to interferences. This disadvantage is somewhat offset by the greater richness of spectral detail observed in these regions, which enhances their usefulness for certain qualitative applications. For example, some colorless drugs give characteristic absorption patterns in the UV. Just knowing whether or not the drug is present in the sample along with an approximate idea of its concentration may be the key information desired in an emergency drug analysis.

Spectrophotometers

A spectrophotometer is an instrument used to measure the absorbance of a solution at one or more wavelengths. When an absorbance measurement is being made for purposes of quantification, the wavelength of choice is generally the one at which the greatest absorbance occurs. This provides the greatest sensitivity since the substance being measured is irradiated with the wavelength of light which it can absorb most strongly. If other light which is less absorbed is passed through the sample at the same time, this light will tend to mask the absorbance already taking place at the optimum wavelength. This is because the device in the spectrophotometer which detects the loss in light intensity (the detector) is simultaneously sensitive to all of the wavelengths of light that are imposed upon it. If only a few of these wavelengths have been absorbed significantly by the sample, the fractional change in light intensity at the detector is much less than if only the highly absorbed wavelengths are being monitored. As an analogy, it is far easier to detect a change in the intensity of a musical sound when it is the only sound rather than when it is part of a chorus or symphony of sounds. For this reason, and also to provide maximum specificity, the spectrophotometer must produce light of limited wavelength (see below) for interaction with the sample. As mentioned already, such light makes it easier for Beer's law to be applicable.

COMPONENTS OF SPECTROPHOTOMETERS

Certain components are common to all spectrophotometers, as illustrated in Figure 4-3. The light, usually provided by a lamp, is resolved (separated) into its component wavelengths by the wavelength selector (monochromator). A small group of adjacent wavelengths, known as a band of light, then is directed at the solution in a transparent container called a *cuvet*. Part of the light is absorbed by the solution. This loss in

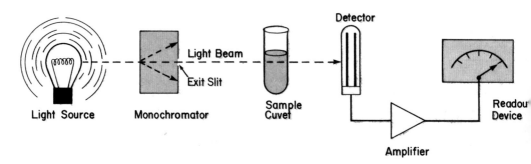

Figure 4-3
Components of a Single-Beam Spectrophotometer

light intensity is measured by the detector because the electrical signal output of the detector depends on the intensity of the light which is incident upon it. This electrical signal is read as an absorbance and is proportional to the concentration of the absorbing substance in the cuvet. Mirrors, slits, and lenses are common devices distributed throughout all spectrophotometers and are used to guide, trim, and focus the light through the major components.

LIGHT SOURCE
A hydrogen- or deuterium-discharge lamp commonly is used for the ultraviolet region while a tungsten-filament lamp is used for the visible region. The actual useful ranges are 200 to 400 nm for the hydrogen- or deuterium-discharge lamps and about 320 to at least 1000 nm for tungsten-filament lamps so that usually either can be used in the 320 to 400 nm interval. The output of the tungsten lamp in the ultraviolet region can be increased greatly by running it at an abnormally high voltage, but this markedly shortens its useful life. Zero absorbance (or $100\%T$) must be set every time the wavelength is changed because the intensity of the light incident on the cuvet changes with wavelength. The double-beam instrument is an exception to this, as is discussed later in this chapter.

WAVELENGTH SELECTOR
The purpose of the wavelength selector, which always involves more than one component, is to isolate specific wavelengths or wavelength bands of light from the source. The principle component is either a filter or a monochromator (Figure 4-3). Filters are of either transmission or interference types, and monochromators are composed of either prisms or gratings.

Filters are mirror or glass devices, each of which passes only a certain band of wavelengths (band pass) of light. Each is inscribed with a particular number that indicates the wavelength region which the filter transmits maximally. For example, a 540 (no. 54) filter absorbs all light except that around 540 nm. The actual band transmitted may be from 520 to 560 nm, with peak transmittance at 540 nm. When a second number appears below the first number, this figure designates the band pass and gives the length of the band in nanometers over which the transmission is at least half as intense as it is at its peak or center wavelength.

Transmission filters are colored glass or colored gelatin (a dyed protein) sandwiched between two plates of glass. Light outside the transmission band is absorbed by the colored material in the filter and thereby removed. It should be clear from the earlier discussion on complementary colors that a red filter is used to transmit red light since it produces this red light by absorbing blue-green light. These filters transmit only a small amount of the incident light and tend to have relatively wide band passes, even up to 50 nm.

Interference filters are composed of two half-silvered pieces of glass with a dielectric material sandwiched between. The light enters the interference filter perpendicular to the silvered surfaces and passes back and forth inside the filter before emerging. Constructive and destructive interferences occur as the light is reflected between the transparent silver films. Only certain wavelength bands of light emerge, depending on the thickness of the dielectric. Other bands cancel due to phase differences during the reflection process. The peak transmittance of these filters can be as high as 90%, and the band pass can be as small as 4 nm.

Prisms and diffraction gratings both function by dispersing (spreading) the light into its component wavelengths (the various colors in the case of visible light). They can provide much narrower wavelength bands than filters can. Another advantage is that a given prism or diffraction grating can provide a whole spectrum of wavelengths, any of which can be isolated and directed at the cuvet.

Glass-prism monochromators, illustrated in Figure 4-4, are used for

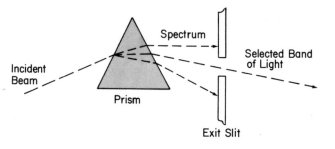

Figure 4-4
Prism Monochromator

the visible spectrum, but quartz prisms are required for the ultraviolet region. (Glass absorbs ultraviolet radiation, which is why it is impossible to get a sun tan through a glass window.) Because shorter wavelengths of light are bent to a greater extent than long ones are when light strikes a prism, the emergent light is dispersed into its component wavelengths. The main disadvantages of this spectrum are that the longer wavelengths are more crowded together than the short ones and that the spectrum is curved. Consequently relatively complex optical and mechanical devices are required with a prism monochromator.

A diffraction grating consists either of a transparent or a reflecting plate whose surface is ruled (scratched) with a large number of closely and equally spaced fine parallel lines, each separated by a fine edge, like the teeth on a saw. The edge of each of these lines reflects the incident light in all directions so that each edge functions as a source of light. With many thousands of adjacent sources, extensive interferences and reinforcements of wavelengths occur, and only certain wavelengths are allowed to leave in certain directions. The result is a uniform display of wavelengths, that is, dispersion of the incident light into a spectrum. Although diffraction gratings produce a spectrum which is not curved and not crowded in the longer wavelength region, they tend to produce more stray (uncontrolled) light than prisms do.

A specific wavelength band is isolated from the spectrum of a prism or grating monochromator by directing the spectrum to a plate which contains a narrow slit. The spectrum or the plate is moved so that only the specific wavelength band desired passes through the slit, leaving the rest of the spectrum masked out. This slit, as well as other slits in the system, also make it more difficult for stray light, which does not possess the desired wavelength characteristics, to enter the sample beam.

CUVET

The transparent container for samples is called a *cuvet*. It is either square or round and made of glass or transparent plastic for the range 320 to 1000 nm and of quartz (silica) for measurements below 320 nm. Square cuvets are recommended for more accurate work since they present flat surfaces to the light. When light encounters a flat rather than a round surface, there is less tendency for the light to be disturbed by reflection, refraction, or lens effects. With square cuvets, it is also easier for the operator to line up the same side of the cuvet toward the light beam. This is important with either type of cuvet in order to keep the effect of the cuvet surface on the light as constant as possible from one sample to the next. This also is the reason why it is important that the cuvet fit firmly into the cuvet holder and, when the cuvet holder itself is movable, that each position of the holder should be free of any variability or wobble.

DETECTOR

The detector measures light intensity by converting the light signal into an electrical signal. The more intense the light, the stronger the electrical signal. Various kinds of devices are used, the two most common of which are *barrier-layer cells* and *photomultiplier tubes*. Phototubes and photo-conductive tubes also are used.

A barrier-layer cell, also called a *photovoltaic cell,* is like a metal sandwich in which a layer of semiconducting material is sandwiched between two dissimilar metals. When light strikes the barrier-layer cell, the cell functions as a battery, and a current flows. This current is proportional to the intensity of the light. The barrier-layer cell requires no external voltage sources; however, its output cannot be readily amplified so that it is not as sensitive as other detectors.

In a photomultiplier tube, the light strikes a light-sensitive metal (cathode), causing electrons to be emitted. These electrons are accelerated by a positive potential in the tube to another electrode, called a dynode. When the electrons strike this dynode, each of them causes several additional electrons to be emitted. The multiplied electrons from the first dynode are accelerated in turn to collide with successive dynodes, each of which is more positively charged than the preceding one. In this way, a large multiplication of electrons is obtained. The electrons are collected at a final positive electrode, the anode.

The advantages of the photomultiplier tubes are high sensitivity, very rapid response times, and low fatigue. It is important not to expose them to room light because they are so sensitive that they may burn out. A high-voltage source is required for their operation. High quality spectro-photometers virtually always use photomultiplier tubes.

Other devices also are used as detectors. Phototubes, for instance, are the same as photomultiplier tubes except they lack the dynodes. The electrons emitted at the cathode go directly to the anode. Because the output of these tubes is readily amplified, they are more sensitive than barrier-layer cells. Also they can be constructed to respond to ultraviolet radiation.

Photoconductive (photoresistive) cells are devices whose electrical resistance decreases as the level of incident light is raised. Cadmium sulfide or cadmium selenide are the light-sensitive materials typically used in the visible region. These cells are about as sensitive as barrier-layer cells, but they do require an external power source.

READOUT DEVICE

The electrical signal from the detector may be read out in terms of $\%T$ or A. The readout itself may be a digital display, a needle reading on a meter or galvanometer scale, or an ink signal on the chart paper of a recorder.

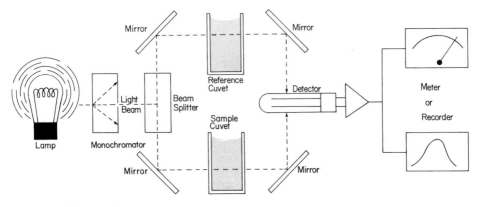

Figure 4-5
Components of a Double-Beam Spectrophotometer

DOUBLE-BEAM SPECTROPHOTOMETER

The double-beam spectrophotometer, illustrated in Figure 4-5, has both
a reference beam and a sample beam. These beams, which both arise
from the same light source, may travel different optical paths through
duplicate components, or they may travel along the same optical path
as an intermingled set of pulses, each pulse of one beam preceded and
followed by a pulse of the other beam. The reference beam is directed
through a second cuvet, called the reference cuvet, which ideally contains the same solution as the sample cuvet except for the absorbing
substance to be measured. The principle value of the reference beam is
in a scanning spectrophotometer. This instrument can plot absorbance
as a function of wavelength or of absorbance as a function of time as
required for kinetic assays of enzymes (see Chapter 20). By constantly
comparing the intensity of the sample beam to the reference beam as the
wavelength is changed, the instrument cancels any intensity or absorbance effects which are common to both beams. Hence the reading at any
given wavelength always represents the *net* absorbance of the solution
in the test cuvet minus the absorbance of the blank. The readout device
for such an instrument is a recorder.

Spectral Absorption Curve

A spectral absorption curve is a plot of absorbance (A) versus wavelength. The curve can be plotted manually from individual readings or
produced automatically using a recorder. An example of such a curve is
given in Figure 4-6. This particular curve has two peaks, one at 300 nm
and one at 475 nm. The beginning of an additional peak is seen as a
steep rise around 260 nm. If the absorption curve were continued into
the ultraviolet region, the third peak which is beginning around 260 nm
might go off scale, since the readout devices on most spectrophotometers

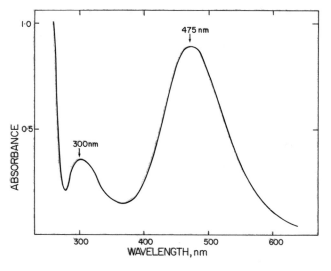

Figure 4-6
Spectral Absorption Curve

provide absorbance readings only up to 2.0. This actually is the point at which almost all of the light at the wavelength of interest is absorbed since the corresponding transmittance value is only 1 $\%\,T$. It is necessary to dilute a solution to obtain reliable absorbance values whenever the absorbance reading at the wavelength of interest is greater than 2.0. Additionally, absorbance readings are more accurate when they are confined (by adjusting the concentration of the solution) to the range 0.1 to 1.0 because values in this range mean that the detector and electrical system are analyzing a moderate rather than a minimal or excessive absorption of light by the sample. These systems perform optimally when this is the case.

The value of the spectral absorption curve is twofold. First of all, its overall shape may serve to identify the absorbing substance. That is, the position, shape, and size of a peak may be unique for a certain substance. Secondly, an optimal wavelength value can be chosen for quantitating the absorbing substance. For the spectrum shown, the wavelength in the visible region where the greatest absorbance occurs is at 475 nm. This wavelength, therefore, would be chosen for measurement of the concentration of this substance in a sample.

CALIBRATIONS

Ideally, a given solution would have the same absorbance value at a given wavelength, in the same size cuvet, in any spectrophotometer. This is not even remotely the case. Due to a variety of discrepancies relating to all of the components in spectrophotometers, absorbance values differ not only from one instrument to another, but even from one cuvet to another of the same type in the same instrument. Therefore frequent calibration and standardization technics are necessary to provide for accurate calculation of test results.

WAVELENGTH CALIBRATION

Didymium or holmium oxide glass plates (filters) have absorption peaks at specified wavelength values. The calibration filter is placed in the path of the beam in the cuvet compartment and readings are made at the wavelength peaks specified for the filters. The monochromator may be adjusted if necessary.

ABSORBANCE CALIBRATION

Although no adjustments are available on spectrophotometers to correct for faulty absorbance readings, these readings can vary significantly from one instrument to another, even when the instruments are of the same make and model. One way to compensate for this is to measure a standard whose absorbance is known and then to correct the absorbances of the tests mathematically. This approach seldom is used in clinical chemistry because it is satisfactory just to compare the relative absorbances of the tests with the observed absorbance of a standard whose absolute concentration (rather than absolute absorbance) is known.

Cuvet Calibration and Care

Colorimeter tubes, or cuvets, are optical glass tubes that are made to suit a particular type of colorimeter. If a cuvet has an imperfection, erroneous readings will be obtained for all solutions put into it. To guard against this source of error, it is advisable to check each cuvet before it is put into use and at frequent intervals thereafter. The following method may be used:

1. Prepare a solution of 0.01% (w/v) potassium ferricyanide. This solution should read between 0.2 and 0.8 absorbance at 420 nm against a Water Blank.
2. Using water for a Blank, set the instrument at 100% T (zero absorbance) at a wavelength of 420 nm.
3. Add a sufficient amount of the solution to the cuvets and read each one, checking the zero setting before each reading. The same solution may be poured from one cuvet to the other if desired.
4. Obtain the average of all the readings. Any cuvet reading outside of ±0.5 %T from the average is unsuitable for use with that set of cuvets.

The surfaces of cuvets must be kept clean and free of scratches. Water spots, dirt, or fingerprints will lead to faulty absorbance readings, especially in the ultraviolet regions. A convenient cleaning procedure for the outside surfaces is a "hot breath" followed by rubbing with a soft tissue paper, such as is used to clean eyeglasses. Strong acids or bases should not be used because they may etch the glass or dissolve the cement in square cuvets. Mild detergent, a 5% (v/v) nitric acid solution, or a 1:1 (v/v) mixture of 5% (v/v) hydrochloric acid and ethanol or methanol are useful cleaning solutions. After having been soaked overnight in one of these solutions, the cuvet should be rinsed thoroughly with water, purified water, and finally with alcohol or acetone before drying.

Calculations

The equation necessary for performing calculations in spectrophotometry is derived from Beer's law. The equation can be written for the standard and for the test as follows:

$$A_s = alC_s \quad \text{and} \quad A_t = alC_t$$

where A_s = absorbance of standard
A_t = absorbance of test
a = absorptivity constant
l = path length of light
C_s = concentration of standard
C_t = concentration of test

Since the absorptivity (a) is a constant for each method, it is the same for both the standard and the test. In addition, the path length (l) is the internal diameter of the cuvet, so that it also is the same for the standard and test. Consequently

$$\frac{A_s}{A_t} = \frac{C_s}{C_t} \quad \text{and} \quad C_t = \frac{C_s}{A_s} \times A_t$$

If %T instead of A values are provided by the instrument, they should be converted to A values using a %T-to-A conversion table before calculations are made.

Calibration Curve

A plot of A or %T versus concentration for a set of standard solutions constitutes a *calibration* or *standard curve*. Three steps are involved: (1) standard solutions are prepared; (2) the A or %T values of these solutions are measured; and (3) a plot of A or %T vs concentration is prepared using graph paper. An example of each kind of plot is shown in Figure 4-7. Since A is related linearly to concentration, the plot of absorbance is made on linear graph paper (Figure 4-7A). Semilog paper

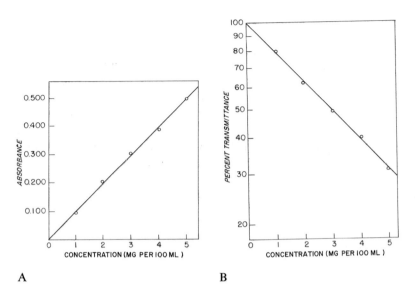

Figure 4-7
Absorbance and Transmittance Curves

(vertical scale is logarithmic, horizontal scale is linear) is used for the $\%T$ plot (Figure 4-7B) since $\%T$ is related logarithmically to concentration.

USE OF CALIBRATION CURVES

Calibration curves are used to replace mathematical calculations to find concentrations of tests. The reading of the test is located on the appropriate graph, and its corresponding concentration is determined from the curve. For example, a solution with $A = 0.35$ or $\%T = 44$ has a concentration of 3.5 mg per dl, as shown in Figure 4-7.

Both plots are linear because the method follows Beer's law. Whenever this is the case, the plot of A against concentration will go through the origin, as shown. This expresses the fact that a concentration of zero yields zero absorbance or 100 $\%T$.

Under these ideal conditions, a constant that represents the slope of the curve can be calculated to relate the concentration of the test (C_t) to its absorbance value (A_t). As is shown in Figure 4-2, the following is true for a standard (s) and a test (t) when Beer's law is followed:

$$C_t = \frac{C_s}{A_s} \times A_t$$

Since the term C_s/A_s will be the same for any point on the curve, this term is a constant. Any A_t reading may be multiplied by this constant to give the corresponding C_t value. Often this is easier than reading the C_t value from the standard curve and can be used whenever the plot is linear and goes through the origin, that is, whenever the plot follows Beer's law.

Calibration curves are not always linear because sometimes a method does not follow Beer's law. In this case more points must be plotted (more standards must be prepared) because it is more difficult to draw a line to fit a set of *curved* points. Nonlinear plots (curves) are used to find concentrations of tests from their readings just as with linear plots. If a calibration curve is not linear, a constant cannot be calculated.

FREQUENCY OF CALIBRATION

With most methods and instruments used in clinical chemistry, readings of standards are variable enough to warrant including standards in every group of tests. With methods following Beer's law, one or two standards may be determined each time, but the entire calibration curve should be checked periodically.

General Rules and Precautions
for Spectrophotometric Measurements

1. A blank should be used for each set of determinations. The primary purpose of the blank is to cancel any absorbance arising from the reagents. Ideally the blank should be prepared in the same fashion and with the same reagents as the specimens to be analyzed.
2. Cuvets should be clean and dry when used. This can be accomplished quickly by rinsing them with volatile solvents such as alcohol or acetone after washing. An alternate method may be used in which the cuvet is rinsed with each colored solution prior to taking the readings, but this is not as desirable as the use of dry cuvets. The outside of each cuvet should be wiped clean before it is placed in the instrument.
3. For each size cuvet there is a specified minimum volume. This prescribed volume is just enough to cover completely the aperture in the cuvet well through which the light beam travels. If a smaller volume is used, the light beam will be deflected by the meniscus and the open area above it, causing serious errors. Thus, the technician should be familiar with the minimum volume for each type of cuvet used and should be certain to have at least that volume in each cuvet.
4. Care should be taken to position the cuvet in the spectrophotometer with the same side facing the light source each time. A mark may be made near the top of the cuvet to guide the alignment.
5. The solution in each cuvet should be free of air bubbles. Inversion after capping with Parafilm will remove air bubbles.
6. Close attention should be paid to the specific details of operating each

instrument and to the method involved in the formation of the colored compound.

7. Light from windows or overhead lights can cause marked errors in readings. The cuvet well should be covered before each reading is made.

Fluorescence[1]

Measurement by fluorescence uses similar principles to those of absorption spectrophotometry except that with fluorescence the molecules which absorb the light energy suddenly reemit some of this energy as light rather than losing it all as heat, as is the case in absorption spectrophotometry. Actually, fluorescent compounds do lose a little of the absorbed energy through collisions with other molecules before the reemission of most of the rest of this energy as light. This means that the light emitted is of lower energy (longer wavelength) than the absorbed light. Since, in addition, the molecules have enough time to spin around between the absorption and emission events, the emitted light is sent out in all directions, even when the excitation occurs only from a single direction. A convenient vantage point for a detector to view or measure this emitted light is at a right angle to the incident beam. In this way, the emitted light is isolated and not superimposed on the incident light that passes through the sample unabsorbed.

The basic components of a fluorometer (Figure 4-8) are essentially the same as those in a spectrophotometer, although there are a few differences. For instance, a higher intensity lamp, such as a mercury-arc discharge or a xenon-arc lamp, is used. Also, as mentioned above, to isolate the emitted light, the detector is mounted to one side of the sample cuvet instead of behind it. In addition to the (primary) filter or monochromator mounted between the light source and the sample, another (secondary) filter or monochromator is mounted between the sample and the detector. This helps to remove stray light that arises in the sample from reflection and refraction of the high-intensity incident beam.

The amount of emitted light, called *fluorescent intensity,* is proportional to the concentration of the fluorescent compound in the cuvet. Instrument settings and emissions can be checked by using stable fluorescent standards such as a solution of quinine sulfate, sodium fluorescein, or a fluorescent block of glass. Some of the substances of clinical interest are fluorescent, or readily made fluorescent, and can be quantitated in this manner. The absorbing (primary) wavelengths generally are in the ultraviolet region, and the corresponding emitted (secondary) wave-

[1] More detailed discussion of fluorescence and related analytic methods may be found elsewhere [3–7, 9].

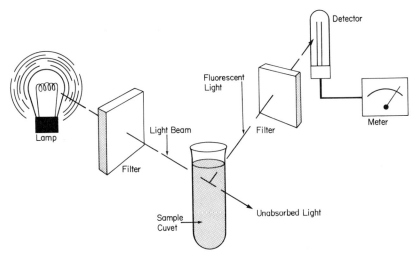

Figure 4-8
Components of a Fluorometer

lengths, which are always longer, are in the visible region. The advantage of fluorescent measurement is its high potential sensitivity, which is 10 to 1000 times that of absorption methods. However, there are precautions which must be observed:

1. The pH and temperature must be controlled. Fluorescence generally is quite sensitive to these factors, especially pH.
2. Precautions must be taken to avoid even trace levels of contaminants. Many materials at low concentrations can contribute extra fluorescence or quench (diminish) the fluorescence of other substances. Consequently all glassware must be cleaned carefully. Fluorescent detergents, chromic acid, stopcock grease, and fingerprints must be avoided.
3. All solvents must be of high purity. Solvents sometimes pick up materials, even from plastics or filter paper, which interfere with fluorescence. Nitric acid is a good cleaning agent for glassware and "fluorescent grade" solvents can be purchased or solvents can be purified by distillation or extraction.
4. Turbidity and air bubbles must be avoided because they cause spurious scattering of light.

Some clinically important substances which sometimes are measured fluorometrically are phenylalanine, catecholamines, porphyrins, cortisol, estrogens, several pharmaceuticals, and some of the metallic ions.

Turbidimetry and Nephelometry[2]

It is possible to measure the concentrations of a substance in a solution by precipitating it as a fine suspension of particulate matter and then measuring the degree to which this suspension blocks light along a given path (turbidimetry) or scatters light to the side (nephelometry). Spectrophotometers are used for turbidity readings, and fluorometers (or nephelometers) are used for nephelometry readings. Within limits, the amount of blocked or scattered light is proportional to the concentration of particles for small-sized particles. Although it can be convenient to quantitate a precipitate without separating it from the solution, this general procedure is associated with certain problems. The most serious one is the difficulty of preparing suspensions in which the particles are always of the same size. Suspensions may block different amounts of light not only because of differences in sizes of particles themselves but also because the settling rates are different for suspensions with different particle sizes. Careful timing and replication of all steps are necessary to minimize these problems.

[2] Turbidimetry and nephelometry are discussed in greater detail elsewhere [10].

References

1. Bender, G. T.   *Chemical Instrumentation: A Laboratory Manual Based on Clinical Chemistry.* Philadelphia: Saunders, 1972.
2. Cresswell, C. J., Runquist, O., and Campbell, M. M.   *Spectral Analysis of Organic Compounds* (2nd ed.). Minneapolis: Burgess, 1972.
3. Elevitch, F. R.   *Fluorometric Techniques in Clinical Chemistry.* Boston: Little, Brown, 1973.
4. Guilbault, G. G. (Ed.).   *Fluorescence.* New York: Dekker, 1967.
5. Guilbault, G. G.   *Practical Fluorescence.* New York: Dekker, 1973.
6. Pesce, A. J., Rosen, C. G., and Pasby, T. L. (Eds.).   *Fluorescence Spectroscopy.* New York: Dekker, 1971.
7. Rubin, M.   Fluorometry and Phosphorimetry in Clinical Chemistry. In Sobotka, H., and Stewart, C. P. (Eds.), *Advances in Clinical Chemistry.* New York: Academic, 1970. Vol. 13.
8. Slayter, E. M.   *Optical Methods in Biology.* New York: Interscience, 1970.
9. Udenfriend, S.   *Fluorescence Assay in Biology and Medicine.* New York: Academic, 1969. Vol. 2.
10. Willard, H. H., Merritt, L. L., Jr., and Dean, J.   *Instrumental Methods of Analysis* (5th ed.). New York: Van Nostrand, 1974.

5. Flame Emission and Atomic Absorption Spectrophotometry

The various metals present in biological fluids are best measured by flame emission or atomic absorption spectrophotometry. The two technics are similar in that they both involve photometric measurements related to energy changes of metals in vaporized samples.

Flame Emission

In flame emission analysis (emission photometry), a solution containing metal ions is sprayed into a flame. The metal ions are energized to emit light of a characteristic color. Sodium is identified by a yellow color, potassium produces a violet color, while lithium imparts a red color to a flame. The intensity of color is proportional to the amount of the element burned in the flame.

Flame photometers are instruments which utilize this principle. These emission instruments are used in clinical chemistry mostly to analyze for sodium, potassium, and sometimes lithium. Most of the salts of the body are either of sodium or potassium; lithium salts are present in detectable quantities only when administered.

The details of operation of a specific flame photometer are best found in the manual provided by the manufacturer. The general system is diagramed in Figure 5-1. Some of the components are discussed in the following sections. More comprehensive accounts of this technic are available [3, 6–8].

NEBULIZER

The purpose of the nebulizer is to create a fine spray of the sample solution and to feed this spray into a burner. Nebulizers operate basically on the same principle commonly used for spraying perfumes and cosmetics. A fine spray is produced by combining a stream of the sample with a stream of air. Two basic kinds of nebulizers have evolved, the premix and the total-consumption types. A premix nebulizer mixes the fuel gases and the sample in a mixing chamber before sending this mixture to the flame. Large droplets go to waste instead of into the flame. Approximately 90% of the sample is discarded in this manner and never gets to the flame. A total-consumption burner feeds the entire sample directly into the flame, large droplets and all. With this system, the gases and the sample are not mixed before entering the flame. This flame can be made hotter, thus potentially providing more sensitivity. However, the large droplets in the flame cause signal noise by light scattering, and the acoustical noise is quite high. Consequently most manufacturers now provide the premix type of nebulizer.

61

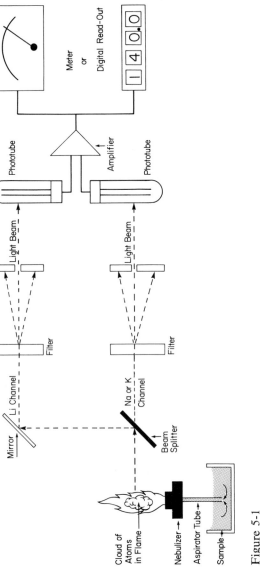

Figure 5-1
Components of an Internal-Standard Flame Emission Spectrophotometer

BURNER

Most types of flame photometers employ a burner of the Meker type that is adapted for the kind of gas to be used with the photometer.

Burners may be classified as premix or total-consumption burners, depending on the nebulizer involved. The combination of nebulization and burning is called *atomization* because the dissolved metal in the sample is converted by these two processes into a cloud of metal atoms.

Emission flame photometers usually use, as fuel, acetylene, propane, or ordinary illuminating gas (mostly methane). These gases burn with decreasing degrees of heat intensity in the order mentioned. A hot flame is advantageous because higher degrees of heat cause greater excitation of molecules and therefore greater intensities of light. This means that greater sensitivity may be achieved by using hotter burning gases. However, at higher flame temperatures, substances other than the ones of interest may emit enough light to interfere with accurate analysis.

DETECTION AND MEASUREMENT

The purpose of the electrical-optical system (Figure 5-1) is to detect and measure the intensity of the emitted light. The technics for admitting only the wavelength of light from the metal being analyzed into the detector are similar to those described for spectrophotometers in Chapter 4. Some photometers are equipped with filters, while others employ monochromators. These latter components spread or disperse the light into its spectral components. The desired wavelength is selected by means of a narrow slit. Photocells or phototubes are used to detect the light by converting it into an electrical current. The amount of current which is generated is proportional to the quantity of light which reaches the detector. The current is measured on a meter or a recorder or both.

INTERNAL-STANDARD METHOD

In certain cases a reference metal is added to both the sample and the standard solution. The instrument measures the ratio between concentrations of the sample metal and the reference metal. This is called the *internal-standard* method (Figure 5-1). The reference metal is added to samples and standards in the same concentration and serves as an internal standard.

Lithium may be used as a reference element for sodium and potassium determinations when it is not already present in the sample. A standard amount of lithium is added to both the sample and standard solutions. A specially designed instrument measures the ratio between the reference and sample metal values by the use of two detectors, one for the sample metal and the other for the reference metal. The *ratio* of the signals from the sample and reference detectors is proportional to the sample concentration. Any change in gas or air pressure, rate of nebulization, flame temperature, or amount of interfering substances affects the signal to each detector simultaneously and to the same extent. Taking the ratio of the two signals, therefore, minimizes these interferences. The resultant signal ratio is amplified and fed to a meter, a digital readout, or a recorder.

The internal-standard method provides more stable and reproducible measurements but with some loss in versatility and convenience. It is used only when large interference effects tend to occur and a convenient reference metal is available, such as lithium for determinations of sodium and potassium in serum or urine.

CALCULATIONS

Under ideal conditions, when the concentration of the metal in the sample is twice as large, the emission of light will be twice as great, and the reading on the meter also will double in value. A series of standards containing various known concentrations of the metal to be analyzed is measured to determine whether this is the case. A calibration curve can be prepared as described in Chapter 4 on spectrophotometry. This curve may be either linear or curved. Many instruments may be calibrated in such a way that the readout of the tests is directly in concentration units.

GENERAL COMMENTS AND COMMON PROBLEMS IN
FLAME EMISSION PHOTOMETRY

1. *Flame.* Proper adjustment of the flame is necessary for the smooth performance of any flame photometer. This requires careful pressure regulation for both the fuel and oxidizing (air or oxygen) gases, a correct ratio between the pressures of these gases, and a clean nebulizer bore. If the flame is not properly adjusted and controlled, instrument performance may be erratic and the readings may drift.
2. *Nebulizer.* After many samples have been analyzed, the nebulizer bore may become partly clogged with protein or metal salt residues. Even air bubbles or dust may cause clogging. This condition results in loss of sensitivity and gross changes in the readings of the instrument. The nebulizer bore should be cleaned at regular intervals with a fine wire and occasionally soaked, if necessary, in 10% (v/v) HCl or a suitable detergent.
3. *Dirt in the gas or air lines.* This situation is easily recognized by the appearance of yellow flecks of light appearing in the flame when no solution is being atomized. Cotton or glass wool filters in these lines will usually remedy the problem.
4. *Dirty burner.* Dirt particles on the burner grid usually exhibit a red or orange glow in the flame. This condition results in unstable performance. The burner grid and orifice should be washed occasionally to prevent the occurrence of this problem.

5. *Static electricity.* Some instruments need to be grounded, although the manuals do not always point out this fact. If static electricity is causing a problem, usually it can be recognized by the fact that the erratic movements of the meter correlate with touching of the instrument by the operator. Static electricity may be eliminated by connecting the chassis of the instrument to a cold-water pipe with a piece of wire.

6. *Electrical or electronic problems.* Instability or decreases in sensitivity may result from gross fluctuations in the line voltage or from faulty electronic parts and electrical connections. The line voltage may be checked with a voltmeter, and a stabilizer installed if necessary. Since photocells and phototubes have a shelf life, it is not advisable to keep spare ones on hand.

7. *Contamination.* Contamination is a common source of error in flame photometry. All glassware, stoppers, water, and standard solutions must be carefully protected from contamination, particularly from sodium or potassium.

8. *Location.* If the instrument is located near a source of variable light (for example, a window), the varying intensity of stray light may strike the detector and cause readings to drift. Drafts from windows, fans, or air-conditioning units may cause erratic performance. Excessive activity in the vicinity of the instrument may cause instability in the readings, especially if the instrument is not equipped with a closed burner system.

Atomic Absorption[1]

In atomic absorption analysis, a cloud of ground-state (unenergized) metal atoms is energized by a beam of light. This results in removal or absorption of some of the light energy. The overall system is like that of an absorption spectrophotometer (see Chapter 4) except that a cloud of metal atoms instead of a cuvet of solution is introduced into the light path. This contrasts with flame emission analysis, which measures light emitted by excited atoms rather than light absorbed by ground-state atoms. Different metals are more suitable for flame emission or atomic absorption analysis based on what happens to them in a flame. If they are atomized *and* energized by the flame, like sodium and potassium, then flame emission is recommended. If they are atomized but not energized by a flame, like zinc and copper, then atomic absorption analysis is preferred. The higher energy necessary to energize these atoms is provided by a light beam. The components of an atomic absorption instrument are illustrated in Figure 5-2.

[1] More comprehensive accounts of this technic are available [1, 2, 4, 5, 9–13].

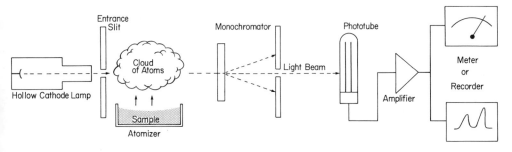

Figure 5-2
Components of a Single-Beam Atomic Absorption Spectrophotometer

Production of atoms or atomization is provided in atomic absorption instruments by flame, flameless, and semiflameless technics.

FLAME ATOMIZER

As in flame emission, both premix and total-consumption nebulizers are used in combination with flames to provide atomization. The basic principles of operation are the same as presented in the discussion of flame emission. The premix type of nebulizer is more commonly used.

FLAMELESS ATOMIZER

Flameless atomizers are small tubes or cups which can be heated to a high temperature by an electric current. A special power supply is required to produce the appropriate voltage. Cups or tubes made of carbon, graphite, or tantalum are used as sample vessels. The sample is placed in the tube or cup. The current is raised either steadily or in steps. At the lower temperature, the sample is dried. At an intermediate temperature the sample is ashed, releasing any organic matter plus any volatile inorganic substances as a cloud of smoke. Finally, the highest temperature is reached, and the metal of the sample is atomized. This process usually is programmed automatically and requires only a few seconds for each step. Throughout the process, the atomizer vessel is purged with an inert gas, like nitrogen or argon, to keep it from burning. A light beam from a lamp is directed through the cloud of atoms. Light absorption then can occur in the same manner as in a flame. Because all of the metal salt of the sample is atomized into a cloud at once and this cloud begins to disperse almost immediately, the absorption signal that results is of very short duration. This signal, therefore, usually is fed to a recorder where a pen deflects on moving chart paper, which results in a peak whose height is proportional to the amount of metal in the sample.

SEMIFLAMELESS ATOMIZER

The semiflameless system involves the use of a flame instead of an electric current to heat the sample in a tube or cup. The Delves cup procedure for lead is a good example of this technic. In this method, a sample of whole blood is placed in a nickel cup and taken to dryness on a hot plate. The cup then is placed in the flame and the metal salt is atomized into the path of the light beam directed over the top of the cup similarly to the flameless method.

HOLLOW CATHODE

The light source for atomic absorption technics is the hollow cathode lamp. It is illustrated as one of the components of an atomic absorption instrument in Figure 5-2. Each of these lamps has a lamplike cathode that contains the element that is being analyzed. For instance, when a sample is analyzed for calcium, a hollow cathode lamp with a calcium cathode is used. Some multielement lamps are available for the analysis of three or four elements. An inert gas like argon fills the lamp.

The hollow cathode lamp produces light which is characteristic of the metal in the cathode. When a potential of 600 to 1000 volts (V) is applied to the lamp, the inert gas is ionized and bombards the cathode. This sputters the metal atoms off the cathode element and creates a cloud of the metal atoms. These atoms are of the same type as those to be analyzed except for one important difference. The atoms in the hollow cathode are energized, and emit light energy as they return to their nonenergized (ground) state. This light energy constitutes the light beam produced by the lamp. It is the appropriate energy to be absorbed by the cloud of unexcited atoms in the flame because corresponding excitation and emission energies are the same for a given metal. The emission of a hollow cathode lamp consists of very narrow bands of light, that is, *line emissions*. This is one of the reasons for the high specificity of atomic absorption analysis.

DETECTION AND MEASUREMENT

For atomic absorption analysis, detection and measurement are accomplished by an electrical and optical system similar to that described earlier in this chapter for emission flame photometry and in Chapter 4 for spectrophotometry. Monochromators or filters are used to isolate the desired wavelength of light, a phototube or photocell converts the light intensity into an electrical signal, and a meter or recorder is the readout device.

INTERFERENCES

Anything which decreases or increases the cloud of atoms from a given sample constitutes a potential interference. The two general types of interferences in atomic absorption spectrophotometry are chemical and physical.

One kind of chemical interference occurs when metal salts are present which are so stable that they are not very efficiently dissociated into atoms at the highest temperature available in the burner. For instance, calcium phosphate will not dissociate into calcium atoms unless a special high-temperature burner is used. This interference can be overcome by introducing a metal salt such as lanthanum or strontium chloride. These added metals bind phosphate stronger than calcium does. Thus calcium is forced to trade its phosphate for chloride. Calcium chloride is dissociated easily into atoms at usual burner temperatures.

Physical interferences derive from physical rather than chemical phenomena. Physical blockage and scattering of light can occur, for instance, when larger droplets are present which deposit solid particles in the flame instead of invisible clouds of atoms. At higher concentrations, these particles constitute a smoke and, therefore, are the equivalent of turbidity or visible cloudiness in a spectrophotometric cuvet. Certain organic solvents like benzene are particularly unsuitable because they produce high concentrations of solid particles. The smoke problem generally is of little concern in flame technics, but is significant in flameless methods. Residual smoke from the ashing step may be present when atomization occurs, or part of the ashing process itself may continue to occur during the atomization step. Background correction may be required to compensate for this problem.

ORGANIC SOLVENTS

Nonsmoky organic solvents tend to produce some very useful physical effects. They can be used to lower the viscosity of the sample, increasing its aspiration rate and thereby generating more ground-state atoms because more sample is delivered to the flame. The result is a stronger signal, that is, increased sensitivity. In addition, such solvents may lower the surface tension of the water droplets, thereby weakening their capacity to remain intact. This increases the efficiency of the nebulization process, which results in a stronger signal. Organic solvents also may provide increased sensitivity by serving as extra fuel to increase the flame temperature.

PRETREATMENT IN ATOMIC ABSORPTION

Many types of biological samples are submitted for metal analysis: whole blood, urine, tissue, hair, and nails, for instance. Those that are not in liquid form must be dissolved prior to analysis. Even samples in liquid form may be too viscous or contain too much salt or protein or interfering substances to allow a direct accurate measurement of a particular metal. Additionally, the metal may be too concentrated or too dilute for a direct analysis. To overcome these problems, pretreatment steps may be required.

1. *Dilution* is the most important and frequently used preparatory procedure. Water or buffer is added to adjust the concentration to a preselected and desirable range. Interfering substances may be sufficiently reduced in concentration so that they no longer interfere and the viscosity is also suitably reduced. The principle disadvantages are that dilutions may introduce errors from pipetting or reagent contamination or may lower the concentration of the substance that is being measured to below the limit of sensitivity of the technic.
2. *Solvent extraction* may be employed to separate and concentrate specific metals from a sample which is unsuitable for direct analysis. One example is the APDC-MIBK system (ammonium pyrolidine dithiocarbamate/methylisobutylketone). The MIBK is immiscible with water and forms a second (organic) phase. The APDC chelates (binds) the sample metals in the water phase but is more soluble, free, or metal complexed, in the organic MIBK phase. Consequently, the metals are extracted into the MIBK phase. When none of the other extracted metals interfere with the metal to be analyzed, the organic solution may be analyzed directly.
3. *Wet ashing* of a solid sample may be used to eliminate the interfering effects of biological macromolecules like proteins. The sample usually is heated vigorously in a mixture of nitric, sulfuric, or perchloric acids, or a mixture of them, until it is dissolved and all the large molecules have been broken down into small ones (digested).
4. *Dry ashing* is similar to the wet-ashing technic. The sample is placed in a heat-resistant container like a platinum dish and heated in a muffle furnace (usually 400 C to 700 C) until all of the organic matter is volatilized. Subsequently the ash of inorganic salts is dissolved (in dilute acid, for example) and analyzed.

COMPARISON OF FLAME AND FLAMELESS ABSORPTION METHODS

The flameless and semiflameless methods usually are 1000 times as sensitive as flame procedures for several reasons. First of all, in the premix burner, a large proportion (approximately 90%) of the sample goes down the drain, whereas the entire sample is atomized in flameless methods. Secondly, all of the sample is taken into the flame at once in the flameless technic, whereas the sample is gradually aspirated when a flame is used. Finally, the sample atoms can remain in the optical path for a second or more in the flameless technic, whereas, with a flame, these atoms are blown away in a fraction of a second.

Both technics have their place in the laboratory. Some commercial instruments currently offer both capabilities. The flameless method offers greater sensitivity, eliminates certain interferences, avoids fuel tanks, and makes a pretreatment procedure unnecessary in some cases. However, the burner retains other advantages. It is easier to use, requires no special power supply, allows a greater speed of analysis, and provides higher precision. The flame method may be used where the highest sensitivity is not required and the sample volume is adequate.

GLASSWARE AND PLASTICWARE

Glassware requires special cleaning for use in atomic absorption analysis. Ordinary washing with detergents may add more metal contamination than it removes. Glassware which is clean in the ordinary sense can be made metal-free by soaking overnight in 10% (v/v) nitric acid. Tap water may be used to thoroughly rinse off the acid, but at least three final rinses with purified water should follow. The glassware may be dried immediately in a dust-free oven, then covered to keep dust out. Faster drying may be achieved by rinsing the glassware with reagent acetone. The use of disposable glassware is highly recommended, but such glassware should be checked for contamination before use. This may be done by adding small amounts of water to some tubes or vials, swirling, then atomizing the water under the appropriate conditions and watching for any instrument response.

Glass pipets may be taken conveniently through the washing and rinsing procedure by the use of a polyethylene pipet-washing container.

Plasticware is less susceptible to metal contamination. Some manufacturers offer disposable plasticware which is metal free and can be used directly. It is worth the expense to buy disposable plastic tubes for metal analyses rather than labor through the acid-soaking procedures. For cleaning nondisposable plasticware, such as bottles and graduated cylinders, the procedure described for glassware can be used.

Calibration Curves

Atomic absorption and flame emission are comparative methods. The response of the instrument to a test solution is compared with its response to several standard solutions of known concentration. At least one standard solution should be less concentrated than the most dilute sample, and at least one standard solution should be more concentrated than the most concentrated sample. A plot of the absorptions versus the standard solution values constitutes the calibration or standard graph or curve. The graph may be straight or curved. It may change slightly day to day or even suddenly during the course of a set of analyses. Consequently, it is advisable to intersperse standards with the test solutions. Any difference in the chemical composition and physical properties of the standards and the sample except the concentration of the metal under analysis is a potential source of error. This includes such variables as protein concentration, salt concentration, pH, metal complexing agents, and viscosity. Any such differences must either be proved to not influence the readings or must be removed by making the chemical and physical composition of the standards and samples the same except for the metal under analysis.

Calculations

The reading on the meter or recorder should double whenever the concentration of metal in the sample is twice as great because twice as much light should be absorbed by the sample under these conditions. This can be checked by measuring a series of standards and constructing a calibration curve as discussed in Chapter 4 on spectrophotometry. As is the case in flame emission, this graph may be linear or curved. Ideally, the solution used for the standards gives the same relative response as the samples. When this is not the case, it may be necessary to use the method of standard additions [13].

General Comments and Common Problems

Almost all of the general comments and common problems raised in the discussion of flame emission in this chapter also apply to atomic absorption. Particular attention must be paid to contamination problems in atomic absorption because of the higher sensitivity of the method. It is important to run standards frequently because of sudden variations which may occur in the nebulization, sample flow, gas flow, and the electrical and optical components of the system.

References

1. Dean, J. A., and Rains, T. C. (Eds.). *Flame Emission and Atomic Absorption Spectrophotometry*. New York: Dekker, 1969. Vol. 1.
2. Elwell, W. T., and Gidley, J. A. F. *Atomic Absorption Spectrophotometry*. New York: Pergamon, 1967.
3. Hald, P. M. Determinations with Flame Photometer. In Visscher, M. B. (Ed.), *Methods in Medical Research*. Chicago: Year Book, 1951. Vol. 4.
4. Hermann, R., and Alkemade, C. T. J. *Chemical Analysis by Flame Photometry* (2nd ed.). New York: Interscience, 1963.
5. Kahn, H. L. Atomic Absorption and Flame Emission Spectroscopy. In White, W. L., Erickson, M. M., and Stevens, S. C. (Eds.), *Practical Automation for the Clinical Laboratory* (2nd ed.). St. Louis: Mosby, 1972. P. 141.
6. MacIntyre, I. Flame Photometry. In Sobotka, H., and Stewart, C. P. (Eds.), *Advances in Clinical Chemistry*. New York: Academic, 1961. Vol, 4.
7. Margoshes, M. Emission flame photometry. *Anal. Chem.* 34:221R, 1962.
8. Margoshes, M., and Vallee, B. L. Flame Photometry and Spectrometry. In Glick, D. (Ed.), *Methods of Biochemical Analysis*. New York: Interscience, 1956. Vol. 3.

9. Ramirez-Munoz, J.   *Atomic Absorption Spectroscopy*. New York: American Elsevier, 1968.
10. Robinson, J. W.   *Atomic Absorption Spectroscopy*. New York: Dekker, 1966.
11. Slavin, W.   *Atomic Absorption Spectroscopy*. New York: Interscience, 1963.
12. Teloh, H. A.   *Clinical Flame Photometry*. Springfield, Ill.: Thomas, 1959.
13. Willard, H. H., Merritt, L. L., Jr., and Dean, J.   *Instrumental Methods of Analysis* (5th ed.). New York: Van Nostrand, 1974. P. 350.

6. Automation and Kits

As the emphasis in the practice of medicine shifts from art to science, the demand for laboratory data continues to grow. It is virtually impossible to perform with manual technics the large numbers of some chemical tests that are ordered daily. Hence automation is essential to the clinical chemistry laboratory. The first practical automation was introduced into the clinical laboratory in 1957 [4]. Since then numerous other types of systems have been devised [2, 3, 5]. Although these instruments may differ from each other in many ways, they all represent *mechanization* of conventional manual technics.

The operating details for each instrument are best found in manuals supplied by the manufacturers. This chapter deals largely with the principles of the various types of automation in use and some of their advantages and disadvantages.

Essentially there are two basic types of automated systems, *continuous flow* and *discrete analysis*. In the continuous flow system, samples follow each other in sequence through a channel. In discrete analysis, each sample occupies a separate container, and the containers are tested in parallel or in sequence.

Continuous Flow Analysis

The heart of a continuous flow system (Figure 6-1), such as the Technicon AutoAnalyzer,[1] is a peristaltic pump composed of steel roller rods which compress several pieces of plastic tubing by rolling on them. By this action, fluids are drawn into the tubing and pushed through the system. Each test requires a manifold of plastic tubes, one for the sample and others for the reagents. The amount of each sample or reagent used in a method is determined by the inner diameter of the tubing employed.

The samples are loaded into a rotating sample tray from which they are picked up in sequence by a probe (a thin metal or plastic tube) attached to a manifold tube. The samples follow each other into the reagent streams, interspersed with a wash solution.

Strange as it may seem, the success of the system depends entirely on small air bubbles injected into the sample and reagent streams at strategic points. These bubbles segment the stream and through a squeegeelike action on the tubing keep the segments intact. This reduces the possibility that a sample may mix with the one following it. At the same time it improves the efficiency of the mixing of reagents in each small segment. Mixing is enhanced by directing the stream through glass coils. This creates a tumbling action in each segment thereby causing the sample and reagents to mix.

The portion of the continuous flow system described thus far is used in all methods for introducing samples and reagents into the system.

[1] Technicon Corp., Ardsley, N.Y.

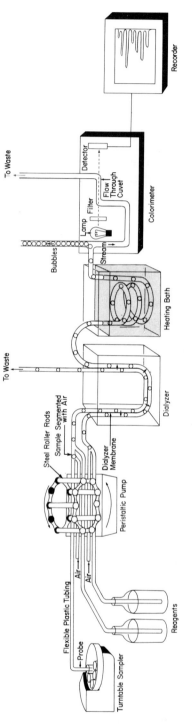

Figure 6-1
Continuous Flow Analysis System

Subsequently, the stream flows through various fittings and modules designed to carry out the required chemical reactions and readings.

If deproteinization is required, the stream containing the sample is directed through a semicircular groove on one side of a dialysis membrane. A stream of recipient solution flows through a similar groove on the other side of the membrane. The solute that is of interest dialyzes into the recipient stream, leaving the proteins behind.

If a period of time is required for a reaction to take place, the stream is directed through a time-delay coil. This is a plastic or glass coil, the length and inner diameter of which determine the amount of time it will take the stream to pass through. If heating is required, a time-delay coil is immersed in a heating bath maintained at a desired temperature by a thermoregulator.

After the reaction has taken place, the stream flows through an appropriate instrument for measurement. This usually is a spectrophotometer, a photofluorometer, or a flame photometer. The signals generated in these instruments activate a strip chart recorder pen to give visual signals, or peaks, the sizes of which are related to the concentrations of the reactant in the samples. A series of standards accompanies each set of tests to give a standard curve. The concentrations of the tests are determined from the standard curve.

Discrete Analysis

As mentioned earlier, *discrete analysis* involves the treatment and measurement of samples in individual containers. Whereas the continuous flow system employs a proportioning system for measuring samples and reagents, discrete systems employ automatic pipetting devices. The samples are loaded into a sampler tray and pipetted by the instrument into reaction tubes. Depending on the type of instrument, the samples may be pipetted in sequence or in parallel. The reaction tubes then move through various stations where more reagents are added and other operations are carried out, for example, mixing, heating, etc.

A sample of the final solution is then removed automatically and presented for measurement to a spectrophotometer, photofluorometer, or flame photometer. Sometimes the reaction vessel itself serves as a cuvet. The readout device may be a recorder or a printout of the final results.

The *centrifugal fast analyzer* [1] is a discrete analyzer which differs from the types described above (see Figure 6-2). The samples and reagents are measured by an automatic pipetter into separate compartments of a Teflon wheel (transfer disk). The disk is placed in the instrument where it rotates at a fixed speed. During the spinning period, the reagent flows by centrifugal force into the sample compartments. The treated samples then flow into cuvets located in a rotor around the outside rim of the disk. The electronics of the instrument are so designed

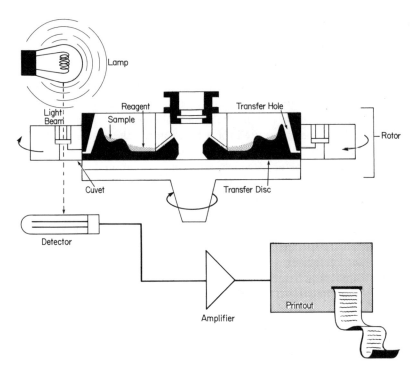

Figure 6-2
Centrifugal Fast Analyzer Concept

that, as these cuvets pass through the vertical light beam of a spectro-photometer, readings are made in rapid sequence and printed out as requested. This permits a single endpoint to be read a number of times or a kinetic reaction to be followed with almost continuous readings. The disk and rotor are diagramed in Figure 6-2.

General Considerations

Automated systems should be recognized as mechanizations of the usual technical steps common to many chemical methods. Table 6-1 lists some common requirements of chemical technics and their automated counter-parts.

Once initial conditions are set, the well-designed automated system maintains or reproduces these conditions with great precision. To ensure *accuracy* of results, frequent standardization of the methods is essential.

Whether sample and reagent measurements are made with automatic pipets or by proportioning (as in the continuous flow system) the important consideration is *repeatability*. The absolute amounts measured are not critical (within certain broad limits) as long as the measurements are precisely reproducible.

Table 6-1. Automation of Manual Technics

Manual Step	Automation
Pipetting samples and reagents	Sampler, Pump, Pipetter
Mixing	Mixing coil, Air, Mechanical mixer
Deproteinization	Dialyzer, Resin column
Heating or incubation	Heating bath or block
Development time	Delay coil, Conveyer travel
Reactant measurement	Flow-through spectrophotometer, Fluorometer, Flame photometer
Reading and calculation	Recorder, Printer, Computer

ADVANTAGES

The advantages of automation are obvious; a few of these are summarized below. These advantages, of course, are stated on the assumption that the instrument or system is basically sound.

1. Large numbers of samples may be processed with minimal technician time.
2. Two or more methods may be performed simultaneously.
3. Precision is superior to that of manual performance.
4. Calculations may not be required.

DISADVANTAGES

Sometimes laboratory personnel are so impressed with the appearance and operation of automated equipment that they fail (or refuse) to note its shortcomings. Some of the problems common to many automated systems are stated below.

1. There are limitations in the type of methodology which can be used. Sometimes a compromise is made in the chemistry which results in less accurate (although possibly more precise) values than with manual technics.
2. The automatic nature of these instruments discourages technicians from making observations and exercising discretion regarding potential problems.
3. The instruments are uniformly objective so that they cannot exercise the judgment expected from an experienced technician in the case of a potentially interfering turbidity or a spurious color, for instance, which can relate to a given sample or to an entire set of tests.
4. Many systems are impractical to use for small numbers of samples. Therefore backup manual methods may be required for individual emergency analyses. Backup methods must also be available in the event of instrumental failures.

5. The systems are expensive to purchase and maintain. Regular mainte- nance schedules require technician time as well as visits from trained service personnel.
6. The relative ease with which automated systems produce results en- courages the accumulation of large amounts of irrelevant data, which must be calculated, recorded, and stored.

Kits

Many methods are available commercially as kits. A kit is a package which contains either some or all of the reagents and components neces- sary to carry out a test. Some kits are well designed and reliable while others are not and should be avoided. The purchase of a kit means that the laboratory buys not only whatever expertise the manufacturer may possess in regard to the test, but also whatever errors, inaccuracies, and general lack of quality that may exist in the kit. The better kits tend to be accompanied by manufacturer's literature which provides:

1. Detailed directions.
2. A statement of the chemical and physical principles involved.
3. Literature references.
4. A description of the calculations and a sample calculation.
5. A typical calibration curve (if applicable).
6. Normal values obtained with the kit.
7. Statements regarding the accuracy and precision to be expected.
8. A statement about the stability of the reagents and recommended storage conditions.
9. A performance evaluation from a reputable laboratory.
10. An assurance that further technical information and assistance will be available from the company if requested. A telephone number and a name are satisfactory, but a sympathetic salesman is not.
11. The number of actual samples which can be analyzed and the actual cost per sample. This information is not always obvious, because it depends on whether the samples are to be performed singly or in duplicate, how many *batches* will be tested, and how many controls and standards will accompany each batch.

Beyond the issue of reliability, the decision as to whether to adopt a kit depends on the cost and convenience involved. Tests which are per- formed infrequently or which are difficult to set up and maintain tend to be more costly and inconvenient. They are the ones for which kits sometimes should be considered.

Sometimes commercial kits are evaluated by professional organiza- tions such as the American Society of Clinical Pathologists or the American Association of Clinical Chemists and the results are published in journals such as the *American Journal of Clinical Pathology* or

Clinical Chemistry. These evaluations are usually objective and informative.

References

1. Anderson, N. G.   The development of automated systems for clinical and research use. *Clin. Chim. Acta* 25:321, 1969.
2. Broughton, P. M. G.   A Guide to Automation in Clinical Chemistry. *Association Clin. Biochem. Tech. Bull.* May 16, 1969.
3. Broughton, P. M. G., and Dawson, J. B.   Instrumentation in Clinical Chemistry. In Sobotka, H., and Stewart, C. P. (Eds.), *Advances in Clinical Chemistry.* New York: Academic, 1972. Vol. 15.
4. Skeggs, L. T., Jr.   An automated method for colorimetric analysis. *Am. J. Clin. Pathol.* 28:311, 1957.
5. White, W. L., Erickson, M. M., and Stevens, S. C.   *Practical Automation for the Clinical Laboratory.* St. Louis: Mosby, 1972.

7. Separation Technics

Biological fluids are extremely complex in composition. For instance, there are literally hundreds of detectable substances in urine [16]. Chemical analysis would be impossible if it were necessary to completely isolate each substance prior to its measurement. An optimal method tests for a specific substance in the presence of all of the others, requiring no isolation of the substance under analysis. A test is specific when none of the other substances present interfere. However, virtually all chemical tests are subject to at least some interferences. This is one of the most important problems in clinical chemistry. In some cases the interferences are sufficiently small or constant so that they do not significantly affect the accuracy or precision of the test. However, in many cases the interferences do affect the results. Therefore some type of separation procedure is required. The substance being analyzed does not have to be isolated from *all* of the other materials in the biological specimen, but at least from the ones which interfere significantly.

It is possible to separate only substances which differ from each other. The more they differ, the easier it is to separate them. Separations in clinical chemistry usually are based on differences in the size, solubility, or charge of the substances involved. The usual basis and a typical description of each of the five kinds of separation commonly employed are given in Table 7-1. Each kind of separation listed is discussed in detail in this chapter. Further discussion of these and other separation procedures is available elsewhere [1, 2, 6, 9, 10, 13].

Precipitation

Many separations are accomplished by *precipitation*. Either the substance being analyzed or the interfering materials are precipitated. The precipitate is isolated by filtration or centrifugation. Paper and scintered glass are useful filtration materials. Both gravity and vacuum (or pressure) flow are sometimes used to aid filtration. When the precipitate contains the substance being tested, the analysis usually is completed either by a turbidimetric measurement of the precipitate (based on the capacity of the precipitate, when suspended, to scatter light) or by a subsequent reaction and measurement after it has been dissolved. Gravimetric methods requiring drying and weighing of precipitates are seldom used in clinical chemistry because the amounts of precipitate usually are very small and the procedures are time consuming.

A number of tests are subject to interferences from proteins. Because proteins are easily precipitated, precipitation technics employing heat, acids, bases, organic solvents, salts, metal ions (or a combination of these technics) frequently are used for their removal.

Three common methods for the precipitation of proteins, providing protein-free filtrates, are the Folin-Wu, trichloroacetic acid (TCA) and

Table 7-1. Separation Technics

Technic	Property	Description
Precipitation	Solubility	Some of the substances precipitate while the others remain dissolved.
Ultrafiltration or dialysis	Molecular size	Some of the substances pass through a layer or sheet of porous material, while the other substances are retained.
Extraction	Solubility	Some of the substances dissolve (partition) more in water, while other substances dissolve more in organic solvent in contact with the water.
Chromatography		
Thin-layer or column	Solubility	Some of the substances dissolve (partition) more in the immobile film of water on a solid supporting medium (or stick more to the exposed areas of the solid supporting medium) while the other substances dissolve more in the surrounding film of flowing organic solvent.
Gas-liquid	Solubility	Some of the substances dissolve more in the immobile film of wax- or oil-like material on a solid supporting medium, while the others dissolve more in the surrounding stream of flowing gas.
Gel filtration	Molecular size	Some of the substances diffuse into the pores in a porous, solid material while others remain outside in the surrounding stream of flowing water.
Ion-exchange	Electrical charge	Some of the substances are bound by immobile charges on the solid supporting medium while others are not bound.
Electrophoresis	Electrical charge	The substances with more charge move faster and, therefore, further. Substances with opposite charges move in opposite directions.

Somogyi-Nelson methods. The first two methods are presented in detail on page 153 for nitrogenous compounds, and the third is presented on page 147 for glucose.

In the Folin-Wu method [5], the proteins are precipitated with tungstic acid (H_2WO_4) produced by successive additions of exact quantities of sulfuric acid (H_2SO_4) and sodium tungstate (Na_2WO_4) to the blood or serum sample. The complete mode of action of this precipitating agent is not understood, but it usually provides a clear filtrate that is suitable for the analysis of a number of substances.

Trichloroacetic acid (CCl_3COOH) causes proteins to precipitate because the positively charged groups on the proteins apparently are neutralized by $CCl_3CO_2^-$ groups, and the negatively charged groups on the proteins are neutralized by the H^+ from TCA. Because the proteins then have less charge, they are less able to remain in solution. So they precipitate. Although this reagent is simpler to use than the Folin-Wu reagents, it may give filtrates which are not as clear. In addition, the acid pH of the filtrate may be less desirable in some methods.

The Somogyi-Nelson method [12, 17] involves successive addition of aqueous solutions of barium hydroxide and zinc sulfate to the sample. The proteins are precipitated as zinc salts, particularly so because these salts stick (adsorb) to the barium sulfate precipitate which forms. This method is comparable to the Folin-Wu tungstic acid method in providing a very clear, protein-free filtrate. Also it removes some other substances besides proteins (sulfhydryl compounds, uric acid, and some creatinine) which might otherwise interfere in the copper-reduction methods for glucose. Thus, the Somogyi-Nelson filtrate is particularly recommended for use with these latter methods.

It is important to examine the filtrate carefully after a protein precipitation since any cloudiness or foaming from shaking indicates incomplete removal of protein. Sometimes complete precipitation may be enhanced by placing the solution in ice water for a few minutes after the precipitant is added.

Ultrafiltration and Dialysis

The filtration process can be used to separate materials when all are dissolved, as long as they are not of the same size. Scintered glass or ordinary paper cannot be used in this process because the pores in these media are much too big to hold back even the largest of most dissolved molecules. Special materials with much smaller pores called *membranes* must be used. The two basic processes which employ membranes to separate large and small dissolved molecules are *ultrafiltration* and *dialysis*. Whereas dialysis emphasizes only the separation of dissolved molecules, ultrafiltration commonly is used to remove particulate matter as well as to separate dissolved materials. Both methods involve the selective passage of dissolved substances through semipermeable membranes. In dialysis, these molecules diffuse through the pores more or less freely. In ultrafiltration, they are driven through by pressure or pulled through by vacuum.

Dialysis conventionally employs cellophane (regenerated cellulose) membrane, which comes as sheets or tubing. A classical application is in the removal of low-molecular-weight substances (for example, salts) from a protein solution. The solution is loaded into dialysis tubing tied off at one end. The tubing then is tied off at the other end to form a bag. This bag of dissolved protein and salt then is suspended in a container filled with water or buffer. The salts diffuse out, but the protein remains inside the bag. This works quite well for large proteins, but small proteins (molecular weight less than 15,000) will slowly be lost to the outside solution because some of the pores in the bag will be large enough to pass materials of this size unless a special smaller pore-size dialysis membrane is used. Dialysis is employed in continuous flow automated systems (discussed in Chapter 4) and to purify or concentrate samples prior to manual analysis.

Ultrafiltration membranes are made from a variety of materials, most of which are cellulose derivatives (for example, mixed cellulose esters, cellulose acetate, cellulose nitrate) or synthetic polymers (for example, polyamide, polyvinyl chloride). A wide range of pore sizes is available with diameters as small as 0.001 to 0.005 μm. (A human hair has a diameter of about 80 μm.) Such ultrafiltration membranes will retain substances in the range of 500 to 10,000 mol wt. However, the exact degree of retention depends on the specific charge, size, and shape of the material being ultrafiltrated. The 0.2-μm membranes may be used to sterilize solutions because they retain bacteria.

Ultrafiltration membranes have been formed into sheets, disks, hair-thin hollow fibers, and conical filter cones. In some cases, the sheets are backed with absorption pads to cause passage of solvent and small solutes through the membrane based on the capillary uptake of the solution by the absorption pad. The disks can be mounted in syringes or on filter cones. The hollow fibers are arranged in bundles, like a handful of straw, and solutions either are filtered from the inside-out or outside-in. The conical filter cones are loaded with solution, placed in a centrifuge tube, and spun. Most or all of the solvent and small substances travel through to the bottom of the tube. Some typical applications of utrafiltration in clinical chemistry include desalting, concentration, fractionation of protein solutions, and the preparation of protein-free filtrates. For instance, urine and cerebrospinal fluid usually do not contain protein in high enough concentration for electrophoretic analysis, but the specimens may be concentrated sufficiently using ultrafiltration.

The general advantage of ultrafiltration over dialysis is speed, but loss of material, particularly proteins, through adsorption to the membrane can be a problem. When it is not essential to recover all of the protein, and rapid concentration is important, the ultrafiltration materials can be very useful.

Extraction

Few of the material components of biological fluids are more soluble in organic solvents than in water. Therefore extraction of a urine or serum sample with an organic solvent isolates a limited number of components in the organic phase, leaving many interfering substances behind in the aqueous phase. This technic also finds application in the separation of many drugs because drugs tend to be soluble in organic solvents. It may be necessary to adjust the pH of urine or serum to a level where the drug is not ionized (has no charge) in order to make it soluble in the organic phase. Sometimes a salt is added to reduce the solubility of the drug in the aqueous phase and help drive it into the organic solvent. A particular advantage and often essential feature of the extraction process is the opportunity to concentrate the substance under analysis. For instance, all of a drug may be extracted in a separatory funnel from 10 ml of urine with 20 ml of a volatile organic solvent. The organic solution then may be concentrated to a volume of 1 drop in the bottom of a conical tube by placing the tube in a warm water bath under a gentle stream of air to rapidly evaporate the solvent. This more concentrated solution often is required for detection of small amounts of drugs.

Chromatography

Chromatography is similar to extraction except one phase is a coated or uncoated solid support, while the second phase is a flowing gas or liquid.

The use of a noncoated solid support defines *adsorption chromatography*. The molecules to be separated are adsorbed part of the time on the surface of the solid, and part of the time are dissolved in the flowing (mobile) part.

Another general kind of chromatography is *partition chromatography*. Here, the solid support (powdered, granular, bead, or sheetlike in form) is coated with a film of water or *nonvolatile* organic liquid (mobilized film or stationary phase). The molecules to be separated are partitioned between the immobilized film and the flowing (mobile) liquid or gas. (This is illustrated later for gas chromatography in Figure 7-4.) Thin-layer chromatography and gas-liquid chromatography are common examples of partition chromatography. Some adsorption effects may be present to at least a small degree, however, because some of the solid support usually is exposed (not covered with the immobilized film) and, therefore, can interact directly with the sample. It is important that the immobilized liquid film essentially be completely insoluble in the flowing phase or the flowing phase will wash the film off the solid support. This is similar to the requirement for immiscible solvents in an extraction process. In either case, partitioning the molecules being separated requires that two phases be available. When both phases are liquids, usually the stationary phase is water and the mobile phase is a volatile organic solvent. However, the use of a gas as the mobile phase in gas chromatography does allow a nonvolatile organic liquid (for example, an oil or a wax) to be employed as a stationary phase.

Chromatographic separation is based on the different amounts of time the components of a mixture spend in the flowing phase versus the immobilized or stationary phase. The more time a given component spends in the flowing phase, the faster it will flow or travel and emerge (elute) from the system. Conversely, the more time a given component spends in the stationary phase, the slower it will travel through the system, and the later it will elute.

PAPER CHROMATOGRAPHY

Paper, which is composed of cellulose, is a very popular chromatographic medium. The semirigid, polymeric matrix of cellulose is the solid portion or solid support. Although paper feels dry, it always has some water (up to 20% under ordinary conditions) tightly adsorbed as a film or coating over its cellulose framework. This water sticks tightly because it is hydrogen bonded to all of the alcoholic (OH) groups on the cellulose. This is why paper can become sticky in very humid weather and why water will wet it so rapidly.

In order to function satisfactorily as a chromatographic medium, paper must be specially processed into a uniform structure and thickness. The cellulose framework of paper provides necessary pores which any chromatographic medium must possess so that molecules can travel through it. The paper used in chromatography is a high quality filter paper.

Thus the stationary (immobilized) phase in paper chromatography is the film of water bound to the cellulose. An organic solvent or mixture of such solvents is used as the mobile phase. The substances being separated will partition between the bound water and the flowing organic solvent. The more soluble a component is in the flowing solvent, the faster and further it will move along the paper. Substances which are more soluble in water do not move very far. The sample mixture is spotted onto the paper, dried, then transported across the paper by a flowing solvent. This process is called *development*. The solvent flows either because of the force of gravity (*descending chromatography*) or because of capillary action (*ascending chromatography*). The solvent front is marked with a pencil and, after the paper is dried, the positions of the compounds present in the mixture are made visible by a staining reaction. The entire procedure is represented in Figure 7-1. The ratio of the distance moved by a compound to that moved by the solvent front is known as the R_f value. This is illustrated in Figure 7-2.

$$R_f = \frac{\text{distance from the origin to the front of the spot}}{\text{distance from the origin to the solvent front}}$$

The origin is the middle point of application of the sample. The R_f value is only approximately constant for a particular compound. It may vary with the composition of the solvent system, the batch of paper, the concentration of the sample, the size of the chamber, humidity, temperature, and pH. Consequently, it always is best to simultaneously include standards next to or even added to a second application of the test, thereby providing exact reference information each time.

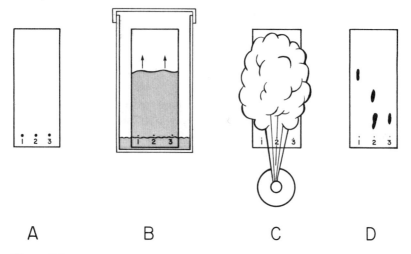

A   B   C   D

Figure 7-1
Paper or Thin-Layer Chromatography. (A) Samples (1, 2, 3) are applied
to the sheet. (B) The sheet is placed in a tank containing a small amount
of organic solvent. Capillary action draws the solvent up the sheet. The
solvent carries compounds with it to different extents. (C) After being dried,
the sheet is sprayed with a staining reagent. (D) The stained compounds
appear as spots. Samples 1 and 3 show one compound each, while sample
2 shows two compounds.

Quantitation of spots, that is, of developed compounds, is achieved by
cutting the spots out and eluting the material with a solvent. In some
cases, the piece of paper containing the spot is dropped into a cuvet
filled with solvent, and a spectrophotometric reading is made after the
substance has been eluted into the solvent and the paper has sunk to the
bottom of the cuvet.

Paper mostly is used for separation of polar molecules with molecular
weights less than a few thousand. An important application along these
lines in clinical chemistry is the separation and identification of amino
acids and drugs in biological samples.

THIN-LAYER CHROMATOGRAPHY [14, 18, 21]

Thin-layer chromatography (TLC) is similar to paper chromatography.
A flat sheet of chromatographic material is used in each technic, and the
basic steps are the same as those illustrated in Figure 7-1. TLC has the
advantage that a variety of supporting media (solid supports) can be
used. The method generally is more rapid and provides more compact
spots than paper chromatography does. The more concentrated spots,
of course, are easier to detect and discriminate. A wider range of detection reagents and conditions can be used because some of the thin-layer

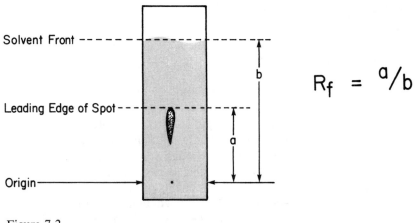

$$R_f = {}^a/b$$

Figure 7-2
Determining the Value for R_f

materials will tolerate corrosive sprays (for example, a mist of an acidic oxidizing agent) and elevated temperatures which would destroy paper.

Typical solid supports used in TLC are silica gel (silica or silicic acid, a polymer of H_2SiO_3) and alumina (aluminum oxide + aluminum hydroxide matrix), both of which are polymeric. Each is cast as a thin continuous film of powder usually on a backing sheet of glass, plastic, or aluminum. The chromatographic separation takes place entirely in the powder layer, the inert backing sheet providing mechanical support. The thin layer of powder is bonded together and to the backing sheet by the addition of a binding agent like plaster of paris ($CaSO_4$), starch, or some inert organic glue. Both silica and alumina tend to maintain a film of water on their surfaces, just like cellulose. Organic solvents are employed as the mobile phase in the development step. Qualitative identification is based on the resultant colors and relative positions of the developed spots after having been made visible (color formation) with a spray reagent. The section of powder that contains the spot of interest can be removed and extracted with a solvent for quantitation.

TLC also is used for the same types of compounds as is paper chromatography. Because of the fundamental differences which exist between paper, alumina, and silica gel media in regard to acidity, basicity, adsorption effects, nature and extent of bound water, and other peculiarities, each of these chromatographic media has unique separatory powers. Failure to separate certain compounds in one of these media does not mean that the others also will be ineffective. Although certain general principles aid in the choice of which chromatographic medium and solvent system to employ, trial and error frequently must be used.

TLC media which provide separation by ion exchange or gel filtration are also available. The principles of these technics are discussed later in this chapter.

COLUMN CHROMATOGRAPHY

Column chromatography is used to scale up (run at a preparative level) those separations which are achieved by TLC. As in TLC, silica gel or alumina usually are the chromatographic media, except that they are packed into a glass tube and usually saturated thoroughly with the developing solvent before the sample is introduced into the system. At the start of the procedure, the sample, in solution, is placed on the media at the top of the column. The developing solvent then is placed in a reservoir over the column and allowed to pass through the column by the force of gravity. Ideally the components of the sample will separate into narrow bands as they move down the column at different rates. The desired components may be eluted (removed) from the column selectively with a special solvent. Sometimes the compound of interest may appear only in a certain segment of the eluate. In this case, the segment desired must be determined in advance by calibrating the column. Then only this fraction is saved for analysis.

This technic finds limited use in the clinical chemistry laboratory because it is slow and tedious and many of the substances in biological samples are too polar to be eluted from the column. Furthermore, there is not a great deal of need for preparative procedures in clinical chemistry because usually a sufficient quantity of material can be separated by TLC and quantitated as previously described in this chapter. Sometimes, however, the column material (for example, silica or alumina) simply can be added to the sample to adsorb either the component being measured or the interfering substances. This is called a *batch* method. If the substance being measured is adsorbed, the solid adsorbent is centrifuged, washed, and then treated with a solvent which dissolves (elutes) the substance from the solid. Alternatively, and somewhat more easily, when the interfering substances are adsorbed, the supernatant (after centrifugation or just standing) or the filtrate (after filtration) can be used directly for subsequent analysis. Removal of interfering substances by means of charcoal is carried out in this manner.

GAS CHROMATOGRAPHY [19, 20, 22]

Most gas chromatography (GC) is gas liquid chromatography (separation based on partition) rather than gas solid chromatography (separation based on adsorption). Unless specifically qualified, the abbreviation GC, therefore, refers to gas liquid chromatography, sometimes abbreviated GLC.

The principles of GC are analogous to column chromatography except that a steady flow of an inert gas (carrier gas) like helium or nitrogen is used instead of a solvent. Although GC requires more

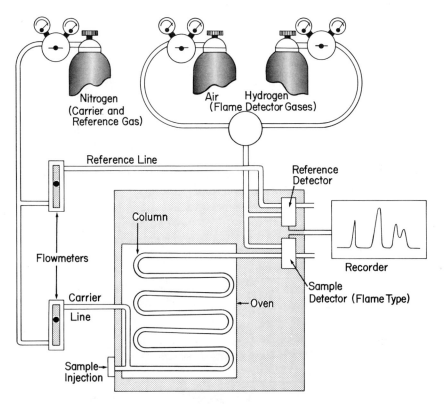

Figure 7-3
Components of a Gas Chromatograph

expensive instrumentation than the other chromatographic methods, it can be particularly advantageous in providing fast, high-resolution analyses coupled with sophisticated quantitation. A compound must be volatile to be measurable by GC; this is an important limitation. Chemical pretreatment can convert small, nonvolatile substances into volatile derivatives, but this adds an extra process to the overall procedure.

Most GC instruments consist of the following components, as illustrated in Figure 7-3:

1. Tank of inert carrier gas.
2. Heated injection port for volatilizing the sample when it is introduced into the system.
3. Oven with variable heat control.
4. Metal or glass column, packed with appropriate material, where the separation takes place. The column is located in the oven.
5. Detector (with fuel if required).

6. Amplifier.
7. Recorder.

The column, which may be straight or coiled, is packed with an inert granular material. A crude form of silica, called *diatomaceous earth,* often is used. It is derived from the geological beds of the silicons, the skeletal remains of tiny sea creatures or diatoms. This solid support is coated with a nonvolatile organic liquid, like a silicone polymer or alcoholic wax, called the *stationary phase.* After appropriate purification and formation of a derivative, if necessary, the sample is drawn into a syringe, then injected into the system through a rubber septum (disk) at the heated injection port, where it is immediately volatilized. A moving stream of inert gas (carrier gas) carries the sample into the heated column. Those components that stay entirely in the gas are swept quickly through the column and emerge immediately. Those components that dissolve in the stationary liquid phase on the solid support are retarded and emerge later. This is illustrated in Figure 7-4. The squares represent molecules of a rapidly eluting component because these molecules mostly avoid the stationary phase. The triangles represent mole-

Step I· Sample Injected into Column

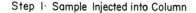

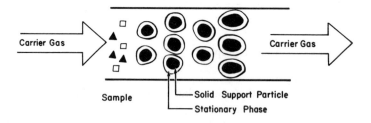

Step 2· Separation Based on Unequal
Solubility in Stationary Phase

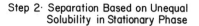

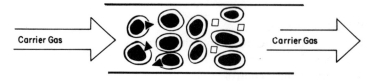

Figure 7-4
Gas Liquid Chromatography

cules of a more slowly eluting component. The molecules of this slowly eluting component spend most of their time dissolved in the stationary phase and, therefore, have a longer *retention time.*

As the separated components elute from the column, they pass through a detector. Several kinds of detectors are available, all of which use some property of each eluting substance to influence an electric current. For instance, a *thermal conductivity detector* reacts to the change in conductivity which occurs across the exit tube when the gas passing through contains a certain compound. A *flame ionization detector* reacts to the ions which are produced when a sample passes through a hydrogen flame and is burned. The electrical signal from each response is fed to a strip chart recorder, where the signal is recorded as a peak on moving chart paper.

Components of a sample are identified based on how much time is required to elute them from the column, that is, their retention times. As with the other chromatographic methods, standards must be used for comparison since the retention times depend on the exact conditions of the system (for example, age of the column or the nature of the previous sample on the column) and the overall composition of the sample. The areas under the peaks are proportional to the amounts of the components and therefore can be used to quantitate the sample. Quantitation also can be based on peak heights when the peaks are narrow and symmetrical. The peak heights are reasonably proportional to the amounts of the compounds under these circumstances. Because the magnitude of responses vary greatly in GC, an *internal standard* often is used to aid quantitation. This is a compound not likely to be found in the sample which is added in the same amount to both the sample and the standard. For each injection, peaks will result, one for the internal standard and one for the compound being measured. In each case a ratio is calculated, and these ratios are used to calculate concentration.

EXAMPLE
Injection 1. Standard peak = 15 mm; internal-standard peak = 30 mm
Injection 2. Test peak = 22 mm; internal-standard peak = 38 mm
Concentration of standard = 20 μg/ml

$$\frac{15}{30} = 0.5 \quad \text{standard ratio} \qquad \frac{22}{38} = 0.58 \quad \text{test ratio}$$

$$\frac{0.5}{20} = \frac{0.58}{x}$$

$$x = 23.2 \; \mu\text{g/ml} \qquad \text{concentration of test}$$

Most biological materials require chemical pretreatment before GC to make the materials volatile. The very type of functional groups, like carboxylic acid and amino groups, which make the substances soluble in water also make them nonvolatile. Chemical modification of these groups, as by esterification or alkylation, results in compounds which are sufficiently volatile for GC analysis. A major application of GC has been drug analysis. Some drugs can be injected directly after organic extraction from serum or urine without requiring any chemical pretreatment. Alcohol can be analyzed even without extraction because of its volatility.

Gel Filtration [4]

This technic involves separating molecules of different size by passing them through a column of gel filtration beads. Beads commonly employed are composed of a polymeric material like cross-linked polyacrylamide (a plastic) or Sephadex (a cross-linked polysaccharide). Even porous beads made of glass sometimes are used. The pores in the beads are too small to admit large molecules, but small molecules flow in and out easily. This allows large molecules to be separated from small ones as shown in Figure 7-5. The sample is represented as a mixture of large and small black dots, representing large and small molecules, respectively. The bigger open circles represent the gel filtration beads. The large molecules or dots cannot penetrate the beads, so they pass through the column in the liquid phase outside the beads and are eluted first. The small molecules move into and out of the beads. This movement retards their passage through the column. Thus they are eluted last.

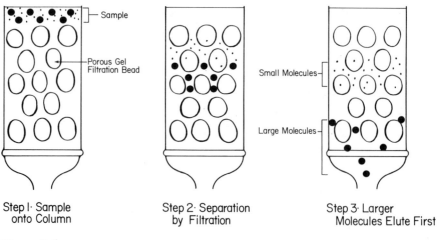

Step 1· Sample onto Column

Step 2· Separation by Filtration

Step 3· Larger Molecules Elute First

Figure 7-5
Gel Filtration

A typical use of this technic in clinical chemistry is the separation of small molecules from proteins. The proteins, being very large molecules, are eluted first.

Ion Exchange [3, 11, 15]

An ion exchanger is an insoluble material containing fixed charged groups of one electrical charge and mobile groups of the opposite charge. The insoluble material can be organic or inorganic, in either natural or synthetic form. Aluminum silicates, polysaccharides, and synthetic resins like polystyrene are used. Frequently these materials are fabricated into beads like the materials for gel filtration. Also ion exchange chromatography usually is carried out in a column. The mobile phase must be aqueous in nature because charged molecules (salts which are soluble only in aqueous media) are involved.

The basic purpose of an ion exchange column is to capture one or more of either the positively or the negatively charged ions in the sample. Two types of ion exchangers are available: anion exchangers, which capture anions, and cation exchangers, which capture cations. The mechanism of this capture involves exchange reactions and is illustrated in Figure 7-6. The exchanger used in the illustration is a cation exchanger because its mobile or exchangeable groups are cations. The sample is represented as a mixture of two anions and two cations. As the sample travels through the column, the cations of the sample exchange with the cations on the column. Hence the anions of the sample start out with one set of cations but end up with another set. The shaded

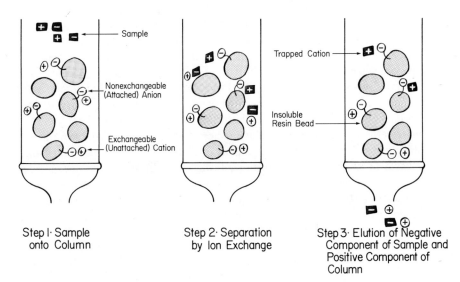

Step 1· Sample onto Column

Step 2· Separation by Ion Exchange

Step 3· Elution of Negative Component of Sample and Positive Component of Column

Figure 7-6
Ion Exchange Separation

cations are captured by the column and will remain on the column until another set of cations is applied to take their place. The mechanism of an anion exchange column is exactly the same except that anions instead of cations are captured by the ion exchange process. Mixed resin columns may be used to remove both anions and cations simultaneously.

Ion exchange chromatography is used both to remove unwanted ions and to concentrate wanted ions. The removal process either involves capture of the unwanted ions so that the immediate eluate contains the wanted species. Or the removal process captures the wanted species, followed by its elution with some other, noninterfering ionic material.

Electrophoresis [8]

Charged molecules in solution migrate to the electrodes when an electric field is applied. This is the principle involved in electrophoresis. Because opposite charges attract, the positively charged molecules (cations) migrate toward the negatively charged electrode (cathode), and the negatively charged molecules (anions) migrate toward the positively charged electrode (anode).

The separations which occur in electrophoresis are based on a variety of factors, although the principal factor is charge. Molecules with net charges of different sign will move in opposite directions. Molecules with different amounts of charge of the same sign will move in the same direction but at different rates. Both of these effects are illustrated in Figure 7-7. Note that the positively charged molecule moves toward the cathode and the negatively charged molecules move toward the anode. Furthermore, the molecule with four negative charges moves further toward the anode than the one with three negative charges. Depending on the medium, most of the previously discussed chromatographic effects may be present, such as adsorption, partition, gel filtration, and ion exchange. Consequently, molecules with the same net charge sometimes also are separated.

Many types of supports are used in electrophoresis [7], such as paper, starch gel, polyacrylamide gel, agar, and cellulose acetate. A most important one in the clinical chemistry laboratory is cellulose acetate. This material is available as thin plastic sheets (strips) with a highly uniform structure of pores and is particularly suitable for large molecules like plasma proteins, catalytic proteins (enzymes), lipoproteins, and special proteins like hemoglobin. Electrophoresis of a biological sample is followed by a staining reaction in which the different zones of proteins are made visible with dyes which bind to the proteins. The opaque cellulose acetate strips then can be cleared (rendered transparent except for the stained proteins) by washing with an organic solvent followed by drying in an oven. A densitometer then can be used to quantitate the stained zones by scanning them photometrically and recording their absorbances. (This technic is described on page 188.) Depending on the

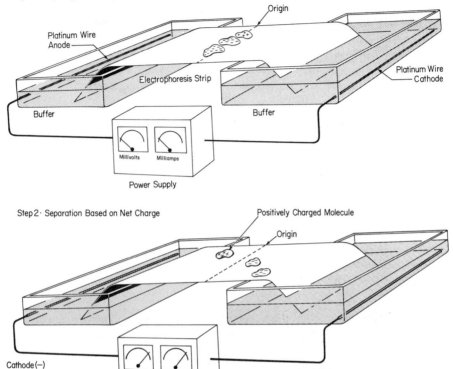

Figure 7-7
Electrophoresis

medium, quantitation can be performed also by cutting out each stained zone (prior to clearing in the case of cellulose acetate) and eluting the stained material into a cuvet with solvent for a spectrophotometric reading.

Control of pH and ionic strength often is quite important in electrophoresis. For this reason, and also to carry the current, buffers are used. The ionic strength of the buffer usually lies in the range $0.05M$ to $0.1M$. A low ionic strength results in a faster analysis, but the zones tend to spread out. A high ionic strength results in sharper zones, but the system heats up because the extra buffer carries extra current through the medium. If the strip gets too hot, the buffer solution will evaporate from the strip and distortion of the patterns will result. Consequently an intermediate ionic strength is chosen to moderate all of these effects.

Electrophoresis is used mostly for the separation of large molecules (molecular weight greater than 1000) such as proteins (see Chapter

15). Proteins exhibit a wide variety of net charge and are not as subject to diffusion problems (spreading out to the point of becoming invisible) as are small molecules like amino acids. Electrophoresis of small molecules generally must be carried out at high voltage. This causes the molecules to migrate so rapidly that they have little time to diffuse excessively. With this technic, considerable extra cooling is necessary because of the heat generated by the high electrical current.

References

1. Bender, G. T.   *Chemical Instrumentation: A Laboratory Manual Based on Clinical Chemistry.* Philadelphia: Saunders, 1972.
2. Colowick, S. P., and Kaplan, N. O. (Eds.).   *Methods in Enzymology.* New York: Academic, 1955–1975. Vol. 1–35.
3. Dorfner, K.   *Ion Exchangers* (3rd ed.). Ann Arbor, Mich.: Ann Arbor Scientific, 1972.
4. Fischer, L.   *An Introduction to Gel Chromatography.* New York: American Elsevier, 1969.
5. Folin, O., and Wu, H.   A system of blood analysis. (Suppl. I.) A simplified and improved method for the determination of sugar. *J. Biol. Chem.* 41:367, 1920.
6. Gaucher, G. M.   An introduction to chromatography. *J. Chem. Educ.* 46:729, 1969.
7. Gordon, A. H.   *Electrophoresis of Proteins in Polyacrylamide and Starch Gels.* New York: American Elsevier, 1969.
8. Gray, G. W.   Electrophoresis. *Sci. Am.* 185:45, 1951.
9. Heftmann, E. (Ed.).   *Chromatography* (2nd ed.). New York: Reinhold, 1967.
10. Karger, B. L., Snyder, L. R., and Horvath, C.   *An Introduction to Separation Science.* New York: Wiley, 1973.
11. Khym, J. X.   *Analytical Ion-Exchange Procedures in Chemistry and Biology: Theory, Equipment, Techniques.* Englewood Cliffs, N.J.: Prentice-Hall, 1974.
12. Nelson, N.   A photometric adaptation of the Somogyi method for the determination of glucose, *J. Biol. Chem.* 153:375, 1944.
13. Plummer, D. T.   *An Introduction to Practical Biochemistry.* New York: McGraw-Hill, 1971.
14. Randerath, K.   *Thin Layer Chromatography* (2nd ed.). New York: Academic, 1966.
15. Rieman, W., and Walter, H. F.   *Ion Exchange in Analytical Chemistry.* New York: Pergamon, 1970.
16. Scott, C. D.   Automated High Resolution Analyses for the Clinical Laboratory by Liquid Column Chromatography. In Sobotka, H., and Stewart, C. P. (Eds.), *Advances in Clinical Chemistry.* New York: Academic, 1972. Vol. 15.
17. Somogyi, M.   A new reagent for the determination of sugars. *J. Biol. Chem.* 160:61, 1945.
18. Stahl, E. C. (Ed.).   *Thin Layer Chromatography: A Laboratory Handbook* (2nd ed.). New York: Springer, 1969.
19. Street, H. V.   The Use of Gas-Liquid Chromatography in Clinical Chemistry. In Sobotka, H., and Stewart, C. P. (Eds.), *Advances in Clinical Chemistry.* New York: Academic, 1969. Vol. 12.

20. Szymanski, H. A. (Ed.).   *Biomedical Applications of Gas Chromatography*. New York: Plenum, 1968. Vol. 2.
21. Touchstone, J. C. (Ed.).   *Quantitative Thin Layer Chromatography*. New York: Wiley, 1973.
22. Tranchant, J. (Ed.).   *Practical Manual of Gas Chromatography*. New York: Elsevier, 1969.

8. Competitive Binding Radioassay

A number of important substances in biological specimens, particularly the hormones, are present in extremely low concentrations, tend to be unstable, and possess few unique chemical properties which can be used for analysis. Complicated separation procedures sometimes allow isolation of these materials, but these procedures tend to be time consuming, destructive, inaccurate, and costly.

One analytical technic which is well suited for the measurement of many of these problem substances is competitive binding radioassay. Although somewhat indirect in its means of measurement, the technic, once it is set up and operating properly, is reasonably simple, accurate, and specific.

Some of the substances more commonly measured by radioassay technics are cortisol, insulin, growth hormone, renin, thyroxine, aldosterone, testosterone, vitamin B_{12} and the cardiac glycosides digoxin and digitoxin.

Principle of Competitive Binding[1]

Competition is the basis of the measurement in competitive binding radioassays. In each case, two closely related substances (one of which is the substance being measured) compete for binding to a third substance. It is useful to let P represent the material to be measured, $*P$ the radioisotopically labeled form of the same material (P and $*P$ are closely related), and Q a binding substance for P and $*P$. The binder, Q, which usually is a protein, has a limited number of binding sites, which are highly specific for P or $*P$. Consequently the competition is between P and $*P$ for the binding sites on Q. The outcome depends on the relative amounts of P and $*P$ which are present. This is illustrated in Figure 8-1. In other words, the outcome of the competition of P and $*P$ for binding to Q always reflects the relative amounts of P and $*P$ which are present. Theoretically, if the absolute amounts of $*P$ and Q are known, the absolute amount of P can be determined. In practice, the competition is monitored by a measurement of the degree of binding of $*P$ since $*P$ is readily quantitated based on its radioactivity. The degree of binding of $*P$ to Q depends on the amount of P which is present since P competes with $*P$. The sample (or some extract of the sample) which contains an unknown quantity of material is mixed with a *known, constant amount* of $*P$ and Q. After equilibration of $*P$ and P with Q, the amount of $*P$ which is bound to Q is measured, and the amount of P is calculated from a standard curve (a plot of the degree of binding of $*P$ as a function of the concentration of P in each of a series of standards).

---

[1] Other descriptions of this principle may be found in the references cited at the end of this chapter.

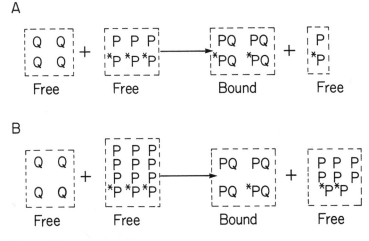

Figure 8-1
Competitive Binding Reaction. In Part A, on the top, equal amounts of P and $*P$ are present, so that equal amounts of P and $*P$ are bound by the binding agent, Q, and equal amounts of P and $*P$ also remain free, that is, unbound by Q. In Part B, the same amount of $*P$ is present, but the amount of P is increased threefold. This changes the amounts of $*P$ which are free and bound relative to the corresponding amounts of free and bound $*P$ in Part A. The presence of extra P in Part B causes the fraction of free $*P$ to increase because the ratios of PQ to $*PQ$ and free P to $*P$ always (after equilibrium is established) must be equivalent to the ratio of total P to total $*P$. These ratios all are $1:1$ in Part A, and $3:1$ in Part B.

This competition often is allowed enough time to go to equilibrium. Therefore it can also be discussed using equilibria equations. In the absence of any P, the equilibrium that is present involves only $*P$ and Q,

$$*P + Q \overset{K}{\rightleftarrows} *PQ \tag{8-1}$$

where $*PQ$ represents the bound form of $*P$, and K is the equilibrium or affinity constant. The greater the affinity of $*P$ for Q, the larger the value of the affinity constant, K. Quantitatively, the equilibrium can be expressed as:

$$K = \frac{[*PQ]}{[*P]\,[Q]}$$

where the brackets indicate molar concentrations.

In general, whenever one or more of the reactants in an equilibrium is lowered in concentration, the equilibrium will shift to compensate for the loss. This is what happens to the equilibrium of equation (8-1) when

P is added. The P reacts with Q to form PQ. This lowers the amount of Q which is free to react with $*P$. The equilibrium of equation (8-1) responds by shifting to the left to replace some of the free Q lost to P. This lowers the concentration of $*PQ$ and raises the concentration of free $*P$. A new equilibrium is reached:

$$*P + Q \rightleftarrows *PQ$$
$$+$$
$$P$$
$$\Updownarrow$$
$$PQ$$

In this equilibrium, the concentration of $*P$ is higher and that of $*PQ$ is lower than in equation (8-1). Since the degree to which the concentration of either $*PQ$ or $*P$ changes depends on how much P is present, a measurement of the degree of this change serves to determine the concentration of P.

The assay is completed by separating $*P$ and $*PQ$ (free $*P$ and bound $*P$) and measuring the amount of one of them (or both, as a double check), based on their radioactivity.

Radioisotopic Labels

A substance is labeled radioisotopically whenever one or more of its atoms is radioactive. Such atoms are called radioisotopes. They can be produced by bombardment of nonradioactive atoms either with neutrons in a nuclear reactor or with charged particles in a cyclotron. Table 8-1 presents the radioisotopes commonly used in competitive binding radioassays in clinical chemistry along with their respective half-lives. Several criteria are used to decide which radioisotopic label to use in each competitive binding radioassay.

Table 8-1. Isotopes Commonly Used in Clinical Chemistry

Isotope	Half-Life
^{131}I	8 days
^{125}I	60 days
^{57}Co	270 days
^{3}H	12 years
^{14}C	5,600 years

1. *Convenience of Preparation.* Frequently the radioisotopic label chosen is the one which allows the most convenient preparation of $*P$. For instance, ^{57}Co is chosen for vitamin B_{12} because simply exposing this vitamin to ^{57}Co results in the uptake of this radioactive

label. This occurs because vitamin B_{12} has an exchangeable cobalt atom. The radioactive cobalt simply substitutes for the nonradioactive cobalt normally present. Similarly, the hormone thyroxine contains exchangeable iodine atoms. Thus exposure of this hormone to ^{125}I or ^{131}I makes it radioactive. Actually ^{125}I and ^{131}I are convenient labels for many of the other hormones as well because these hormones contain tyrosine, or can be substituted by tyrosine, an amino acid which is easily labeled with iodine atoms. With other substances, ^{14}C or ^{3}H (tritium) may be chosen because it is more convenient to prepare the compound with these labels.

2. *Level of Radioactivity.* In some competitive binding assays, *P must be very highly radioactive. This is the case whenever the concentration of P, the substance to be measured, is extremely low because in these cases the concentration of *P also must be kept very low. Too much of an excess of *P over P results in negligible and therefore immeasurable changes in the concentrations of *PQ and free *P when P is added. This is because a meaningful (measurable) competition between P and *P for binding to Q can occur only when comparable amounts of P and *P are present. Consequently the assay will be insensitive to P unless the concentration of *P is comparable to that of P. This means that *P must be intensely radioactive. Otherwise it cannot be measured, based on its radioactivity, at very low concentrations. The most radioactive atoms (that is, the ones which give off the most pulses of energy or "counts" per unit time, like counts per minute) are used in these cases.

   For instance, ^{125}I and ^{131}I are much more radioactive than ^{14}C or ^{3}H. The use of ^{125}I or ^{131}I, therefore, allows more sensitive assays than can be made with ^{14}C or ^{3}H. This greater sensitivity is required in many cases because the concentrations of many of the hormones are very low.

3. *Half-Life of the Label.* The half-life of a radioisotope is the time after which half of the radioactive atoms originally present have decayed. Thus after one half-life the counts per minute are one-half the original value (unless the decay products also are radioactive). The disadvantage of using radioisotopes with short half-lives is that they lose their radioactivity rapidly and therefore must be replaced on a regular basis. This is the case with ^{131}I (half-life, 8 days) and ^{125}I (half-life, 60 days). There is no way around this problem because it is the short half-life in the first place which gives these iodine atoms their intense radioactivity. The radioisotopes ^{14}C and ^{3}H have half-lives of 5,600 years and 12 years, respectively and therefore need not be replaced regularly.

4. *Stability of the Radioactive Material.* Some compounds slowly break down after they are radioactively labeled because of radiation damage or inherent instability. Although the radioactivity level of the prepa-

ration may not change, the capacity of the preparation to provide a useful competitive binding assay gradually is lost.

5. *Instrumentation Available.* Radioisotopes ^{125}I, ^{131}I, and ^{57}Co emit primarily gamma radiation, while ^{14}C and ^3H emit beta radiation. If a laboratory is equipped to measure only one type of radiation, then it is limited to using isotopes which emit the corresponding radiation.

Binding Substances

The key ingredient in the competitive binding radioassay is the binding substance Q because it is this substance which determines the sensitivity and specificity of the assay. An assay is specific only when the binding sites on Q are specific for P and $*P$. If other substances besides P can displace $*P$ from these binding sites, the assay will measure these substances as if they were P. Generally, interferences like this arise only when substances are present which are very similar to P in structure.

Four types of materials have been employed as binding substances: antibodies, blood-transport or binding proteins, tissue receptors, and special binding substances. In each case there is a biological basis for the specificity and affinity that is observed for each binding substance Q and the particular substance or substances P which it binds. In other words, at least part of the actual biological function of each of these binding substances, Q, involves binding of the particular substance P. For instance, antibodies against insulin are used to test for insulin; also, the blood-transport protein, thyroxine-binding globulin, is used to test for thyroxine.

ANTIBODIES

Antibodies have been used widely in competitive binding radioassays as the binding agents because of their unique specificity. They can be prepared against a wide variety of substances. The first step is to prepare a pure form of P, suitable for injection. If P itself is not antigenic, it must be attached to a larger molecule, such as gamma globulin, to render it antigenic. When a carefully prepared solution of P is injected into an animal, the animal will make anti-P antibodies. The animal's blood is collected, and the serum is isolated. Such serum is called *antiserum,* because of the antibodies against P which are present. Sometimes this antiserum is used essentially directly (after dilution) in the competitive binding radioassay. Sometimes only the antibody fraction is used. In either case, whenever antibodies are used as the binding agents in competitive binding radioassays, these assays are called *radioimmunoassays* (RIA). Further discussion of antibodies may be found on page 191.

Separations

Another important requirement for a good competitive binding radioassay is a suitable method for the complete and rapid separation of the

free and bound forms of the radioactive material (*P and *PQ) so that
the radioactivity of either or both of these can be accurately measured.
Regardless of the method of separation chosen, it must be efficient and
relatively simple to perform. A wide variety of separation procedures
have been used. Each method has its own advantages and problems.
Some of the more common and representative categories of procedures
are discussed here. In all cases, it is desirable to have the concentrations
of serum proteins and other constituents of all of the assay mixtures as
uniform as possible. Minor alterations in the amount of *P, Q, the ionic
strength, pH, or serum concentrations can greatly affect the completeness of any of the separation methods.

1. *Chromatography.* One general technic employed is chromatography,
   including any of the specific technics from electrophoresis to gel
   filtration. Major differences in size and charge between *P and *PQ
   generally will be present because of the large size and charge of Q.
   Some type of chromatography always can be relied upon, but these
   technics are complex and time consuming.
2. *Precipitation.* The addition of various protein-precipitating agents,
   whenever Q is a protein and *P is not, is a rapid and simple way to
   separate *P from *PQ. Among the agents used are ammonium sulfate, sodium sulfate, acetone, ethanol, and polyethylene glycol. The
   precipitation is followed by centrifugation or filtration to separate the
   soluble *P from the *PQ precipitate. Although this technic can be
   quite satisfactory, the separation of *P from *PQ generally is not as
   sharp as in other separation methods. This means that some free *P
   is trapped in the *PQ precipitate, and some *PQ remains in solution.
   However, it is difficult to rival the speed and simplicity of this general
   technic.
3. *Solid-Phase Adsorption.* A number of separation methods are based
   on the removal of *P by adsorption onto an insoluble material.
   Adsorbents include charcoal, dextran- or hemoglobin-coated charcoal, Florisil (a magnesium silicate), kaolin (aluminum silicate),
   QUSO (microgranules of silica), cellulose powder, and ion-exchange
   resins. In this case, it is *PQ which is left in solution instead of *P,
   as in the previous precipitation procedure. The adsorption is followed by filtration or centrifugation. This general separation method
   is attractive because of its speed and ease of separation, although
   sometimes the separation again is not as sharp as one would like.
   Sometimes it is critical that exactly the same amount of solid adsorbent be added to each tube, while in other cases this is less important.
   The conditions, such as pH, ionic strength, temperature, and time,
   can be very critical.
4. *Antibody Precipitation.* Another method for the separation of *P
   and *PQ is the use of an antibody to precipitate *PQ. When Q itself

is an antibody, this procedure is called the *double-antibody* technic. The precipitating or second antibody, then, is an anti-Q or anti-antibody. This procedure is based on the fact that antibody-antigen complexes usually precipitate when the antigen is a protein. Antibodies themselves can be antigenic, of course, because they are proteins. The second antibodies are prepared in the usual way, which in this case involves the use of Q as an antigen. A large animal usually is chosen for the production of precipitating antibodies because a large amount of these antibodies is required to thoroughly precipitate the *PQ complex, even though the *PQ complex is present in quite low concentration. Because of the cost involved in producing antibodies and the fact that this procedure uses large amounts of precipitating antibodies, it is more expensive than the other separation methods. Another disadvantage is that usually a longer period of time is required for the complete precipitation of *PQ by antibodies than in other separation methods. However, the advantages of the method are that it can be used for practically any competitive binding radioassay, the separation is complete, and it may be used with large volumes of incubating solutions.

5. *Insolubilized Binder*. One of the more recent developments in separation technics is the insolubilization of the binding substance throughout the entire assay. This can be done without destroying the capacity of Q to form *PQ complexes. After P is added and competes with *P for insolubilized Q and the system comes to equilibrium, *PQ (insoluble) and *P (soluble) are separated by simple decantation or centrifugation.

Insoluble antibodies have been produced in a number of ways. They may be covalently bonded or fixed to insoluble polymers, or physically adsorbed to some type of plastic or glass which binds them quite strongly. Although this technic is generally more expensive to use, it does offer the advantage of great convenience.

Counting

The radioisotopes commonly used in competitive binding radioassays are ^3H, ^{14}C, ^{57}Co, ^{125}I, and ^{131}I. The first two are beta emitters. (Beta particles or rays are electrons.) The last three, ^{57}Co, ^{125}I, and ^{131}I, are principally gamma emitters. (Gamma rays are high-energy photons.) All of these radioisotopes are most conveniently measured or counted by using instruments called *scintillation counters,* as illustrated in Figure 8-2. These instruments employ scintillators, which are substances that release small flashes of light whenever they are exposed to pulses of radiation like beta or gamma rays. The light flashes are detected by a photomultiplier tube, which converts them into pulses of electricity. These electrical or voltage pulses are amplified and fed to a scaler, an

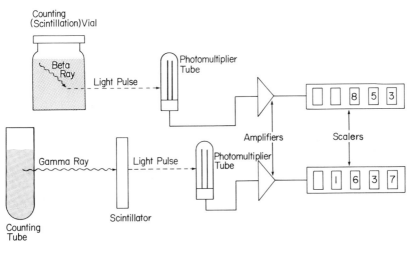

Figure 8-2
Components of Radiation Counters

electrical device which counts the number of voltage pulses which reach it. The number of counts per unit of time then indicates how much radioactivity is present in the sample which is being counted.

Scintillators give off pulses of light when struck by high-energy particles because these particles collide with the molecules of the scintillator, imparting energy to these molecules and thereby exciting them to a higher energy state. As each scintillation molecule returns to its starting or ground state, it gives off a pulse of light. Scintillators usually are solutions or solids. Usually an organic solution of a scintillating substance is used for counting the beta emitters ^{3}H and ^{14}C, while a solid scintillator is used to count the gamma emitters ^{57}Co, ^{125}I, and ^{131}I.

For liquid scintillation counting (Figure 8-2, top), the radioactive sample is dissolved in an organic solution of a scintillating substance in a vial. The vial is placed, manually or automatically, in front of a photomultiplier tube for detection of the light pulses which are emitted from the vial. (These pulses of light are much too weak to be seen by a human eye.)

For solid scintillation counting (Figure 8-2, bottom), a crystal of sodium iodide activated with 1% thallium is most frequently used as the scintillator. In this case, the instrument is called a gamma counter. There is no direct contact of the radioactive sample, which is located in a plastic or glass tube, and the sodium iodide scintillator, although the sample is positioned immediately adjacent to the sodium iodide crystal when it is counted. The sample emits gamma rays, which strike the sodium iodide crystal. The crystal then emits small flashes of light, and a photomultiplier tube detects the flashes of light, converting them into voltage signals.

Gamma counting is more convenient than beta counting because the sample can be counted directly after the separation step without the need to dissolve the sample in a scintillation fluid. This advantage is offset somewhat, however, by the problems associated with the short half-lives and radiation damage (to the labeled compound) from the ^{125}I and ^{131}I radioisotopes.

The number of counts obtained for a given sample depends on the amount of radioactivity in the sample and the length of time it is counted. Precision of radioactivity counting increases with the number of counts obtained. For instance, 3% precision requires 4,000 counts, 2% precision requires 10,000 counts, and 1% precision requires 40,000 counts. This means that the counts-per-minute (cpm) value calculated from the time required to obtain the total counts (for example, 4000) is within the specified percentage (for example, 3%) of the counts-per-minute value which would be calculated from an "infinite" number of counts over an "infinite" period of time under the same instrumental and sample conditions. Most counting equipment may be set either to obtain a predetermined number of counts (preset precision or error), in which case time is read as a variable, or to count for a set amount of time, in which case the number of counts (and hence the counting precision or error) is the variable.

Calculations

As illustrated in Figure 8-3, the relation between counts and concentration is not linear. This is true in both theory and practice. Although a wide variety of plots have been introduced (including a logarithmic type of plot called a *logit plot* which will linearize the data from certain assays), for practical purposes many of these plots are satisfactory. For instance, it is acceptable to plot counts or counts per unit time (usually counts per minute, cpm) against concentration on linear paper and draw a line connecting the points.

Sample Radioimmunoassay Method[2]

The concepts of competitive binding and radioimmunoassay as applied to specific methods can be illustrated with a method outline for the hypothetical substance, human beauty hormone (HBH), in serum. HBH is P in the designation for binding substances as outlined in the section on binding substances earlier in this chapter.

---

[2] Actual examples of radioassays may be found on page 312 (thyroxine) and on page 357 (digoxin).

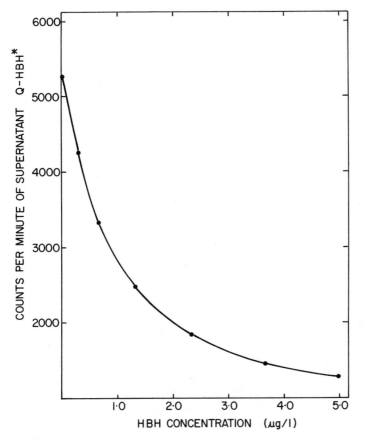

Figure 8-3
Radioassay Calibration Curve

REAGENTS

1. Stock standard made with pure HBH.
2. Working standards made by diluting the stock standard.
3. Buffer of appropriate pH.
4. HBH tagged with [125]I. This is *HBH, that is, *P in the binding designations.
5. Antiserum containing antibodies to HBH. This is Q.
6. Charcoal suspended in buffer. This is the separating agent.

PROCEDURE

1. Into each of a series of 12 × 75 mm polystyrene tubes measure 100 μl of each test serum. Include several control sera containing previously measured amounts of HBH.

2. Into each of another such series of tubes measure 100 μl of each working standard, including 100 μl of buffer for a zero standard.
3. Add 50 μl of the solution of *HBH to each tube and mix.
4. Add 100 μl of antiserum to each tube, mix, and let stand for 15 minutes at room temperature. (HBH and *HBH compete for binding to the antibody during this incubation.)
5. Add 0.5 ml of charcoal suspension to each tube, mix, and let stand 15 minutes. (The charcoal binds free *HBH but not antibody-bound *HBH.)
6. Centrifuge the tubes for 5 minutes.
7. Decant the supernatant fluids into appropriate counting tubes or vials and count the radioactivity (antibody-bound *HBH) in a gamma counter set at 2% counting error (precision).

CALCULATIONS
1. Plot the counts per minute (cpm) of the standards against their respective concentrations on linear graph paper (see Figure 8-3).
2. Determine the concentrations of the tests from their cpm readings using the standard curve.

Practical Considerations

1. *Equipment.* There are three pieces of equipment universally required for performing competitive binding radioassays: a scintillation counter, a centrifuge, and a pipetting device. The scintillation counter is the largest investment, although the costs range widely, depending on the level of sophistication of the instrument. Generally speaking, the less expensive instruments lack the precision and sample-handling capacity necessary for convenient operation with large test volumes. Since neither gamma- nor beta-emitting isotopes have completely dominated the field, the ability to count both is desirable.

   Almost all competitive binding radioassays call for centrifugation at some point. Although most such procedures do not require centrifugation in the cold, none of them suggest the use of an elevated temperature, which can occur in an old, overloaded centrifuge when the motor overheats and gradually raises the temperature of the samples. A refrigerated centrifuge is advantageous because some separation steps work best in the cold. As an alternative, a centrifuge can be placed in a cold room or refrigerator. Centrifugations are best carried out in horizontal or swinging-bucket heads. Special attention should be given to the size and convenience of the carrier head, that it have the capacity to accommodate all of the tubes in any given procedure because of the importance of exact timing in competitive binding radioassays.

   It is of the utmost importance that the pipetting equipment be of high quality. A poor pipet system will ruin any competitive binding radioassay procedure. Pipets fall into three categories: manual, semi-automatic, and automatic. When manual pipets are used, the radio-active solutions should never be pipetted by mouth. Because the semiautomatic and automatic pipets are considerably faster than the manual pipets, they are preferable for high-volume work. The performance of a pipetting system being considered should be evaluated carefully before it is put into use. Performance should not be compromised with cost.

2. *Technic*. Some technologists are not familiar with the technics involved in the use of micropipets. The ability of each person to measure small samples accurately and precisely must be checked before he or she is given the responsibility to run competitive binding radioassays.

3. *Timing*. Precise timing is essential. Once an assay is underway, the timing sequence must be followed precisely. There are certain steps in the assay, like the mixing and separation steps, which must be carried out on all of the samples with no interruptions whatsoever, not even to answer the phone.

4. *Kits vs. Assembly*. One issue is whether to buy commercial kits, which have everything in them and can be convenient, or to assemble the various components independently. There are advantages and disadvantages with both approaches. For the large-volume laboratory, assembly of components can be quite advantageous. There is more control available over the quality of each of the reagents and certain costs can be cut, although more time and therefore money must be spent on inventory, ordering, assembly, and testing the various components. For the low-volume laboratory the kit approach can save time and money. Kits, however, do not always perform according to the manufacturer's claims. Some points worth considering before purchasing kits are presented in Chapter 6.

5. *Reagents*. Whether one opts for kits or assembly of components or both, a general problem is that different lots of reagents, sometimes even different bottles from the same lot, may vary in their quality and properties. Therefore minor alterations in conditions and procedures often are necessary to obtain optimum performance from the assays. This means that each new batch of reagents must be checked. A particular problem is the quality of the binding agent. Different batches of a given material may have different properties, particularly in regard to the important properties of binding affinity and specificity in the case of the antibodies. A reliable supplier of high-quality, pretested binding agents must be found, or a reliable program for the development of these agents by the laboratory itself must be established. Unreliable kits and components are not uncommon. Therefore the choice of a supplier should be made critically.

This problem extends as well to the labeled substances. In the case of the iodine radioisotopes in particular, $*P$ may be impure as a result of radiation damage from the radioisotopic atoms it contains or from the reagents used in the preparation of *P from P. The substance P or *P may be somewhat unstable in any case. This requires checks on the purity of *P (for example, chromatography, gel filtration, ability to bind completely to Q). Such tests should provide assurance that all of the radioisotopic label still is bound to *P.

6. *Sensitivity and Range.* The two opposing factors which are compromised in competitive binding radioassays are sensitivity and range. A very sensitive assay is achieved by using a minimum amount of antibody. However, this results in a small analytic range. The use of a more concentrated antibody solution results in a wider analytic range and provides better measurements of higher concentrations of the substance being analyzed. On the other hand, a loss in sensitivity results.

7. *Samples.* It is not sufficient that the assay perform well for the standards. The next step is to test actual samples. This may not be a small step since samples can behave differently than the standards do. Recoveries of pure standards from patients' samples and from serum pools can be checked. Sometimes an independent method may be used as a reference method. Quality-control materials are commercially available for some radioassay tests.

8. *Safety.* Precautions in handling radioisotopes cannot be overemphasized. One should follow recognized safety procedures and not eat, drink, or smoke in the immediate area where isotopes are handled. A trained technologist with good laboratory sense and skills will be exposed to no significant radiation hazards from the amounts of radioactive materials usually on hand in clinical chemistry. However, each analyst should be familiar with the various rules and recommendations issued by the Nuclear Regulatory Commission (formerly the Atomic Energy Commission) and should observe them faithfully.

References

1. Aubert, M. L.   Critical study of the radioimmunological assay for the dosage of the polypeptide hormones in plasma. *J. Nucl. Biol. Med.* 14: (1, 2, 3), 1970.

2. Hayes, R. L., Goswitz, F. A., and Murphy, B. E. P.   *Radioisotopes in Medicine: In Vitro Studies.* Oak Ridge, Tenn.: U.S. Atomic Energy Commission, 1968.

3. Jaffe, B. M., and Behrman, H. R. (Eds.).   *Methods of Hormone Radioimmunoassay.* New York: Academic, 1974.

4. Kirkham, K. E., and Hunter, W. M.   *Radioimmunoassay Methods.* Edinburgh: Churchill/Livingstone, 1971.

5. Odell, W. D., and Daughaday, W. H. (Ed.).   *Principles of Competitive Protein-Binding Assays.* Philadelphia: Lippincott, 1971.

6. Peron, F. G., and Caldwell, B. V.   *Immunologic Methods in Steroid Determination.* New York: Appleton-Century-Crofts, 1970.

7. Skelley, D. S., Brown, L. P., and Besch, P. K.   Radioimmunoassay. *Clin. Chem.* 19:146, 1973.
8. Sönksen, P. H. (Ed.).   Radioimmunoassay and saturation analysis. *Br. Med. Bull.* 30:(1) 1974. Entire issue.
9. Zettner, A.   Principles of competitive binding assays (saturation analyses): I. Equilibrium techniques. *Clin. Chem.* 19:699, 1973.
10. Zettner, A., and Duly, P. E.   Principles of competitive binding assays (saturation analyses): II. Sequential saturation. *Clin. Chem.* 20:5, 1974.

II. Methods

9. Blood Gases, pH, and Bicarbonate

The blood gases most commonly measured are carbon dioxide (CO_2) and oxygen (O_2). Although small amounts of these substances exist as dissolved gases in the blood, they are, for the most part, chemically combined with other materials. Carbon dioxide exists chiefly as bicarbonate (HCO_3^-) while most of the oxygen is bound to hemoglobin to form oxyhemoglobin. The pH, which is defined in the section on pH and Buffers in Chapter 1, measures the acidity of the sample and is dependent upon the relative amounts of bicarbonate and carbonic acid (H_2CO_3) which are present.

Carbon Dioxide

Carbon dioxide exists in the blood in several forms. Most of it is present as bicarbonate ion (HCO_3^-), but there also is some carbonic acid (H_2CO_3), dissolved CO_2, and CO_2 that is bound by proteins.

P_{CO_2}

Gases occur both in gaseous states and dissolved in liquids. In each case, they exert a pressure. When more than one gas is present, the total pressure depends on the fractional pressure or partial pressure of each contributing gas. The partial pressure (P) of a gas therefore is a measure of its concentration either in a gaseous or liquid phase. Dissolved CO_2 in blood is measured by its partial pressure, P_{CO_2}. This is most easily accomplished with a P_{CO_2} electrode [5, 6] which can be constructed starting with a pH electrode. A small amount of a bicarbonate solution is trapped around the glass bulb of a pH electrode with a thin Teflon film. This film passes dissolved CO_2 (a neutral molecule) but not H^+ or HCO_3^- (charged molecules). The more CO_2 present in the sample, the more that diffuses through the Teflon film into the trapped bicarbonate solution. This CO_2 changes the pH of the solution, and the change is detected by the pH electrode. Such P_{CO_2} electrodes are available commercially. The P_{CO_2} also can be calculated from the pH and CO_2 content of the blood using a modified form of the Henderson-Hasselbalch equation, which is given in one of the following sections of this chapter.

Normal values for P_{CO_2} in arterial blood are 33–48 mm of mercury (Hg).

CO_2 Content

The CO_2 content (also referred to as total CO_2 or total CO_2 content) includes all of the forms of CO_2 that are present (dissolved CO_2, H_2CO_3, HCO_3^-, and protein-bound). When acid is added to a physiological sample, all of these forms are converted to dissolved CO_2. A measure of this dissolved CO_2 after acid addition therefore provides the CO_2 content. Although a P_{CO_2} electrode can be used, the reference procedure for measuring CO_2 content is the gasometric method of Van Slyke [13, 14]. This technic is described and discussed below.

A commonly used automated method for measuring CO_2 content [11] takes advantage of the fact that the dissolved CO_2 is evolved from an acidified sample. Dilute sulfuric acid is used to acidify the sample, and the released CO_2 gas is reabsorbed in a weak bicarbonate buffer solution containing the pH indicator, phenolphthalein. The absorbed CO_2 causes the pH to decrease so that some of the phenolphthalein is converted from its basic form (red) to its acid form (colorless). The loss of red color therefore is proportional to the CO_2 content of the sample.

The CO_2 content can be calculated also from the pH and P_{CO_2} values (see page 125), analogous to the calculation of P_{CO_2} from the pH and CO_2 content values.

The reference method for measuring CO_2 content (total CO_2) is the gasometric method of Van Slyke. The method requires the use of an apparatus designed to measure gas pressures at constant volume. The modern version of such an apparatus is the Natelson Microgasometer, a micro adaptation of the more cumbersome Van Slyke manometric gasometer.

CARBON DIOXIDE CONTENT METHOD

METHOD
Van Slyke and Neill [13, 14]; Natelson [8].

MATERIAL
Anaerobic serum.

REAGENTS
Lactic acid 1N. With water, dilute 9 ml of concd lactic acid to 100 ml. Make only enough to last 2 weeks since molds tend to grow in the solution, which increases the CO_2 concentration.

Sodium hydroxide 3N. Dissolve 12 g of sodium hydroxide in water and dilute to 100 ml.

Low-Foam Detergent. Dilute 10 ml of stock Low-Foam Detergent (reagent no. 810)[1] to 100 ml with water.

Anti-Foam Reagent (reagent no. 820).[1]

Clean mercury.

APPARATUS
Natelson Microgasometer (see Figure 9-1).

[1] Scientific Industries, Springfield, Mass.

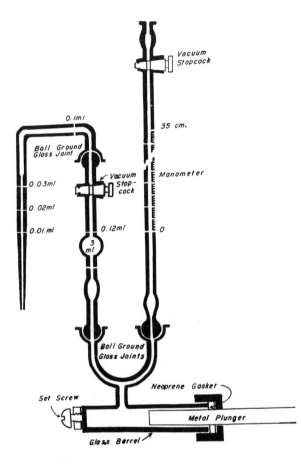

Figure 9-1
Natelson Microgasometer. Cross sectional schematic diagram of the microgasometer. Changes in pressure are produced by movement of the plunger.
The set screw forces the barrel against a gasket for a seal. Thus the plunger
is not close fitting to the barrel. The level of the Hg in the manometer is a
measure of the pressure in the chamber since they are connected as in a
U-tube. (From Natelson, Samuel, *Techniques of Clinical Chemistry* [3rd
ed.], 1971. Courtesy of Charles C Thomas, Publisher, Springfield, Illinois.)

PROCEDURE
Expel any water from the chamber and from the pipet tip.
Draw into the pipet in order:
0.03 ml of serum
0.01 ml of mercury
0.03 ml of lactic acid
0.01 ml of mercury
0.01 ml of Anti-Foam Reagent
0.01 ml of mercury

0.10 ml of Low-Foam Detergent
Mercury to the 0.12 ml mark on the chamber

Close the stopcock and draw the mercury meniscus almost to the bottom
   of the reaction chamber.
Mix the solution for 1 minute.
Advance the piston until the aqueous meniscus is at the 0.12 ml mark.
Read the mercury manometer in mm (P_1).
Advance the piston until the mercury reaches the top of the manometer.
Immerse the pipet tip in the sodium hydroxide reagent, open the stop-
   cock, and express a small drop of mercury from the tip.
Draw in 0.03 ml of sodium hydroxide.
Draw in mercury to the 0.12 ml mark.
Close the stopcock and draw the mercury meniscus almost to the bottom
   of the reaction chamber.
Mix for 10 seconds.
Advance the aqueous meniscus to the 0.12 ml mark.
Read the mercury manometer (P_2).
Advance the piston until the mercury meniscus is at the top of the
   manometer, discharge the solution from the instrument, then rinse the
   chamber with water, lactic acid, then water again.

BLANK
Perform an analysis as described above, substituting water for the
sample. If a significant Blank is obtained it should be used in the calculations.

CALCULATIONS
Since P_1 = the initial gas pressure in mm and since P_2 = the initial
pressure minus that of carbon dioxide, then $P_1 - P_2$ = mm of carbon
dioxide pressure. Subtract the Blank (if any) from this value and
multiply the resulting pressure by the appropriate factor for the
ambient temperature (Table 9-1). The formula for the calculation is

$$[(P_1 - P_2) - \text{blank}] \times \text{factor} = CO_2 \text{ mmol/liter}$$

The factors are derived from the basic gas laws, but they include also
certain empirical corrections. Further discussion of their derivation is
available [9, 15].

STANDARDIZATION
Since the gasometric apparatus measures gases directly in terms of pressure and volume, calculations are made without reference to standards.
However, sometimes Standards are helpful in checking the technic or
the instrument. It is advisable to use sodium carbonate to make the
standard solution, since this salt may be obtained in a pure, anhydrous
form.

Table 9-1. Factors for Calculating CO_2 Content

Temperature (°C)	Factor for calculating (mmol/liter)
17	0.242
18	0.240
19	0.238
20	0.237
21	0.236
22	0.235
23	0.234
24	0.233
25	0.232
26	0.231
27	0.230
28	0.229
29	0.228
30	0.227
31	0.225
32	0.224

Carbon dioxide standard, 25 mmol/liter. The calculations give the carbon dioxide concentration in millimols per liter but, since the carbon dioxide concentration is expressed as bicarbonate ion, which is monovalent, the millimols are also milliequivalents. It is desirable to apply the same calculations to the standard sodium carbonate solution and to have the results compare with those of 25 meq per liter serum carbon dioxide concentration. Since sodium carbonate (Na_2CO_3) has a valence of two, a 25-millimol-per-liter solution contains 50 milliequivalents per liter.

One mole of Na_2CO_3 = 106 g, and 1 millimol = 0.106 g. Therefore, to make 100 ml of this 25 millimol-per-liter standard solution requires

$$0.106 \times 2.5 = 0.2650 \text{ g } Na_2CO_3/100 \text{ ml}$$

Store the solution in a plastic bottle and use in the same manner as serum. This solution keeps for a few weeks if refrigerated.

NORMAL VALUES

The normal range for CO_2 content in the serum of healthy adults is 25–32 meq per liter.

DISCUSSION

When the carbon dioxide method is performed as described here, the accuracy is approximately ±0.5 meq per liter.

Acidification of the serum with lactic acid results in the formation of carbonic acid from bicarbonate with the subsequent release of carbon dioxide:

$$H^+ + HCO_3^- \rightarrow H_2CO_3 \rightarrow H_2O + CO_2\uparrow$$

The serum gases are released by the combined effects of the acid, vacuum, and agitation. The first manometer reading is the total pressure of the gases released from the sample plus the aqueous vapor pressure in the chamber. The sodium hydroxide solution then absorbs carbon dioxide rapidly and completely:

$$NaOH + CO_2 \rightarrow NaHCO_3$$

Because the carbon dioxide has been removed, the second manometer reading represents the pressure of the gases minus the carbon dioxide.

During the mixing process, when there is a vacuum in the chamber, the mercury meniscus in the manometer should not move. If the mercury rises appreciably, a leak at the reaction-chamber stopcock is indicated whereas, if the mercury falls, the leak is at the manometer stopcock. The apparatus should be checked for leaks prior to use by creating a vacuum and watching for a change in the mercury level in the manometer. Stopcocks should be lubricated frequently with a thin coating of a high-vacuum silicon lubricant.

When not in use, the chamber may be left filled with water or mercury. In either case, the chamber stopcock should be left open to avoid unnecessary pressure on the stopcocks which may occur with changes in room temperature.

With some instruments, mixing is accomplished by manual manipulation; other instruments are equipped with magnetic stirrers. With the latter type care must be taken to position the stirring bar so that it is in the solution when rotating.

When the apparatus is filled with mercury, the piston support should be ½ to 1 inch from the center support. If the distance exceeds this, some mercury should be expressed from the system.

The apparatus should be dismantled and cleaned periodically. Hot detergent is satisfactory for removing grease from the system. After the glassware has been cleaned, it should be rinsed thoroughly and dried before reassembly. The ball joints and stopcocks must be cleaned and lubricated as well.

It is important that the mercury be kept clean; mercury cleaning filters are commercially available for this purpose. Since mercury vapors are toxic, mercury should be kept covered. Any spills should be cleaned up promptly. Dirty mercury may be stored under water until cleaned or sold. Also, mercury amalgamates with other metals. Therefore metallic apparatus (or even jewelry) should be kept away from mercury.

If whole blood comes into contact with air, carbon dioxide escapes from the blood. As carbon dioxide leaves the serum, chloride diffuses from the cells to replace it. Thus contact between whole blood and air has the effect of lowering the serum carbon dioxide concentration and elevating the serum chloride concentration. If blood is not kept anaerobically, the serum usually will lose 2 to 5 meg per liter of carbon dioxide over a very short period of time. The use of vacuum tubes for collecting blood for carbon dioxide analysis is recommended [4]. The tube should be well filled, tightly stoppered, and protected from agitation.

If the determination cannot be performed the same day the sample is received, the unopened tube may be centrifuged and stored in the refrigerator for at least 24 hours with virtually no change (0 to 1 meq per liter) in the carbon dioxide content. If other determinations are to be made with the same sample, the serum should be removed with a Pasteur pipet and carefully transferred to another vacuum tube for storage to avoid contamination of the serum by intracellular substances.

The physiological role of carbon dioxide in the blood is discussed in Chapter 10.

pH

The best way to measure pH is with a pH meter. The pH meter basically is composed of a glass bulb electrode, a reference electrode (usually a calomel electrode), and a sensitive meter. These two electrodes are illustrated in Figure 9-2.

The principle of this system is that an electrical potential develops across the glass-bulb electrode when the solution in which it is immersed has a different H^+ concentration than the solution of fixed pH which is inside the bulb. This potential is proportional to the H^+ concentration in the test solution and is measured with the aid of the reference electrode, which puts out a reference potential, and a meter, which measures the potential of the pH electrode by comparing it with the potential of the reference electrode. When the glass bulb and reference components are housed together, the resulting unit is called a *combination electrode*.

More detailed discussions of pH and its measurement are available [1, 16].

USE OF pH METER

A number of considerations and precautions are important in pH measurements.

1. *Standardization.* Regularly pH meters must be standardized with at least two standard buffers with different known pH values. One buffer

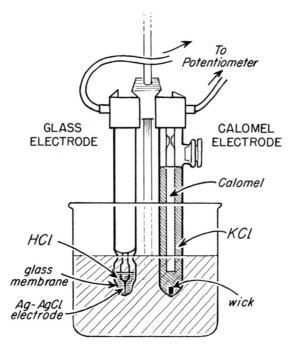

Figure 9-2
pH Electrode

is used to set a correct pH reading on the meter (for example, pH 7.0), and the second buffer (for example, pH 4.0) is used to set the correct slope of the pH meter readings so that all values between pH 4.0 and pH 7.0 will be correct, not just the ones near pH 7.0. This is accomplished by using different control knobs to set the pH 7.0 and pH 4.0 readings when the electrode is immersed in the corresponding solutions. Standard buffer solutions or powders to prepare these solutions are commercially available. Instructions for the preparation of standard pH buffers are available from the National Bureau of Standards. The preparation date or date of opening of a commercial buffer should be recorded, and these solutions should be capped whenever they are not being poured since their pH gradually changes because of absorption of CO_2 from the air. Bacterial growth also results in pH changes.

2. *Care of pH Electrodes.* The manufacturer's directions in regard to activation and storage of the pH electrode should be followed carefully. Sometimes water is the recommended storage solution, sometimes saline or buffer.

3. *Use.* The liquid and salt levels inside the electrode should be checked before use. As with any measurement system involving liquids,

checks for air bubbles should be made whenever spurious readings occur. The pH electrode should be rinsed thoroughly before and after each sample or buffer measurement. Since most glass electrodes are fragile, careful handling is necessary.
4. *Temperature.* At the time of measurement, the temperature should be controlled to less than ±0.5 C for both standard buffers and samples since pH changes with temperature. The electrode must be fully equilibrated to the sample temperature when the reading is taken.

In health, the pH of arterial blood is remarkably constant at 7.38–7.42. Values below 7.30 or above 7.50 almost always require treatment for the underlying disturbance, and values as low as 7.0 or as high as 7.8 are incompatible with life.

Bicarbonate

The methods which measure only bicarbonate are based on the alkaline property of bicarbonate. They are simply acid-base titrations [10, 12]:

$$HCl + NaHCO_3 \rightarrow H_2O + CO_2 \uparrow + NaCl$$

In these methods, a known excess of acid is added. The amount that escapes neutralization by the bicarbonate in the sample is determined by titration with standard base.

The indicator for these titrations must be carefully chosen to ensure detection of the proper equivalence point. Detection of the endpoint is best made with a pH meter. Since the pH of normal plasma is 7.4, this pH is used as the endpoint for the titrations.

In practice, this test is not very reliable. It is preferable to determine CO_2 content instead, which usually is within 1 mmol per liter of the bicarbonate value, or to calculate bicarbonate from CO_2 and pH values, as discussed below.

Normal values for bicarbonate in serum are 24–30 meq per liter.

Henderson-Hasselbalch Equation

The relationship among the various parameters discussed above may be derived starting with the following equilibrium:

$$H_2CO_3 \rightleftarrows H^+ + HCO_3^-$$

Letting K be the equilibrium constant,

$$K = \frac{[H^+] [HCO_3^-]}{[H_2CO_3]}$$

Therefore,

$$[H^+] = K \times \frac{[H_2CO_3]}{[HCO_3^-]}$$

Taking the logarithm of both sides,

$$\log [H^+] = \log K + \log \frac{[H_2CO_3]}{[HCO_3^-]}$$

Multiply both sides by -1:

$$-\log [H^+] = -\log K - \log \frac{[H_2CO_3]}{[HCO_3^-]}$$

By definition, $-\log [H^+] = pH$, and $-\log K = pK$. Therefore,

$$pH = pK + \log \frac{[HCO_3^-]}{[H_2CO_3]}$$

For blood serum at 37 C, the pK of $H_2CO_3 = 6.1$. Thus,

$$pH = 6.1 + \log \frac{[HCO_3^-]}{[H_2CO_3]}$$

This is the Henderson-Hasselbalch equation. The H_2CO_3 includes the actual H_2CO_3 which is present along with dissolved CO_2.

Normally the bicarbonate concentration in blood is approximately 26 mmol per liter while the concentration of carbonic acid is approximately 1.3 mmol per liter. The blood pH may be calculated by inserting these values into the above formula:

$$pH = 6.1 + \log \frac{26}{1.3}$$
$$pH = 6.1 + \log 20$$
$$pH = 6.1 + 1.30 = 7.40 \qquad \text{(normal blood pH)}$$

The calculations and correction factors used in determinations of blood pH and blood gases have been discussed in other publications [2].

P_{CO_2} and CO_2 Content Equations

The greater the partial pressure of CO_2 (P_{CO_2}), the greater the amount of H_2CO_3 (actual H_2CO_3 plus dissolved CO_2) present in the sample. Actually,

$$[H_2CO_3] \approx 0.03\, P_{CO_2}$$

where the factor 0.03 is essentially a solubility factor for a given partial pressure of CO_2. Also, by definition,

$$CO_2 \text{ content} = [HCO_3^-] + [H_2CO_3]$$

or, using $[H_2CO_3] = 0.03 \, P_{CO_2}$,

$$CO_2 \text{ content} = [HCO_3^-] + 0.03 \, P_{CO_2}$$

The following three equations allow calculation of all the parameters once certain pairs of these parameters are known:

$$pH = 6.1 + \log \frac{[HCO_3^-]}{[H_2CO_3]}$$
$$[H_2CO_3] = 0.03 \, P_{CO_2}$$
$$CO_2 \text{ content} = [HCO_3^-] + 0.03 \, P_{CO_2}$$

Oxygen

About 95% of the oxygen in blood is in the oxyhemoglobin molecule. The other 5% is present as dissolved oxygen. The oxygen status of a blood sample can be expressed in various ways:

1. P_{O_2}, *the partial pressure of oxygen,* is the fraction of the total gas pressure in the sample which is contributed by oxygen. This parameter is measured with an oxygen electrode. In this electrode (Figure 9-3), a voltage is applied across a platinum wire cathode and a silver wire anode in a dilute salt solution, all of which is separated from the

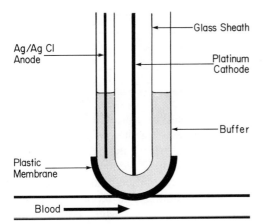

Figure 9-3
Oxygen Electrode

blood sample by a thin plastic (for example, polypropylene) membrane. Some of the oxygen in the blood diffuses through the membrane and is reduced to water at the cathode. The magnitude of the electron current at this cathode is a measure of the amount of diffusible oxygen in the blood sample. Although this technic measures only the dissolved oxygen in the sample, the amount of dissolved oxygen depends on the amount of oxyhemoglobin present.

2. *O$_2$ capacity* is the amount of oxygen which can be combined with all of the hemoglobin present in the sample. When blood is saturated with oxygen before measurement of the oxygen concentration, the subsequent measurement gives the *oxygen capacity* of the blood. Although spectrophotometric [7] and gasometric [12] technics can be used to measure O$_2$ capacity, usually this parameter is calculated from the hemoglobin concentration since, at full saturation, 1 g of hemoglobin binds 1.36 ml of oxygen (37 C, pH 7.4).

3. *O$_2$ content* is the sum of the dissolved and the combined (oxyhemoglobin) O$_2$ in the circulating blood. This is derived from the Po$_2$ and O$_2$ capacity or is measured gasometrically. In the gasometric technic, the sample is treated with reagents which cause hemolysis and also liberate the O$_2$ from the oxyhemoglobin. The amount of liberated oxygen is measured based on its pressure (manometrically).

4. *Percent O$_2$ saturation* is the O$_2$ content divided by the O$_2$ capacity, expressed as a percent. This parameter gives the amount of O$_2$ relative to the full saturation value. It also can be determined from the Po$_2$.

Normal values for the various oxygen parameters are given in Table 9-2.

Table 9-2. Normal Blood Oxygen Values

	Content (vol %)	Capacity (vol %)	Saturation (%)
Arterial	17–22	18–23	95–97
Venous	10–16	—	60–85

Sample Handling

A sample for blood gas or pH analysis should be collected in a heparinized syringe. The syringe should be capped immediately after the collection. If the analysis cannot be performed within the next 5 minutes, the filled syringe should be cooled immediately in ice and stored at this temperature until just before the analysis is carried out. Samples may be kept in this manner for approximately 4 hours. The blood-gas values change with time because of the metabolic processes which continue to occur in freshly drawn blood.

Physiological Significance of CO_2 and HCO_3^- and of pH

Acidic materials (for example, CO_2, lactic acid, phosphoric acid) are produced during the metabolism of foods for energy. Because these acidic materials are constantly excreted into the blood, it is necessary for the body to dispose of them on a regular basis. Consequently both breath and urine are acidic.

Prior to disposal, the effect of these acidic materials on the pH of the blood is controlled by various buffers, the principal one of which is the bicarbonate and carbonic acid system.

Two types of acid-base disorders occur in the body; acidosis and alkalosis. By definition, in acidosis the pH of the blood is lower than normal. In alkalosis the blood pH is higher than normal.

Acid-base disorders can arise in various ways. Typical sources are lung and kidney problems because these organs are responsible for the acidic breath and urine, respectively. Hypofunction of either of these organs results in insufficient acid disposal, which leads to acidosis because the H^+ concentration in the blood increases. Alternately, when these organs are not the cause of a pH disturbance, they will combat the disturbance by either increasing or decreasing their disposal rate of acid, whichever tends to bring the pH back to normal.

Acid-base or pH disturbances which arise from breathing (lung, trachea) problems are called respiratory disturbances, while problems which arise elsewhere in the body (for example, kidney, intestines) are metabolic disturbances. The four basic types of acid-base disorders, therefore, are respiratory acidosis, respiratory alkalosis, metabolic acidosis, and metabolic alkalosis. Although knowing just the pH value determines whether the disorder is an acidosis or alkalosis, additional knowledge of the components of the bicarbonate buffer system is necessary in order to determine whether the origin of the disorder is metabolic or respiratory. This should be clear from an examination of Table 9-3, in which the directions of change of the important parameters in acid-base disorders are given. In general, at least two of the parameters must be known before a primary acid-base disorder can be defined. Note, for instance, that PCO_2 is decreased in both metabolic acidosis and respiratory alkalosis.

Table 9-3. Changes in Acid-Base Disorders

Primary Disorder	pH	PCO_2	CO_2 content or HCO_3^-
Metabolic acidosis	↓	↓	↓
Metabolic alkalosis	↑	↑	↑
Respiratory acidosis	↓	↑	↑
Respiratory alkalosis	↑	↓	↓

METABOLIC ACIDOSIS

Some typical causes of metabolic acidosis are diabetes (ketoacids from excess lipid metabolism accumulate), diarrhea (loss of bicarbonate-rich fluids from the lower part of the intestine), renal tubular acidosis (excess bicarbonate is lost to the urine), and accidental ingestion of acid.

METABOLIC ALKALOSIS

Two of the most common causes of metabolic alkalosis are vomiting (loss of stomach fluids, which are rich in acid) and potassium deficiency secondary to diuretic or steroid therapy. (Potassium deficiency triggers excessive renal secretion of acid into the urine.)

RESPIRATORY ACIDOSIS

Any process which interferes with breathing causes respiratory acidosis, because a reduced breathing rate raises the CO_2 level of the blood. (CO_2 is acidic by virtue of its capacity to react with the water in the blood to form H_2CO_3.) Thus trapped food in the trachea (obstruction of the upper breathing tube), narrowed bronchioles (narrowing of the passages of the lower breathing tubes, as in asthma), damaged lung membranes (destruction of the alveoli or tiny air sacs in the lung, as in emphysema), and fluid in the lung (fluid in the alveoli and bronchioles, that is, pulmonary edema) all cause respiratory acidosis because they block CO_2 escape in the breath from the body.

RESPIRATORY ALKALOSIS

Respiratory alkalosis results from hyperventilation (overbreathing) because of the excessive loss of CO_2 acidity which is involved. The dizziness and tingling of the lips which occurs when a person voluntarily hyperventilates are typical onset symptoms of mild alkalosis. Some persons feel these symptoms before a traumatic event (like a speech to a large crowd) because their emotional state causes them to hyperventilate. A more serious cause of respiratory alkalosis is salicylate intoxication from an overdose of aspirin. The salicylate, at least initially, causes hyperventilation by chemically stimulating the respiratory center in the brain.

Physiological Significance of Blood Oxygen Concentration

Decreased blood oxygenation may produce cyanosis (blueness of the skin), which is found in many disorders involving the heart and lungs. In severe diffuse lung diseases such as emphysema, the disease process in the lungs prevents blood from coming in contact with adequate amounts of oxygen. In certain forms of congenital heart disease, shunts are present whereby venous blood bypasses the lungs, allowing unoxygenated blood to mix with oxygenated blood and resulting in unsaturated blood in the peripheral circulation. Determination of blood oxygen saturation is used in establishing a diagnosis during special examinations such as cardiac catherization and respiratory function tests and thus may contribute to the planning of appropriate surgical or medical therapy. The test is also of great value in monitoring the oxygen saturation of blood during and immediately following open-heart surgery.

References

1. Bates, R. G. *Determination of pH*. New York: Wiley, 1964.
2. Burnett, R. W., and Noonan, D. C.   Calculations and correction factors used in determinations of blood pH and blood gases. *Clin. Chem.* 20: 1499, 1974.
3. Davenport, H. W. *The ABC of Acid-Base Chemistry* (5th ed., rev.). Chicago: University of Chicago Press, 1969.
4. Gambino, S. R.   Heparinized vacuum tubes for determination of plasma pH, plasma CO_2 content, and blood oxygen. *Am. J. Clin. Pathol.* 32:285, 1959.
5. Gambino, S. R.   Determination of blood P_{CO_2}. *Clin. Chem.* 7:236, 1961.
6. Gambino, S. R.   Oxygen, partial pressure (P_{O_2}) electrode method. In S. Meites (Ed.), *Standard Methods of Clinical Chemistry*. New York: Academic, 1965. Vol. 5.
7. Gordy, E., and Drabkin, D. L.   Spectrophotometric studies: XVI. Determination of the oxygen saturation of blood by a simplified technique, applicable to standard equipment. *J. Biol. Chem.* 227:285, 1957.
8. Natelson, S.   Estimation of sodium, potassium, chloride, protein, hematocrit value, sugar, urea, and nonprotein nitrogen in fingertip blood. Construction of ultramicro pipets. A practical microgasometer for estimation of carbon dioxide. *Am. J. Clin. Pathol.* 21:1153, 1951.
9. Peters, J. P., and Van Slyke, D. D.   Methods. *Quantitative Clinical Chemistry*. Baltimore: Williams & Wilkins, 1932. Vol. II, Chap. 7.
10. Scribner, B. H.   The bedside determination of bicarbonate in serum. *Proc. Staff Meet. Mayo Clin.* 25:641, 1950.
11. Skeggs, L. T., Jr.   An automatic method for the determination of carbon dioxide in blood plasma. *Am. J. Clin. Pathol.* 33:181, 1960.
12. Van Slyke, D. D., Stillman, E., and Cullen, G. E.   Studies of acidosis: XIII. A method for titrating the bicarbonate content of the plasma. *J. Biol. Chem.* 38:167, 1919.
13. Van Slyke, D. D., and Neill, J. M.   The determination of gases in blood and other solutions by vacuum extraction and manometric measurement. *J. Biol. Chem.* 61:523, 1924.
14. Van Slyke, D. D.   Note on a portable form of the manometric gas apparatus, and on certain points in the technique of its use. *J. Biol. Chem.* 73:121, 1927.
15. Van Slyke, D. D., and Sendroy, J., Jr.   Carbon dioxide factors for the manometric blood gas apparatus. *J. Biol. Chem.* 73:127, 1927.
16. Willard, H. H., Merritt, L. L., Jr., and Dean, J. *Instrumental Methods of Analysis* (5th ed.). New York: Van Nostrand, 1974.
17. Winters, R. W., Engel, K., and Dell, R. B. *Acid-Base Physiology in Medicine* (2nd ed.). Cleveland: London Company, 1969.

10. Electrolytes

In general, electrolytes are substances which form or exist as ions (charged particles) when dissolved in water. Sodium chloride is an electrolyte, because it forms Na^+ and Cl^- in water. These two ions also are called *electrolytes*. The sodium ion is called a *cation* because it is attracted by a negatively charged electrode or cathode, and the chloride ion is called an *anion* because it is attracted by a positively charged electrode or anode. All charged particles either are cations (positively charged) or anions (negatively charged).

As used in clinical chemistry, the term electrolytes refers primarily to sodium, potassium, chloride, and bicarbonate because these substances constitute the major ions in the body. They often are considered together because changes in the concentration of one of them almost always is accompanied by changes in the concentration of one or more of the others.

Carbon dioxide methodology and the importance of carbon dioxide and bicarbonate in acid-base chemistry are presented in Chapter 9. Normal values for electrolytes in serum, cerebrospinal fluid (CSF), and urine are shown in Table 10-1.

Sodium and Potassium

Measurement of sodium and potassium is almost always made by emission flame photometry. However, colorimetric methods also have been employed. Sodium can be precipitated as sodium uranyl acetate [4]. This yellow complex can be redissolved and measured spectrophotometrically [13]. Potassium can be precipitated as potassium sodium cobaltinitrite [3]. This complex can be dissolved, then quantitated based on the capacity of its cobalt to reduce the Folin-Ciocalteu phenol reagent to a blue color [8].

Ion-selective electrodes for measuring sodium and potassium are available. These electrodes employ the same principle as pH electrodes (see the section on pH in Chapter 9) except that they are so constructed as to respond only to the ion being measured (that is, Na^+ or K^+). Electrode methods are not widely used in clinical laboratories for analysis of sodium or potassium because they are much less efficient to use than flame emission photometry.

EMISSION FLAME PHOTOMETRY

The general principles of emission flame photometry are described in Chapter 5. Sodium produces a yellow color in a flame and potassium a violet color. The amount of each color is proportional to the amount of each of these elements in the sample. The color is measured with a detector and read with a meter. Lithium, which produces a red color in the flame, is used as an internal standard with some instruments. An

130

Table 10-1. Normal Values for Electrolytes

Electrolyte	Serum (meq/liter)	CSF (meq/liter)	Urine (meq/d)
Bicarbonate	24–30	25–28	None
Chloride	100–108	120–130	75–200
Sodium	138–146	142–150	75–200
Potassium	3.8–5.0	2.2–3.4	40–80

exact amount of lithium is added to each sample. The intensity of the yellow (sodium) and violet (potassium) colors in the flame are compared by the instrument with the intensity of the red color from the lithium internal standard. This helps to compensate for changes in the condition of the flame and in the levels of interfering substituents in the samples. The absolute intensity levels of the colors being measured are more easily changed by these effects than are their relative or ratio values. For example, a slight change in gas pressure might result in a change in the sodium or potassium signal, but the lithium signal would change to approximately the same degree, so the *ratio* would not be affected significantly.

Automated methods for sodium and potassium analyses are available in both continuous flow and discrete analysis systems in which the samples are diluted automatically and presented to a flame photometer equipped with a recorder or printout device.

Most flame photometric methods have a coefficient of variation of approximately 1.0% for sodium and 2.0% for potassium.

PREPARATION OF SAMPLES AND STANDARDS
The glassware, pipets, and stoppers used in preparing dilute samples for flame photometry must be carefully cleaned and rinsed well with purified water. The purity of the water may be checked by aspirating some water into the photometer and watching for any color change in the flame. A light yellow color suggests sodium contamination. For the direct measurement procedure, the sample is diluted quantitatively with water and mixed well. Because of the large difference between the concentrations of sodium and potassium in serum (the ratio of sodium to potassium is approximately 30:1), some methods require that different dilutions of the sample be made for determination of each element. Although the dilution requirements vary from one instrument to another, potassium usually is determined with a 1:25 or a 1:50 dilution and sodium with a dilution ranging from 1:100 to 1:500. However, some instruments permit determination of both sodium and potassium with a single dilution.

When the internal-standard method is employed, a certain amount of lithium solution must be added to the sample before dilution. Since this method measures the ratio of lithium to sodium or potassium, measurement of the lithium solution must be as exact as that of the sample. However, since it is a *ratio* rather than an absolute concentration which is measured, the final dilution of the sample with water need not be as accurate as when the direct method is used.

In practice, dilution errors of 5%, or even more in some instances, may be tolerated without effect on instrument readings. However, analysts are urged to check the response of their own instrument to differences in dilution by measuring several portions of the same sample or standard, including the appropriate amount of lithium, and diluting each to a slightly different total volume. Aspirating these samples into the instrument in succession will show whether or not there is a response to dilution differences.

Emission photometry, like spectrophotometry, requires that the sample be compared with standard solutions. Proper preparation and use of standards are fundamentally important to the accuracy of the procedure. Stock standards of sodium and potassium should be prepared from reagent-grade sodium chloride and potassium chloride. The salts should be dried in an oven at 110 C and cooled in a desiccator before use.

REAGENTS

Stock sodium chloride 100 meq/liter. Weigh 5.8454 g of dry sodium chloride (NaCl).[1] Dissolve in purified water, transfer quantitatively to a 1 liter volumetric flask, and dilute to volume.

Stock potassium chloride 10 meq/liter. Weigh 0.7455 g of dry potassium chloride (KCl).[1] Dissolve in purified water, transfer quantitatively to a 1 liter volumetric flask, and dilute to volume.

Stock lithium solution. The optimum concentration of lithium to be used with an internal-standard method is usually specified by the manufacturer of the instrument. The chloride, nitrate, carbonate, or sulfate salts of lithium may be used in preparation of lithium solutions. Lithium sulfate ($Li_2SO_4 \cdot H_2O$) is recommended for this purpose, since it has a relatively high molecular weight, is quite soluble, and is not significantly deliquescent.

The stock lithium solution should be prepared in such concentration that some easily measured quantity may be used in preparing the samples. Thus, if the final working concentration of lithium is to be 15 meq per liter, and 5 ml of stock lithium solution are to be used to prepare a

[1] Available as a Standard Reference Material from the U.S. National Bureau of Standards.

1:100 dilution of the sample, the concentration of the stock lithium solution must be

$$\frac{100}{5} \times 15 = 300 \text{ meq/liter}$$

If lithium sulfate ($Li_2SO_4 \cdot H_2O$) is used (mol wt = 128, eq wt = 64),[2] a 300 meq per liter solution would require 64×0.3 (eq) = 19.2 g of the salt dissolved and diluted to 1 liter.

Because of the wide application of flame photometry in industry, the term *parts per million* (ppm) often is used as an expression of lithium concentration. This term is defined in Tables 1-1 and 1-2 in Chapter 1. Since the equivalent weight of lithium is approximately 7, the approximate relationship of parts per million and milliequivalents per liter for lithium is as follows:

meq/liter $\times$ 7 = ppm

Thus the 300 meq/liter solution of lithium described above is approximately 2100 ppm in lithium.

The exact concentration of lithium used is not as important as the fact that the lithium concentration in all samples and standards be the *same*. Some analysts prefer to make the stock lithium solution to *approximately* the desired concentration and to prepare new standards each time a new stock lithium solution is made. However, if each new lithium solution is carefully prepared and tested against the old one, preparation of new standards is unnecessary.

All stock solutions should be stored in plastic containers because a significant amount of sodium may be leached from some types of glass bottles.

CALCULATIONS

At least two standards should accompany each series of sodium or potassium analyses. Ideally one standard should have a higher concentration than the tests and another a lower concentration. Calculations are made by plotting a standard curve as described for spectrophotometric methods (see page 55) and determining the concentrations of the tests from the standard curve. In emission photometry the readings are directly and linearly related to concentration. So curves are plotted on linear graph paper.

Some flame photometers permit calibration in concentration units and therefore provide direct readout of test results.

---

[2] Molecular weight = mol wt and equivalent weight = eq wt.

SODIUM AND POTASSIUM IN URINE

Because of the wide variation in sodium and potassium concentrations in urine, sometimes samples must be prepared separately for sodium and potassium analyses. If urine first is diluted 1:10 with water and this solution is then rediluted in the same manner as serum, the readings for potassium usually will fall somewhere in the range of potassium concentrations in serum.

For sodium analysis, the urine first should be treated the same as serum. If the resultant readings are outside the range of the available standards, a more suitable dilution should be made. With the internal-standard method, the lithium concentration remains constant even when the size of the sample is altered.

Some flame photometers have sufficient range to accommodate, with a single dilution, any concentration of sodium or potassium likely to be found in urine.

PRECAUTIONS

Since sodium is such a common element, contamination can be a serious problem. All glassware and stoppers should be rinsed carefully with purified water after washing. Use of chemically clean disposable glassware is recommended. Care should be taken not to handle stoppers or containers in such a way as to introduce sodium from sweat on the skin. That the fingers are a common source of contamination can be demonstrated by dipping a finger into some pure water, nebulizing the water, and noting the color change in the flame.

The measuring instrument itself should be kept clean. There should be a regular schedule for cleaning the nebulizer, the burner, and the optical windows.

The compressed air line should have a filter, and the filter element should be changed periodically. The fuel (usually propane) should be free of contaminants (usually evident as yellow flecks in the flame).

Tobacco smoke is a source of potassium contamination. Even though a closed burner system minimizes potassium contamination, smoking should not be permitted in the immediate vicinity of the instrument while potassium analyses are being performed.

The potassium concentration inside the red blood cells is approximately 25 times that in the serum. Therefore hemolysis will cause serious positive errors in the serum potassium level. Gross increases in the serum potassium concentration, with equivalent decreases in the sodium concentration, also have been noted when the serum is allowed to remain in contact with the cells for a period of time. Thus spurious potassium values may occur without visible evidence of hemoglobin in the serum. These errors, due to exchange of intracellular and extracellular material, are accelerated by refrigeration. For this reason, serum should be separated from the cells as soon as possible if accurate sodium and potassium results are to be expected. Whole blood should never be refrigerated prior to separation.

Normal values for sodium and potassium in serum, urine, and CSF are given in Table 10-1.

Chloride

There are several methods in common use for the quantitative determination of chloride. Some are based on the principle of precipitating chloride as an insoluble salt and quantitating either the precipitate or the excess precipitant. One such method [9] employs an excess quantity of a standard solution of silver nitrate in nitric acid. Chloride in the sample combines with silver to form insoluble silver chloride. After digestion of the protein,[3] the excess silver is titrated with potassium thiocyanate solution, resulting in the precipitation of silver thiocyanate. Ferric ammonium sulfate (ferric alum) is included as an indicator since it combines with free thiocyanate, after the free silver is used up, to form orange-colored ferric thiocyanate. The reactions are as follows:

1. $Cl^- + AgNo_3 \rightarrow AgCl\downarrow + $ excess Ag^+

2. Excess $Ag^+ + KCNS \rightarrow AgCNS\downarrow$

3. Excess $CNS^- + Fe^{3+} \rightarrow Fe(CNS)_3$ (orange color)

This is an excellent reference method that may be used for determining chloride concentration in biological materials. However, it is somewhat involved to use for multiple analyses in the clinical laboratory.

Chloride may also be determined by shaking a solution containing chloride with solid silver iodate [14]. Since silver chloride is less soluble than silver iodate, the following reaction takes place:

$$Cl^- + AgIO_3 \rightarrow AgCl\downarrow + IO_3^-$$

Iodide is added to the solution, and the free iodine formed by the interaction of the iodate (IO_3^-) and the iodide (I^-) is titrated with thiosulfate solution. Protein should be removed at the outset by precipitation and centrifugation. This is a satisfactory method in which the titrated amounts of thiosulfate are large and the endpoint is sharp and permanent. The method has also been adapted so that the iodine con-

---

[3] Since protein combines with heavy metals (like silver), it should be removed before final titration.

centration may be determined colorimetrically, but the titrimetric procedure is more sensitive and accurate. A modification of the iodate method has been suggested using silver iodate in solution rather than in solid form [7].

The method commonly used with continuous flow automated systems employs a mercuric thiocyanate reagent [12, 16]. In this method, chloride forms a soluble, nonionized complex with mercury thereby releasing thiocyanate ions which combine with ferric ions to form the colored complex ferric thiocyanate. The reactions are similar to those shown on page 135.

$$Hg(SCN)_2 + 2Cl^- \rightarrow HgCl_2 + 2(SCN^-)$$

$$3(SCN^-) + Fe^{3+} \rightarrow Fe(SCN)_3$$

A potentiometric method is in common use which requires a special instrument called a *chloridometer* [5]. This instrument has two sets of electrodes, one set serves as a sensor (endpoint detector) and the anode of the other pair is a spool of silver wire. The ends of all the electrodes are immersed in the test solution (for example, serum diluted with nitric acid). When the titration is begun a constant current is generated which effects the release of silver ions from the silver wire into the solution. These silver ions combine with chloride in the sample to form insoluble silver chloride. When all the chloride is used, the presence of free silver ions in solution results in an abrupt change in potential between the sensor electrodes. This change activates a microswitch, automatically stopping the titration. The elapsed time of each titration is automatically recorded on a timer. Since the rate of release of silver ions is constant, the amount of time during which they are released is directly proportional to the amount of chloride in the test solution. That is, the more chloride present in the solution, the longer it takes to generate enough silver ions to combine with all of it. The instrument is standardized with a standard chloride solution. Calculations are made comparing the titration time of the sample to the titration time of the standard. This simple, accurate method is highly recommended.

The mercurimetric titration method is presented here in detail. It is a simple, reliable method which may be used when automated systems or chloridometers are not available.

CHLORIDE METHOD

METHOD

Schales and Schales [11]; Rice [10].

MATERIAL

Serum, cerebrospinal fluid, urine.

REAGENTS

Mercuric nitrate. Weigh 3.2 g of mercuric nitrate, $Hg(NO_3)_2$, and dissolve in a solution of 3 ml of concd nitric acid plus 20 ml of water. Transfer to a 1 liter volumetric flask and dilute to 1 liter with water.

Although the exact concentration of this solution is not important, the calculations are simplified if the solution is tested and adjusted so that the 100 meq per liter standard titrates 1.00 ml of this solution.

Nitric acid, approximately 1N. Dilute 6 ml of concd nitric acid to 100 ml with water.

Diphenylcarbazone, 0.1% (indicator). Dissolve 100 mg of diphenyl-carbazone, $C_6H_5N:NCO(NH)_2C_6H_5$, in 100 ml of 95% ethyl alcohol. Store in a refrigerator and protect from the light.

Ethyl ether.

APPARATUS

Microburet of 2 ml capacity calibrated in increments of 0.01 ml.

PROCEDURE

A. *Serum and cerebrospinal fluid*

   Into a 25 ml Erlenmeyer flask measure:

   2 ml of water
   0.2 ml of sample
   1 drop of 1N HNO_3
   3 drops of indicator
   1 ml of ethyl ether

   Titrate carefully to the first trace of purple[4] color that persists in the ether layer. A standard should be titrated in the same fashion (see Standardization).

B. *Urine*

   The chloride concentration in urine is quite variable, and an amount of sample should be used that will result in a suitable titration. If a very low or high volume of titrant is used, the determination should be repeated using a more suitable amount of urine.

   Occasionally a very alkaline urine will produce a pink color as soon as the titration is begun [2]. If this happens, add 1N nitric acid dropwise until the color disappears. Then proceed with the titration to the usual endpoint.

CALCULATIONS

$$\frac{\text{titration of sample (ml)}}{\text{titration of standard (ml)}} \times 100 = Cl \text{ (meq/liter)}$$

---

[4] With very icteric serum the endpoint color may be more red than purple.

If an amount other than 0.2 ml of sample is used, the results must be multiplied by

$$\frac{0.2}{\text{amount used (ml)}}$$

Sometimes it is desirable to express the concentration of urine chloride in terms of grams per liter of chloride, or sodium chloride. These may be calculated as follows:

$$\frac{\text{meq/liter of Cl}}{1000} = \text{eq/liter of Cl}$$

Cl (eq/liter) $\times$ eq wt of Cl (35.5) = Cl (g/liter)

Cl (eq/liter) $\times$ eq wt of NaCl (58.5) = NaCl (g/liter)

Grams per 24 hours may be obtained by multiplying grams per liter by

$$\frac{\text{volume (ml)}}{1000}$$

STANDARDIZATION

Either of the following solutions is suitable to use as a chloride standard:

1. Exactly $0.100N$ HCl, which has a chloride concentration of 100 meq per liter
2. Sodium chloride 100 meq per liter
   Dissolve 5.8454 g of sodium chloride[5] in water and dilute to 1 liter. Before weighing, the salt should be dried in an oven at 110 C for several hours and allowed to cool in a desiccator. This solution may be used also for preparing sodium standards. (See "Preparation of Samples and Standards" in the section on Sodium and Potassium of this chapter.)

NORMAL VALUES

Normal values for chloride in serum, urine, and CSF are shown in Table 10-1.

DISCUSSION

The method described above has an accuracy of approximately $\pm 2\%$.

When mercuric nitrate is added to an acid solution containing chloride ions, undissociated mercuric chloride is formed:

$$2Cl^- + Hg(NO_3)_2 \rightarrow HgCl_2 + 2NO_3^-$$

[5] Available as a Standard Reference Material from the U.S. National Bureau of Standards.

When the chloride is used up, free mercuric ions appear in the solution. These combine with diphenylcarbazone to form a purple-colored complex. This complex is extracted into the ether layer where its color is more distinctive. Appearance of the endpoint color is quite obvious if the flask is tilted at an angle and viewed from the side at eye level.

Heavy metals bind protein; therefore, in methods employing silver or mercury, protein inadvertently may be measured as chloride. It has been shown that in the original, Schales-and-Schales method, failure to deproteinize yields spuriously high results [1]. In the present method, slight excess of mercuric ions added after the chloride is used up combine with indicator in the ether phase and, therefore, are not available to the protein in the aqueous phase.

Blood for chloride determination preferably is drawn in vacuum tubes. The serum should be separated from the cells as soon as possible since prolonged contact between serum and cells may result in falsely high chloride values for the serum. Moderate hemolysis does not significantly affect the chloride concentration of serum.

Glassware and pipets used for chloride analysis should be clean. Sources of chloride contamination in the laboratory are many; even tap water may contain a significant amount of chloride. For this reason, all apparatus should be rinsed well with purified water after washing. The use of clean disposable glassware is recommended.

If bromide is present in the sample, mercuric bromide will be formed during the titration and it will be calculated as chloride. The amount of bromide normally present in blood is not detectable by this method; however, in bromide therapy or poisoning, the quantity may be significant. Although this represents an analytical error in the chloride method, clinically it is not important since bromide behaves as chloride in electrolyte physiology. If the bromide (Br) concentration is determined (see Chapter 26), the true chloride (Cl) may be calculated as follows:

apparent Cl − [Br, mg/100 ml × 0.125] = true Cl (meq per liter)

Physiological Significance of Electrolyte Concentrations in Serum

Electrical neutrality is maintained at all times in the body fluids. That is, the sum of all the cations (positively charged substances) equals the sum of all the anions (negatively charged substances). This is illustrated in Table 10-2 using some average data for plasma. Two types of changes in plasma electrolyte concentrations occur in disease. One is a change in the *total* electrolyte concentration to a value other than 155 meq per liter. The patient then either is hypertonic (electrolytes > 155 meq per liter) or hypotonic (electrolytes < 155 meq per liter). Electrical neutrality still is maintained, however. If the new total cation value is 130 meq per liter, for instance, then the new anion value also is 130 meq per

Table 10-2. Average Electrolyte Composition of Human Serum

Cation	Concentration (meq/liter)	Anion	Concentration (meq/liter)
Na+	143	Cl⁻	104
K+	4.5	HCO_3^-	29
Ca++	5	Protein	16
Mg++	2.5	HPO_4^{--}	2
Total	155	SO_4^{--}	1
		Organic Acids	3
		Total	155

liter. The other type of change is a change in the *relative* amounts of the electrolytes. For instance, an increase in chloride of 5 meq per liter may accompany a decrease in bicarbonate of 5 meq per liter. The resulting total anion value still is 155 meq per liter. Of course, simultaneous changes in the total amount as well as relative amounts of the individual anions or cations also can occur.

Electrolyte values often are requested for the purpose of defining the electrolyte status of the patient. Medications and fluids then can be chosen which will help restore the electrolyte values to normal. For instance, if the potassium level is low, and there are no complicating factors, the patient may be given potassium. If the patient is hypertonic, he may be given extra water.

A second and simultaneous use of electrolyte values is to aid in diagnosis. Primary value usually is placed on other findings (history, physical examination, other laboratory values), however, because there are numerous combinations of normal and abnormal electrolyte concentrations which accompany various physiological disorders. It must be emphasized that proper clinical interpretation of electrolyte values requires clinical experience as well as a working knowledge of electrolyte metabolism. The significance of an abnormal concentration of a given electrolyte frequently depends on the concentration of the other serum electrolytes as well as the state of hydration of the patient. Furthermore, patients may have, at a given moment, more than one disorder influencing electrolyte concentrations in the serum. For these reasons, it is impossible to relate abnormal concentrations of serum electrolytes to specific disorders in a simple manner. Table 10-3 contains an outline of some of the more common situations that are encountered. This subject is treated in greater depth elsewhere [15].

RELATIONSHIP OF SODIUM TO CHLORIDE AND BICARBONATE

It is not the prerogative of the technician to evaluate results or to suggest therapeutic measures, but an elementary knowledge of the relationships

Table 10-3. Some Serum Electrolyte Patterns

Clinical Condition	Na	K	HCO₃	Cl
Dehydration	↑	N[a]	N or ↓	↑
Diarrhea	↓	↓	↓	↓
Congestive heart failure	N or ↓	N	N	↓
Malabsorption syndrome	↓	↓	N or ↓	N
Pyloric obstruction	↓	↓	↑	↓
Acute renal failure	↓	↑	↓	↑
Pulmonary emphysema	N	N	↑	↓
Diabetic acidosis	↓	N or ↑	↓	↓
Excessive sweating	↓	N	N	↓
Starvation	N	↓	↓	N
Ammonium chloride administration	↓	↓	↓	↑

[a] N means normal.

between serum electrolytes is helpful in determining the probable accuracy of laboratory data before they are reported.

Sodium is the cation that is present in highest concentration in serum, while chloride and bicarbonate are the two anions of highest concentration. Table 10-2 shows that the sum of bicarbonate and chloride concentrations is about 10 meq per liter less than the sodium value. Unless pathological variations occur, this general relationship holds true to within about 5 meq [6].

If the difference between the sum of the carbon dioxide plus chloride concentration and the sodium concentration is significantly greater than 10 (15 or more), an increase in some unmeasured anion is indicated. The three most common causes of this occurrence are: (1) Uremia or kidney damage, resulting in the retention of endogenous acids. These acids may be either inorganic (phosphate or sulfate) or organic (amino acids). This situation usually is verified by an elevated blood urea nitrogen. (2) Diabetic acidosis with retention of ketonic acids in the blood. These acids act in the same way as those present in uremia. Their presence may be verified by high blood glucose and blood acetone concentrations accompanied by a positive urine acetone. (3) Types of poisoning resulting in the formation of unmeasured acids in the blood. For example, ingested methyl alcohol is converted to formic acid.

An unusually small difference (less than 5) between the sum of the common anion concentrations and the sodium value is not very common. If the serum albumin concentration is low, such a small difference between the measured anion and cation concentrations may be observed.

One general rule that technicians can apply is that the sum of the carbon dioxide[6] plus chloride concentrations must always be *less* than the sodium concentration. Each set of electrolyte analyses should be added and evaluated with respect to other analyses (urea nitrogen, proteins, glucose) as discussed above, and to previous chemical results for the patient.

It should be evident that at least carbon dioxide, chloride, sodium, and potassium determinations are necessary in order adequately to interpret a complicated problem of electrolyte disturbance. In such cases, physicians should be encouraged to request all the electrolyte analyses simultaneously to facilitate interpretation and to obviate the need for single analyses which later may be found necessary for interpretation of the results. However, there are many instances where one or two of the electrolyte concentrations provide adequate information.

When initial results or changes do not coincide as described herein, the validity of the results should be checked in the laboratory beginning with the most suspicious-looking results.

EXAMPLES
1. Admission determinations:

$$\left.\begin{array}{l} CO_2 = 29 \\ Cl = 112 \end{array}\right\} = 141$$

$$Na = 144$$

$$K = 4.5$$

$$144 - 141 = 3 \qquad \text{(small difference)}$$

Since the carbon dioxide and the sodium values are normal while the chloride concentration is high, the chloride determination should be repeated.
2. Serum electrolyte concentrations for one patient on two successive days:

   a. $CO_2 = 22$, $Cl = 98$, $Na = 132$, $K = 3.8$

   b. $CO_2 = 17$, $Cl = 97$, $Na = 133$, $K = 3.9$

The unmeasured anion value has increased by 7 meq per liter. Check urea nitrogen for rise; chloride for possible negative error; carbon dioxide for possible negative error; and sodium for possible positive error.

Occasionally a set of results will not coincide with the above-stated conditions even after verification by repeat determinations with the same

[6] Either CO_2 content or bicarbonate values may be used.

sample. In these cases the patient's physician should be informed personally regarding the problem. Besides the usual laboratory errors caused by contamination or faulty technic, there are two main causes of erroneous electrolyte results which are common enough to warrant mention here: (1) occasionally, because the proper equipment is not at hand, a wet or contaminated syringe or tube is used; and (2) sometimes blood may be drawn during the administration of an intravenous infusion. In the latter case, if the infusion contains electrolytes there may be gross contamination of the blood sample. In any other infusion (for example, glucose), the dilution factor may be large enough to invalidate the results.

Electrolytes in Urine

The physiological significance of chloride, sodium, and potassium concentrations in urine is an intricate problem because urinary electrolyte concentrations are grossly influenced by many factors such as intake of the various ions and water, the state of relative body depletion or repletion, acid-base balance of the body, or functional integrity of the renal tubules.

Determination of urinary electrolytes is of value only when the physician is able to evaluate all of the relevant factors. Since electrolytes are not excreted in uniform concentration throughout the day, the total twenty-four-hour excretion value usually is more important than expressions of concentration in random samples. Due to the many influencing factors, it is difficult to cite normal values for electrolyte concentrations in urine. However, Table 10-1 gives approximate normal values for healthy persons under average conditions of diet and activity.

References

1. Annino, J. S.   The effect of serum proteins on the determination of chloride by mercurimetric titration. *J. Lab. Clin. Med.* 38:161, 1951.
2. Asper, S. P., Jr., Schales, O., and Schales, S. S.   Importance of controlling pH in the Schales and Schales method of chloride determination. *J. Biol. Chem.* 168:779, 1947.
3. Breh, F., and Gaebler, O. H.   The determination of potassium in blood serum. *J. Biol. Chem.* 87:81, 1930.
4. Butler, A. M., and Tuthill, E.   An application of the uranyl zinc acetate method for determination of sodium in biological material. *J. Biol. Chem.* 93:171, 1931.
5. Cotlove, E., Trantham, H. V., and Bowman, R. L.   An instrument and method for automatic, rapid, accurate, and sensitive titration of chloride in biological samples. *J. Lab. Clin. Med.* 51:461, 1958.
6. Hald, P. M., Heinsen, A. J., and Peters, J. P.   The estimation of serum sodium from bicarbonate plus chloride. *J. Clin. Invest.* 26:983, 1947.
7. King, E. J., and Bain, D. S.   A simplified silver iodate method for the determination of chloride. *Biochem. J.* 48:51, 1951.
8. Looney, J. M., and Dyer, C. G.   A photoelectric method for the determination of potassium in blood serum. *J. Lab. Clin. Med.* 28:355, 1942.

9. Peters, J. P., and Van Slyke, D. D. *Methods.* In *Quantitative Clinical Chemistry.* Baltimore: Williams & Wilkins, 1932. Vol. II, p. 829.
10. Rice, E. W. Improved direct mercurimetric titration of chloride in biologic fluids. *Am. J. Clin. Pathol.* 27:252, 1957.
11. Schales, O., and Schales, S. S. A simple and accurate method for determination of chloride in biological fluids. *J. Biol. Chem.* 140:879, 1941.
12. Skeggs, L. T., Jr. *AutoAnalyzer Manual.* Ardsley, N.Y.: Technicon Instruments.
13. Stone, G. C. H., and Goldzeiher, J. W. A rapid colorimetric method for determination of sodium in biological fluids and particularly in serum. *J. Biol. Chem.* 181:511, 1949.
14. Van Slyke, D. D., and Hiller, A. Application of Sendroy's iodometric chloride titration to protein-containing fluids. *J. Biol. Chem.* 167:107, 1947.
15. Welt, L. G. *Clinical Disorders of Hydration and Acid-Base Equilibrium* (2nd ed.). Boston: Little, Brown, 1959.
16. Zall, D. M., Fisher, D., and Garner, M. O. Photometric determination of chlorides in water. *Anal. Chem.* 28:1665, 1956.

11. Glucose

Many methods are available for the determination of glucose. Basically, they fall into three categories: aromatic amine methods, enzymatic methods, and oxidation methods.

A number of aromatic amines, such as ortho-toluidine (o-toluidine), aniline, and benzidine react with glucose in a hot acid solution to form colored products. These methods involve a reaction between the aldehyde group

$$\overset{\displaystyle H}{(-\overset{|}{C}=O)}$$

of glucose and the amino group ($-NH_2$) of the aromatic amine. Currently, o-toluidine [3], shown below, is the most widely used aromatic amine for glucose determinations.

o-Toluidine

The two most widely used enzymatic methods are based either on the enzymes glucose oxidase [8] or hexokinase [7]. In the former case, glucose and oxygen react in the presence of glucose oxidase to form gluconic acid and hydrogen peroxide.

$$
\begin{array}{c}
\text{H}-\text{C}=\text{O} \\
\text{HCOH} \\
\text{HOCH} \\
\text{HCOH} \\
\text{HCOH} \\
\text{CH}_2\text{OH} \\
\text{Glucose}
\end{array}
\;+\; \text{O}_2 + \text{H}_2\text{O} \;\xrightarrow{\text{glucose oxidase}}\;
\begin{array}{c}
\text{CO}_2\text{H} \\
\text{HCOH} \\
\text{HOCH} \\
\text{HCOH} \\
\text{HCOH} \\
\text{CH}_2\text{OH} \\
\text{Gluconic Acid}
\end{array}
\;+\; \text{H}_2\text{O}_2
$$

The change in concentration of either the O_2 or the H_2O_2 may be monitored by an electrode process, or a second enzyme, peroxidase, may be used to catalyze the oxidation of a substance like o-toluidine [11] to a colored product.

$$\text{H}_2\text{O}_2 + o\text{-toluidine} \xrightarrow{\text{peroxidase}} \begin{array}{c}\text{oxidized}\\ o\text{-toluidine}\end{array} + \text{H}_2\text{O}$$
$$(\text{color})$$

Although the first step in the colorimetric method (involving glucose oxidase) is quite specific, the second step (involving peroxidase) is subject to interference from a number of substances such as ascorbic acid, uric acid, bilirubin, glutathione, or hemoglobin. In each case, less color development occurs, probably because the hydrogen peroxide is being consumed by the interfering materials. These interferences can be eliminated by the use of a Somogyi-Nelson filtrate [9], a deproteinization procedure in which these substances and proteins are precipitated as zinc and barium salts in an alkaline solution (see the following section).

The other enzymatic method involves hexokinase. This method is less subject to interferences than the glucose oxidase-peroxidase method and involves the following two reactions in sequence:

$$\text{glucose} \xrightarrow[\text{ATP}]{\text{hexokinase}} \text{glucose 6-phosphate}$$

$$\text{glucose 6-phosphate} \xrightarrow[\text{NADP}]{\substack{\text{glucose 6-phosphate} \\ \text{dehydrogenase}}} \text{6-phosphogluconate} + \text{NADPH}$$

The latter reaction is followed by noting the increase in absorbance at 340 nm.

The two most widely used oxidizing agents in the oxidation methods for glucose are cupric ion, Cu^{2+} and ferricyanide ion, $Fe(CN)_6^{3-}$. Both are used in alkaline solution and involve a change in color. In the former case, the light blue cupric ion is reduced by glucose to the cuprous ion, Cu^+, which can occur as yellow cuprous hydroxide (CuOH) or be converted by heat to red cuprous oxide (Cu_2O). However, this reaction usually is not quantitated by this color change, but by the further reaction of the cuprous ion with a reagent like phosphomolybdate [4], arsenomolybdate [9], or neocuproin [1] to form a colored product. The former two substances are reduced by cuprous ion, while neocuproin is a complexing agent which is specific for Cu^+. Because the intermediate cuprous ion can be oxidized back to cupric ion by air, it is important to take precautions in any cupric-ion method to minimize the contact of the reaction mixture with air. Time, temperature, salt concentration, and other variables must be controlled carefully because the glucose is oxidatively split into a number of fragments by the cupric ion. The exact nature and extent of the products formed depends on all of these factors. In order to obtain reliable glucose values with any of the cupric-ion methods, it is necessary to prepare a protein-free filtrate, such as the Somogyi-Nelson filtrate [9, 10] (see following section), since this technic removes most of the noncarbohydrate-reducing substances from biological samples (along with protein). This includes glutathione, creatinine, creatine, uric acid, ascorbic acid, phenols, and certain amino acids. When they are not removed, these substances may cause spuriously high results.

Ferricyanide, which is yellow, is reduced by glucose to ferrocyanide, $Fe(CN)_6^{4-}$, which is almost colorless. The reaction may be measured, for example, either by the loss in yellow color [6] (at 420 nm) or by the addition of ferric ions [5], which react with the ferrocyanide to form Prussian blue (ferric ferrocyanide), as shown below:

$$3Fe(CN)_6^{4-} + 4Fe^{3+} \rightarrow \underset{\text{Prussian blue}}{Fe_4[Fe(CN)_6]_3}$$

One of the advantages of the ferricyanide methods over the cupric methods is that ferrocyanide is not as readily oxidized by air as is cuprous ion. However, this method is subject to the same interferences as the cupric method and therefore requires an appropriate protein-free filtrate for reliable glucose values.

Somogyi-Nelson Filtrate [9, 10]

Although the Somogyi-Nelson filtrate is not required in the glucose method presented in this chapter, directions for its preparation are included because of its importance in other glucose methods.

REAGENTS

Barium hydroxide, 0.3N. Weigh 28.4 g of barium hydroxide mono-hydrate, $Ba(OH)_2 \cdot H_2O$, or 47.2 g of barium hydroxide octahydrate ($8H_2O$). Dissolve the salt in 1 liter of water. Let it stand for several days in a covered container. Then decant. Store the solution in a plastic bottle and protect it from air.

Zinc sulfate 5% (w/v). Dissolve 50 g of zinc sulfate heptahydrate ($ZnSO_4 \cdot 7H_2O$) in water, and dilute to 1 liter.

These two solutions should be tested as follows:
To a 50 ml Erlenmeyer flask add 5 ml of zinc sulfate solution, 10 ml of water, and one drop of 0.1% phenolphthalein. Titrate the barium hydroxide from a 10 ml buret until a faint pink endpoint is reached. This should require 4.7 to 4.8 ml; if the titration is outside these limits the *stronger* solution should be adjusted. If the titration is *less* than 4.7 ml, the barium solution is stronger, and the concentration factor is

$$\left(\frac{4.75}{\text{titration}} \right)$$

This factor is multiplied times the total volume of the barium solution to be adjusted. The difference between this calculated result and the actual volume of the barium solution represents the amount of water to be added.

If the titration is *greater* than 4.8, the concentration factor is

$$\left(\frac{\text{titration}}{4.75}\right)$$

and this is applied to the total volume of the zinc solution to calculate the amount of water to be added to it.

PROCEDURE
Filtrate. Into a 50 ml Erlenmeyer flask measure:

1.0 ml of blood, serum, or CSF
15 ml of water
2.0 ml of barium hydroxide solution

Mix and allow to stand at least 30 seconds.
Add 2.0 ml of zinc solution, mix well, and after 2 minutes filter.

The *o*-Toluidine Method for Glucose

The *o*-toluidine method is presented here because it is essentially free of interferences, simple, and widely used. Although serum can be used without deproteinization, additional specificity is provided with deproteinized samples. For this reason, the method presented here includes the preparation of a protein-free filtrate.

GLUCOSE METHOD
METHOD
Dubowski [3].

MATERIAL
Serum or CSF.

REAGENTS
Benzoic acid, 0.1% (w/v). Dissolve 1 g of benzoic acid in 1 liter of water.

Trichloroacetic acid, 3% (w/v). Dissolve 15 g of trichloroacetic acid in water and dilute to 500 ml.

Ortho-toluidine reagent. Dissolve 0.3 g of thiourea in 188 ml of glacial acetic acid. Add 12 ml of *o*-toluidine, mix, and store in a dark bottle in a refrigerator.

PROCEDURE
To 0.5 ml of serum in a 16 × 100 mm test tube, add 4.5 ml of trichloroacetic acid (Test).
Shake to mix and let stand for 5 minutes.
Centrifuge for 5 minutes.
To 2 other tubes add 4.5 ml of trichloroacetic acid. To one add 0.5 ml

of water (Blank), and to the other add 0.5 ml of working standard (Standard). Mix each tube.

Measure 0.5 ml of the Test, Blank, and Standard into three 16 × 100 mm tubes.

Add 3.5 ml of *o*-toluidine reagent to each and mix.

Cover the tubes with Parafilm and place in a boiling water bath for 12 minutes.

Cool the tubes in ice water. Then bring them to room temperature.

Transfer the solutions to 12 × 75 mm cuvets and read the absorbance of the Test and Standard at 630 nm against the Blank set at zero.

CALCULATIONS

$$\frac{\text{absorbance of Test}}{\text{absorbance of Standard}} \times 100 = \text{glucose (mg/dl)}$$

STANDARDIZATION

Stock standard, 5 g/dl. Dissolve 5 g of glucose (dextrose)[1] in 50 ml of benzoic acid solution, and dilute to 100 ml with the same solvent. Store in the refrigerator. (The benzoic acid is an antibacterial agent.)

Working standard, 100 mg/dl. Dilute 1 ml of stock standard to 50 ml with benzoic acid solution and store in the refrigerator.

This 100 mg/dl glucose standard is processed the same as the serum samples.

NORMAL VALUES

The normal range for glucose concentrations in the serum of fasting adults is 65–100 mg per dl. With some less specific methods the stated normal range may extend to 110 or even 120 mg per dl.

DISCUSSION

The nature of the product(s) involved when *o*-toluidine reacts with glucose has been discussed [2]. The *o*-toluidine need not be highly pure; actually, purified reagent may not work as well as technical-grade material. In addition, older reagents tend to give less color. It is important to always include one or more standards with each set of tests.

Close adherence to the heating times and temperatures is important since the intensity of the color produced depends on these factors and may vary from one lot of *o*-toluidine to another. The intensity of the color is constant for about 0.5 h and then starts to decrease slowly.

Best results are achieved when the glacial acetic acid used to prepare the *o*-toluidine reagent is as anhydrous as possible. The solution is stabilized against air oxidation with thiourea, a reducing agent.

However efficient a glucose method may be, it cannot compensate for

---

[1] Available as a Standard Reference Material from the U.S. National Bureau of Standards.

errors inherent in the blood sample. Glycolysis, or breakdown of glucose, takes place rapidly in collected blood and in cerebrospinal fluid. In blood samples enzymes enhance glycolysis and, in infected cerebrospinal fluid, bacteria destroy glucose. Consequently samples for glucose analysis should be delivered to the laboratory as soon as possible after being obtained. When samples are received in the laboratory, deproteinization should be carried out without delay. In the protein-free supernatant, the glucose concentration is maintained reasonably well, whereas in whole blood or cerebrospinal fluid as much as 10 mg per dl of glucose per hour may be lost at room temperature through glycolysis.

Frequently it is desirable to preserve samples for glucose analysis by means other than the preparation of a protein-free supernatant. Refrigeration retards glycolysis and may maintain the glucose concentration for several hours. Sodium fluoride in a concentration of 1 mg per ml of blood is effective in inhibiting glycolysis. However, because of its anti-enzymatic action, this preservative should not be used in conjunction with enzymatic methods. Since the glycolytic enzymes are located in the red cells, the glucose concentration is much more stable in serum or plasma that has been separated from the cells.

The concentration of glucose in serum is approximately 12 mg per 100 ml higher than in whole blood. However, with methods which do not determine true glucose, this difference may not be apparent since the lower concentration of glucose in whole blood is offset by the higher concentration of nonglucose reducing substances from the intracellular water.

The quantitative determination of glucose in urine is discussed in Chapter 28.

Serum Acetone

Because of the accumulation of ketonic acids in the blood of patients in diabetic acidosis, an approximation of the concentration of acetone and acetoacetic acid in the blood sometimes is of clinical interest. This test may be performed using a mixture of sodium nitroferricyanide (nitroprusside), ammonium sulfate, and sodium carbonate in the proportions of 1:20:20. These salts should be ground to a powder separately, then mixed in proper proportion and kept in a tightly stoppered bottle. For the test, place a small mound of the powder on a piece of filter paper and add one drop of serum. A purple color signifies an abnormal concentration of acetone (10 mg per dl or more). If the test is positive, various dilutions of the serum may be made with water. These dilutions should be tested in the same manner. The final result may be reported in terms of the last dilution that gives a positive test, or an approximate concentration (milligrams per 100 ml) may be calculated by multiplying 10 times the number of times the serum was diluted to give the last positive reaction.

Although this is referred to as an acetone test, the reaction is much more sensitive to acetoacetic acid than to acetone. Both these compounds are present in ketosis, but the positive reaction is due largely to acetoacetic acid.

The reagents may be checked with dilutions of acetone in water. Since the density of acetone is approximately 0.8, the reagent contains approximately 800 mg of acetone per milliliter. The lower limit of detectability with this test is approximately 10 mg per dl. Therefore, a 1:8000 dilution of acetone should show a weakly positive reaction.

This test may also be performed using commercially available tablets.[2] However, if tablets are used for *serum* acetone they should first be crushed and a drop of water applied before the serum. The positive reaction is not as sharp as with the mixed powder described above.

Detection of acetone in urine is discussed in Chapter 28.

Physiological Significance of Glucose Concentrations

Glucose is one of the few chemical constituents of the blood which may change in concentration rapidly and dramatically. Although there are many diseases that cause disturbances in glucose metabolism, the commonest and most important of these is diabetes mellitus. In this condition, the capacity to metabolize glucose is reduced because of an effective deficiency of insulin. The result is a high blood glucose level, with glucose subsequently appearing in the urine. Insulin injections generally are prescribed for the more severe cases, dosage being dependent on conditions such as the intensity of the disease, the amount of exercise and the quality and quantity of diet.

The glucose concentration in the blood sometimes attains a very high or a very low level. At either of these extremes a comatose state may result, and it is important that the physician know whether the patient is in diabetic coma (high glucose) or hypoglycemic shock (low glucose). This is not always obvious from the physical appearance of the patient. The responsibility of supplying the vital information falls upon the laboratory technician. In these cases it is essential that treatment be instituted as soon as possible. Thus, emergency glucose analyses should be performed rapidly but meticulously, always with the realization that error or delay may have serious consequences.

Blood glucose levels, particularly in emergency cases, may range from an undetectable amount to over 2000 mg per dl. During treatment the level may change very rapidly.

Blood samples obtained during the administration of glucose infusions will have high glucose concentrations that may be misleading to the physician.

---

2 Acetest tablets, Ames Co., Elkhart, Indiana.

Glucose and Insulin Tolerance Tests

Knowledge of the rate of utilization of glucose or insulin is frequently of diagnostic value to the physician. Glucose tolerance and insulin tolerance tests are performed in much the same manner. With the patient in a fasting state, a control sample of blood is obtained; glucose or insulin is then administered, and blood samples are taken at intervals thereafter. The amount of insulin or glucose administered, the number of blood samples required, and the intervals at which they are to be obtained all are determined by the physician on the basis of the information which he is seeking.

A simple and satisfactory glucose-tolerance test, used in many laboratories for the diagnosis of diabetes mellitus, is as follows:

Obtain a fasting blood sample and administer either 100 g of glucose orally or administer 0.5 g of glucose per kg of body weight intravenously as a 25% solution. In exactly 2 hours, obtain another blood sample for glucose analysis. In the absence of diabetes, the 2-hour level should be back to normal or be slightly below normal.

More complex tests, utilizing greater numbers of blood specimens, and perhaps urine specimens as well, may be selected by the physician to suit special diagnostic purposes.

References

1. Campbell, D. M., and King, E. J.   Colorimetric determination of glucose in 20 microliters of blood. *J. Clin. Pathol.* 16:173, 1963.
2. Cooper, G. R.   Methods for determining the amount of glucose in blood. *CRC Crit. Rev. Clin. Lab. Sci.* 4:101, 1973.
3. Dubowski, K. M.   An *o*-toluidine method for body-fluid glucose determination. *Clin. Chem.* 8:215, 1962.
4. Folin, O., and Wu, H.   A system of blood analysis (Suppl. I). A simplified and improved method for the determination of sugar. *J. Biol. Chem.* 41:367, 1920.
5. Folin, O.   A new blood sugar method. *J. Biol. Chem.* 77:421, 1928.
6. Johnson, J.   Protein free filtrate or dialysate. Some experience with automation in a clinical chemistry laboratory, with special reference to the routine blood glucose determinations. *Am. J. Med. Technol.* 24:271, 1958.
7. Keller, D. M.   An enzymatic method for glucose quantitation in normal urine. *Clin. Chem.* 11:471, 1965.
8. Kingsley, G. R., and Getchell, G.   Direct ultramicro glucose oxidase method for determination of glucose in biological fluids. *Clin. Chem.* 6:466, 1960.
9. Nelson, N.   A photometric adaptation of the Somogyi method for the determination of glucose. *J. Biol. Chem.* 153:375, 1944.
10. Somogyi, M.   A method for the preparation of blood filtrates for the determination of sugar. *J. Biol. Chem.* 86:655, 1930.
11. Washko, M. E., and Rice, E. W.   Determination of glucose by an improved enzymatic procedure. *Clin. Chem.* 7:542, 1961.

12. Nitrogenous Compounds

Aside from protein, the major nitrogen-containing compounds in serum are urea, amino acids, uric acid, creatinine, creatine, and ammonia. These substances collectively are referred to as nonprotein nitrogen (NPN). They may be measured all together as NPN or they may be measured individually. This chapter discusses methods for NPN, urea, amino acids, and ammonia. Creatinine and creatine are presented in Chapter 13; uric acid is discussed in Chapter 14. Indirect determination of protein by nitrogen analysis is discussed in Chapter 15.

Nonprotein Nitrogen (NPN)

The collective quantitative estimation of nitrogen-containing substances that are not proteins (NPN) formerly was a common laboratory procedure. Although NPN still may be of some interest, it has largely been replaced by urea determinations. One requirement common to all nonprotein-nitrogen methods is that all proteins must be removed from the sample before the actual determination. The deproteinization procedure consists of the precipitation of proteins and the preparation of a protein-free filtrate. There are many methods for preparing protein-free filtrates suitable for nonprotein-nitrogen determination, of which two are described here.

Preparation of Protein-Free Filtrates

Folin-Wu Tungstic Acid Method

To 1 volume of blood, add 7 volumes of water. Add 1 volume of $0.67N$ sulfuric acid, mix, and allow to turn dark brown. Add 1 volume of 10% (w/v) sodium tungstate ($Na_2WO_4 \cdot 2H_2O$), mix well, allow to stand for 2 minutes, then filter. This method yields an acid filtrate that is suitable for the analysis of several substances. Any amount of filtrate may be prepared, but the above-stated proportions should be maintained.

Trichloroacetic Acid Method

To 1 volume of blood, add 9 volumes of 5% (w/v) trichloroacetic acid (CCl_3COOH), with mixing, and filter. The filtrate thus formed is strongly acid.

Of the two methods described above, the second is preferred for nonprotein-nitrogen determination, since it requires only one reagent and gives a greater yield of filtrate. The amount of sample to be used depends upon the quantity of filtrate needed for the analysis. It is advisable to prepare enough filtrate to repeat the determination if necessary.

After the removal of proteins, the nitrogenous substances in the filtrate must be broken down and the nitrogen converted into a common mea-

153

surable form. This is accomplished by a wet-digestion method using an acid-digestion mixture and heat. There are many digestion mixtures, but most of these achieve the same results. The major ingredient in all these mixtures is sulfuric acid, which acts as an oxidizing agent. The mixtures differ mostly in the type of catalyst and other digestion aids used.

The digestion mixture described in the method presented here is only one of many suitable mixtures. It includes sulfuric acid as an oxidizing agent, perchloric acid as a catalyst, and phosphoric acid to raise the boiling point.

During the digestion process the various nitrogenous substances are broken down. Their nitrogen is converted into ammonia, which is held by the sulfuric acid in the form of ammonium sulfate.

The ultimate determination of ammonia can be accomplished by one of three general methods: (1) by nesslerization, (2) by distillation and titration, and (3) by the phenol-hypochlorite (Berthelot) reaction. An example of nesslerization is contained in the method presented following this discussion; the phenol-hypochlorite reaction is discussed further in conjunction with the ammonia method (see page 162).

The distillation and titration procedure [17] is part of the Kjeldahl technic [12, 22] for nitrogen analysis. In this method a protein-free filtrate of a whole blood or serum sample is placed in a boiling flask. Copper sulfate, potassium sulfate, and sulfuric acid are added. Boiling this mixture over a micro burner for 12 minutes breaks down all of the nitrogen-containing compounds because of the oxidizing action of hot sulfuric acid. The copper sulfate and potassium sulfate speed up this wet-digestion procedure: copper sulfate serves as a catalyst for the oxidation reactions; potassium sulfate raises the boiling point.

The nitrogen from all of the nitrogen-containing compounds is converted into ammonia, which is held in the form of ammonium sulfate by the sulfuric acid. At this point, aqueous sodium hydroxide is added to convert the ammonium ions into neutral volatile ammonia molecules. A brief distillation of the reaction mixture, following alkalinization, liberates the ammonia, which is trapped in an excess amount of a boric acid [15] solution. Finally, the amount of ammonia is determined by titrating the acid back to its original acidity value (before the ammonia neutralized part of the acid) with standard hydrochloric acid. This micro Kjeldahl technic is a reliable nonprotein-nitrogen method, but it is cumbersome and time consuming for a busy clinical laboratory.

NONPROTEIN-NITROGEN METHOD (NESSLERIZATION)
METHOD
Folin and Wu [6, 11], modified.

MATERIAL
Whole blood or serum.

REAGENTS

Digestion mixture. To 978 ml of water in a 2 liter beaker add 180 ml of
85% phosphoric acid and 630 ml of concd sulfuric acid. Cool thoroughly and add 3.0 ml of perchloric acid slowly, with stirring. Mix
and keep well stoppered.

Nessler's solution. This reagent is an alkaline solution of the double
iodide of mercury and potassium ($HgI_2 \cdot 2KI$). Its preparation from
raw reagents is tedious, involving shaking a potassium iodide solution
with metallic mercury to form the double iodide [13]. However, this
compound is available commercially as "Nessler's powder," which
can be dissolved in a sodium hydroxide solution according to the
manufacturer's instructions. Store at room temperature and observe
the precautions listed on page 157.

PROCEDURE

Prepare a protein-free filtrate by either of the methods described on
page 153.

To a Folin-Wu digestion tube calibrated at 35 ml add:

1 ml of digestion mixture
3 ml of protein-free filtrate
2 glass beads

Boil over a micro burner until digestion is complete.

The solution will blacken. Dense white fumes will fill the tube. Oxidation
is complete when the mixture becomes colorless and the white fumes
lift slightly.[1]

Cool to room temperature, dilute to 35 ml with water, and invert to mix.

To another digestion tube add 1 ml of digestion mixture, dilute to 35 ml
with water, and mix (Blank).

To a third digestion tube add 3 ml of ammonium sulfate standard, 1 ml
of digestion mixture, dilute to 35 ml with water, and mix (Standard).

Pipet 5 ml of Blank, Standard, and Test into 19 mm cuvets and add 2 ml
of Nessler's solution to each. Invert to mix.

Read the absorbances of the Standard and Test at 480 nm, setting the
Blank at zero.

Determinations reading over 0.600 should be repeated using 1 ml of
filtrate.

CALCULATIONS

$$\frac{\text{concentration of Standard (40 mg/dl)}}{\text{absorbance of Standard}} \times \text{absorbance of Test}$$
$$= NPN \ (mg/dl)$$

If 1 ml of filtrate is used, multiply the final result by 3.

---

[1] Occasionally a sample will not turn colorless, even after prolonged heating. If this
happens, remove the tube from the flame, cool slightly, and add 1 drop of 30%
hydrogen peroxide. Resume heating until the fumes lift.

Because Nessler's solution is subject to variations, standards should accompany each group of nonprotein-nitrogen determinations.

The method may be standardized with an ammonium sulfate solution. If a standard is to have a nitrogen concentration equivalent to a blood nonprotein nitrogen of 40 mg per dl, the nitrogen concentration in the standard must be made equal to that in the blood filtrate. Since the filtrate is a 1:10 dilution of the blood, the nitrogen concentration in a filtrate made from a blood sample containing 40 mg/dl of nonprotein nitrogen is $\frac{1}{10} \times 40 = 4$ mg per dl.

Ammonium Sulfate standard. Ammonium sulfate, $(NH_4)_2SO_4$, has a molecular weight of 132. There are two nitrogen atoms, each having an atomic weight of 14. Therefore, $\frac{28}{132}$ of the weight of ammonium sulfate is nitrogen. This means that for every unit of nitrogen desired, $\frac{132}{28}$ units of ammonium sulfate must be used.

To make 1 liter of a 4 mg/dl nitrogen standard requires 40 mg of nitrogen, or $40 \times \frac{132}{28} = 188.6$ mg of ammonium sulfate.

Dissolve the 188.6 mg of ammonium sulfate in approximately 500 ml of water. Add 10 ml of concentrated sulfuric acid. Then dilute to 1 liter with water.

This standard need not be subjected to the digestion procedure. When used as a filtrate, the relative NPN value of this standard is 40 mg per dl.

The normal range for NPN concentrations in the blood of healthy adults is 25–40 mg per dl.

The nesslerization method is neither so accurate nor so reliable as the titrimetric procedure, due to the sensitivity of Nessler's solution to temperature, contamination, and other variable conditions. However, with adequate control, the method is suitable for most purposes. Moreover, it requires far less time, space, and labor than the Kjeldahl technics involving distillation and titration.

A precision of ±8% may be expected from the nesslerization technic as described here.

The principles of this method are the same as for any Kjeldahl method in that ammonium sulfate is formed during digestion. (See preceding section.) At this point an alkaline solution of the double iodide of potassium and mercury (Nessler's solution) is added to the acid ammonium sulfate solution, resulting in the formation of an orange colloid (possibly dimercuric ammonium iodide, $NH_2Hg_2I_3$). The reaction is

shown here in two phases, the first being the alkaline conversion of the
ammonium sulfate to ammonium hydroxide, and the second the formation of the color.

1. $(NH_4)_2SO_4 + 2NaOH \rightarrow 2NH_4OH + Na_2SO_4$

2. $NH_4OH + (KI)_2HgI_2 \rightarrow$ orange colloid

The successful use of this method depends, to a large degree, upon the
control of Nessler's solution. A few suggestions for this control are listed
here.

1. Freshly made solution should be allowed to settle for the prescribed
   amount of time before being used.
2. The solution should be kept stoppered to prevent contamination by
   dirt and fumes.[2]
3. Nessler's solution should not be subjected to large temperature
   changes.
4. In order to avoid stirring up sediment from the bottom of the container, the solution should not be agitated.
5. Blank and Standard should accompany each group of analyses.

Comments pertinent to the physiological significance of nonprotein
nitrogen concentrations are presented on page 160.

Urea Nitrogen

Normally urea nitrogen in blood comprises about 45% of the nonprotein nitrogen. Gross changes in the nonprotein nitrogen usually are
reflections of changes in the urea concentration. Urea determinations
yield essentially the same clinical information about nitrogen retention
as nonprotein-nitrogen determinations do.

A wide variety of methods for the determination of urea have been
devised. Some of these may be performed directly with whole blood or
serum while others require preliminary preparation of a protein-free
filtrate. Generally these methods are of two basic types: those which
determine urea directly by reaction with a colorimetric reagent, and
those which measure ammonia resulting from the action of the enzyme
urease [14] on urea.

Diacetylmonoxime is a widely used colorimetric reagent for urea. It is
employed under acidic conditions, and reacts first with water to give
diacetyl, then with urea to give a yellow diazine derivative.

---

[2] Acetone fumes cause gross turbidity in Nessler's solution.

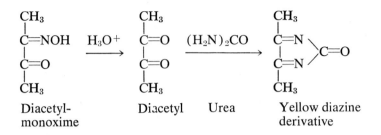

Diacetyl-monoxime   Diacetyl   Urea   Yellow diazine derivative

The method has been adapted to both manual and automated analyses. Methods are available for carrying out the reaction with or without prior deproteinization of the sample. Thiosemicarbazide often is included in the reagent, resulting in a final color which is more intense than the color which results in its absence.

When urease is used to form ammonia and carbon dioxide from urea, the ammonia is measured by one of three common technics:

1. Reaction with phenol and sodium hypochlorite in an alkaline medium to form indophenol, a blue substance. (This reaction is used in the method for ammonia presented later in this chapter in the section on Blood Ammonia.)
2. Reaction with Nessler's reagent [8]. This reaction is used in the method for nonprotein nitrogen presented in the preceding section.
3. Titration with standard acid back to an original acidity level after the ammonia has been aerated into a standard acid solution [20]. (This technic is similar to that discussed for the Kjeldahl method discussed in the preceding section.)

The urea method presented here employs the diacetylmonoxime reaction with a deproteinized sample.

UREA METHOD

METHOD
Evans [3].

MATERIAL
Whole blood or serum.

REAGENTS
Stock diacetylmonoxime. Dissolve 2.5 g of diacetylmonoxime in water and dilute to 100 ml. Store in a dark bottle and renew monthly.

Stock thiosemicarbazide. Dissolve 0.25 g of thiosemicarbazide in water and dilute to 100 ml. Store in a dark bottle and renew monthly.

Diacetylmonoxime-thiosemicarbazide reagent. Mix 24 ml of stock diacetylmonoxime with 10 ml of stock thiosemicarbazide and dilute to 100 ml. Store in a dark bottle in the refrigerator.

Acid reagent. To 1 liter of water add 80 ml of concd sulfuric acid and 10 ml of 85% phosphoric acid. Dissolve 0.5 g of ferric chloride in this solution.

Color reagent. Mix 5 parts of the acid reagent with 1 part of the diacetylmonoxime-thiosemicarbazide reagent.

Trichloroacetic acid, 5%. Dissolve 5 g of trichloroacetic acid in water and dilute to 100 ml.

Trichloroacetic acid, 10%. Dissolve 10 g of trichloroacetic acid in water and dilute to 100 ml.

PROCEDURE

Into a 13 × 100 mm test tube measure 0.8 ml of water, 0.2 ml of sample, and 1 ml of 10% trichloroacetic acid (Test).

Mix and allow to stand at least 5 minutes.

Centrifuge for 10 minutes.

Measure 0.5 ml of the clear supernatant into a 16 × 125 mm tube.

Measure 0.5 ml of 5% trichloroacetic acid into another tube (Blank).

Set up a series of Standards as described under Standardization.

To each tube add 5 ml of color reagent and mix.

Heat in a boiling-water bath for 8 minutes.

Cool to room temperature, transfer to 12 mm cuvets, and read the absorbances at 520 nm against the Blank set at zero.

CALCULATIONS

Plot the absorbances of the Standards against their respective concentrations on linear graph paper. Determine the values of the Tests from the standard curve.

STANDARDIZATION

Stock standard. Dry approximately 2 g of urea[3] in an oven at 100 C for 1 h. Then cool to room temperature in a desiccator. Dissolve 1.712 g of this dried urea in water and dilute to 1 liter, including 5 drops of concd sulfuric acid. The nitrogen content (urea nitrogen) of this solution is 80 mg per dl. (Nitrogen accounts for $28/60$ of the weight of urea.)

Working standard. Dilute 1 ml of the stock urea standard to 10 ml with 5% trichloroacetic acid.

[3] Available as a Standard Reference Material from the U.S. National Bureau of Standards.

To use, measure 0.1, 0.2, 0.3, 0.4, and 0.5 ml of the working standard into 16×125 mm tubes and add 0.4, 0.3, 0.2, 0.1, and 0.0 ml of 5% trichloroacetic acid to each respectively. Then treat the Standards the same as the Test supernatants. The standard values are 16, 32, 48, 64, and 80 mg per dl, respectively.

NORMAL VALUES

The concentration of urea nitrogen in the blood of healthy adults ranges between 10 and 20 mg per dl. This is 32 to 42 mg per dl of urea.

DISCUSSION

The method presented above is precise to ± 2 mg per dl in the normal and moderately elevated ranges.

It is preferable to form diacetyl from diacetylmonoxime in the reaction rather than to use diacetyl as a reagent because diacetyl is unstable. A number of metal ions and organic reagents have been used to intensify and stabilize the color. Ferric ions and thiosemicarbazide are used for these purposes in the present method. Since the color from the reaction is not stable indefinitely, it is important to measure color without a long delay.

The common anticoagulants (heparin, EDTA, and oxalate) do not interfere with the method.

Blood for urea nitrogen determination may be kept in a refrigerator for at least 72 hours without measurable change in the blood urea-nitrogen concentration. The determination may also be carried through the filtrate stage. The filtrate may be stored for long periods of time without alteration.

PHYSIOLOGICAL SIGNIFICANCE OF BLOOD UREA NITROGEN
AND NONPROTEIN NITROGEN CONCENTRATIONS

The nonprotein nitrogen (NPN) determination may be considered an indirect method for determining the urea nitrogen (BUN) concentration in the blood. Both methods yield approximately the same clinical information because nonprotein nitrogenous blood components, other than urea, do not usually undergo changes in concentration large enough to affect the nonprotein-nitrogen concentration significantly. The only probable exceptions to this generalization are the amino acids. If accurate information concerning amino acid concentrations is desired, special methods may be used that determine total amino acid concentration. (See the section later in this chapter on Amino Acids.)

The urea nitrogen concentration in serum or plasma is approximately 2 mg per dl higher than in whole blood. Also, since there is a greater concentration of nonprotein nitrogenous material inside the red cells than in the plasma, the nonprotein nitrogen concentration is approximately 7 mg per dl higher in whole blood than in serum or plasma. Some reports indicate that the use of serum in methods employing nessleriza- tion reduces the incidence of turbidity formation. Generally, however, the choice between whole blood or serum may be made on the basis of convenience to the laboratory and the physician.

If an anticoagulant is used, care should be taken that blood for deter- mination of nonprotein nitrogen or of blood urea nitrogen is not col- lected in tubes that contain the double oxalate of potassium and ammo- nium, commonly used in hematology containers. Ammonia, of course, is detrimental to the accuracy of any nitrogen method involving the ultimate determination of ammonia.

Blood urea nitrogen is an expression of nitrogen and not urea. In order to convert milligrams per 100 ml of urea nitrogen into milligrams per 100 ml of urea the blood urea nitrogen is multiplied by $^{60}\!/_{28}$ or 2.14. Urea is expressed as nitrogen, or blood urea nitrogen, by using the reciprocal of the above expression: $^{28}\!/_{60}$ or 0.467 × milligrams per 100 ml of urea equals milligrams per 100 ml of urea nitrogen.

An expression of nonprotein nitrogen concentration cannot accurately be converted to blood urea nitrogen or urea mathematically. Approxi- mately 45% of a normal nonprotein nitrogen concentration is urea nitrogen, but elevations in nonprotein nitrogen concentration are almost entirely urea nitrogen. At high nonprotein nitrogen concentrations, the blood urea nitrogen may be 70% to 80% of the total.

Urea is the chief end product of protein metabolism in the body. The importance of the urea concentration in blood lies in its value as an indicator of kidney function. Elevation of the urea concentration in blood signifies inadequate kidney function. Invariably there are toxic substances retained in the blood along with the urea. It is for this reason that a high nonprotein nitrogen or blood urea nitrogen level is a matter of concern to the physician. However, the urea concentration in blood is influenced by diet. Thus people who are malnourished or who are on low protein diets may have blood nitrogen levels that are not accurate indi- cators of kidney function. For these and other reasons, creatinine some- times is considered a better kidney function test (see Chapter 13). Dur- ing pregnancy, the blood nonprotein nitrogen or urea level usually is lower than normal.

Blood Ammonia

Ammonia is present in blood as the ammonium ion, $NH_4{}^+$. Some methods for measuring blood ammonia involve addition of a base at some stage in the procedure to liberate the ammonium ion as gaseous ammonia. The ammonia then can be trapped in an acidic solution, which converts it back into the ammonium form. It then is quantitated colori- metrically or by titration.

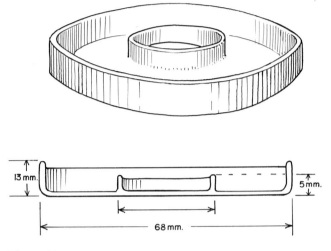

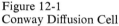

Figure 12-1
Conway Diffusion Cell

In the titration method [18], the amount of ammonia trapped by an acid solution is determined by the volume of a second, standardized, acid solution used to titrate the former solution back to its original acidity value (before it was partly neutralized by the ammonia which it trapped). Standard base also can be used as the titrant. In this case, the amount of acid not neutralized by the ammonia is determined. Knowing the total amount of acid before and after part of it has been neutralized by ammonia allows the amount of ammonia to be calculated.

A simple method for diffusing the ammonia out of a blood sample and into an acid solution is the Conway microdiffusion method, which employs a Conway cell [1] (Figure 12-1). A sample of blood or urine is placed in the compartment between the two walls, and a measured amount of standard acid is pipetted into the center compartment. The sample is alkalinized, and the cell is covered with a glass plate that has been greased in order to ensure an air-tight seal. The dish is left for several hours, during which time the ammonia diffuses out of the alkaline sample and is taken up by the acid. After complete diffusion, the amount of acid neutralized by the ammonia is determined by titration with a standard acid or base. Generally, the microdiffusion methods require relatively long periods of time for complete diffusion.

Probably the most common colorimetric procedure for ammonia is the indophenol reaction, originally discovered by Berthelot. An alkaline solution of phenol-hypochlorite reacts with ammonia to give the blue product, indophenolate.

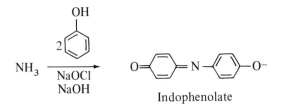

Indophenolate

The reaction usually is accelerated by the addition of sodium nitroprusside, $Na_2Fe(CN)_5NO$.

Nesslerization (previously discussed in the section on Nonprotein Nitrogen) and reaction with ninhydrin also can be used to quantitate ammonia colorimetrically.

Not all methods require gaseous liberation of the ammonia. For instance, an ion-exchange resin [4] or Permutit [5] can be used to adsorb the ammonium ions followed by elution and quantification with a colorimetric reaction. Also, there is an enzymatic method [16] in which α-ketoglutarate is converted to glutamic acid through the action of glutamic dehydrogenase. The reaction is quantitated based on the decrease in absorbance at 340 nm or fluorescence at 460 nm (from 360 nm excitation) associated with conversion of NADH to NAD^+. (See page 242.)

The method presented here involves liberation of the ammonia with base, trapping the ammonia on the ground-glass tip of a rod which is coated with acid and suspended above the sample in a closed bottle (Figure 12-2), eluting the ammonia off the rod into a cuvet, and quantitating with the indophenol reaction.

BLOOD AMMONIA METHOD

METHOD
Seligson and Katsugi [19] (modified).

MATERIAL
Whole blood (heparinized).

REAGENTS
Phenol reagent. Dissolve 5 g of phenol and 0.25 g of sodium nitroprusside in water and dilute to 1 liter.

Hypochlorite reagent. Dissolve 25 g of NaOH and 40 ml of Chlorox in water and dilute to 1 liter.

Sulfuric acid, 1N. Dilute 2.8 ml of concd sulfuric acid to 100 ml.

Potassium carbonate, 45% (w/v). Dissolve 45 g of potassium carbonate in water and dilute to 100 ml.

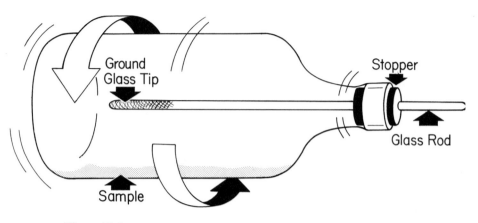

Figure 12-2
Ammonia Diffusion Bottle

APPARATUS
Diffusion bottles with absorption rods (see Figure 12-2).
Rotator.
Bath (56 C).

PROCEDURE
Note: It is advisable to set up the Blank, Standards, and Tests in dupli-
   cate. Rinse out the diffusion bottles with deionized water.
Into four diffusion bottles measure:
   Blank: 1 ml water
   Standard 1: 0.5 ml of working standard (S_1)
   Standard 2: 1 ml of working standard (S_2)
   Test: 1 ml of whole blood
For each bottle proceed as follows:
a. Dip the ground end of an absorption rod into $1N$ H_2SO_4 and set it
   down without touching anything with the acid-dipped end.
b. Add 1 ml of 45% potassium carbonate solution to the bottle.
c. *Quickly* insert the rod and tighten the stopper, taking care not to
   touch the mouth of the bottle or the sample with the rod.
Carefully place the bottles horizontally in the rotator and rotate for 30
   minutes. Avoid any agitation which might splash sample onto the rod.
Remove the rod *carefully* from each bottle and suspend it in a 12 mm
   cuvet.
Wash down each rod with:
a. 1 ml of phenol reagent.
b. 1 ml of hypochlorite reagent.
c. 1 ml of water.

Remove each rod, cover the cuvets with Parafilm, mix, and place them in a 56 C bath for 5 minutes.

Remove the cuvets from the bath and cool them to room temperature. Wipe the outsides of the cuvets and read the absorbances of the Tests and Standard at 630 nm, setting the Blank at zero.

CALCULATIONS

$$\frac{100}{\text{absorbance of } S_1} + \frac{200}{\text{absorbance of } S_2} \times \frac{\text{absorbance of Test}}{2}$$
$$= \text{ammonia nitrogen } (\mu g/dl)$$

STANDARDIZATION

Stock standard, 10 mg/dl of nitrogen. Dissolve 236 mg of ammonium sulfate, $(NH_4)_2SO_4$, in water containing 0.5 ml of concd sulfuric acid and dilute to 500 ml.

Working standard, 200 μg/dl of nitrogen. Dilute 2 ml of the stock standard to 100 ml with water, including 0.1 ml of concd sulfuric acid in the dilution.

When used as designated in the method, 1 ml and 0.5 ml of this standard are equivalent to blood ammonia nitrogen values of 200 and 100 μg per dl, respectively.

NORMAL VALUES

The normal ammonia nitrogen concentration in blood is 50–150 μg/dl. Postprandial levels are higher than fasting levels.

DISCUSSION

The ammonia content of freshly drawn blood rises rapidly at room temperature because of enzymatic hydrolysis of various amino substances in blood. Therefore the analysis should be performed immediately on receipt of the sample. However, the ammonia content can be kept reasonably constant for about 2 hours if the blood is kept refrigerated or on ice.

Because of the low concentration of ammonia in blood and its relative ubiquity, ammonia methods are difficult to set up and control. Glassware must be cleaned meticulously. Reagents must be prepared with ammonia-free water, and then kept tightly covered when not in use to minimize absorption of ammonia from the air. This is particularly true for any acid reagents. Because organic solvents like ether and acetone tend to be contaminated with ammonia, they should not be used to dry the glassware or cuvets involved.

In the method presented here, the ammonium ion in the blood is converted to ammonia (NH_3) by the potassium carbonate solution, which is very alkaline. Rotating the sample in a closed bottle then allows a high rate of diffusion of the volatile ammonia into the air space, where it is irreversibly trapped as ammonium ion by the acid solution on the absorption rod suspended above the sample (Figure 12-2). The exact extent of transfer of the ammonia is not critical because the results are compared with standards run at the same time and under the same conditions. The film of acid solution containing the ammonium ions then is washed into a cuvet and the indophenol reaction (discussed at beginning of this section on Blood Ammonia) is used to quantitate the ammonia colorimetrically.

PHYSIOLOGICAL SIGNIFICANCE OF BLOOD AMMONIA CONCENTRATIONS
Ammonia is one of the byproducts of the metabolism of nitrogen-containing compounds in the body. Because it is a toxic substance, it is converted to urea as it is formed and finally excreted in this less toxic form. This conversion takes place in the liver. Therefore hepatic failure results in an increase in concentration of blood ammonia. A mild increase in the blood ammonia level also can accompany physical exercise.

Amino Acids

Amino acids can be determined either individually or collectively as amino acid nitrogen. Usually, chromatographic methods are used for the determination of individual amino acids. These include high-voltage electrophoresis [21], thin-layer chromatography [2], and the use of a commercial amino-acid analyzer [23] which employs an ion-exchange separation. These amino-acid screening technics can be used with either serum or urine. They are helpful in detecting familial metabolic disorders or certain transient illnesses.

The classic method for the determination of amino acid nitrogen in plasma, serum, or urine is the gasometric Ninhydrin method of Van Slyke [10] in which Ninhydrin liberates CO_2 from the carboxyl groups of the amino acids. The CO_2 is subsequently measured manometrically (based on its volume). However, colorimetric methods are usually used in clinical laboratories. For instance, 1-fluoro-2,4-dinitrobenzene [9] and sodium β-naphthoquinone-4-sulfonate [7] each has been used to quantitate amino acid nitrogen by the formation of a color. The former reagent, introduced more recently than the latter, is more convenient to use and is less subject to interferences.

Quantitation or semiquantitation of individual amino acids is primarily of importance in regard to certain hereditary disorders such as phenylketonuria (PKU). Elevated blood levels of total amino acids, that is, of amino-acid nitrogen, may occur in such pathological conditions as uremia, leukemia, and yellow atrophy of the liver.

Urinary Nitrogenous Substances

The nitrogenous substances discussed below, with the probable exception of urea, are not requested very frequently in clinical laboratories.

For this reason, they have been given only superficial treatment here. Details may be found in the cited references. Creatinine, creatine, and uric acid are also nitrogenous compounds which are present in urine but, since they are determined more frequently and by specific technics, they are considered separately in Chapters 13 and 14.

Most of the nitrogenous compounds in urine are products of protein metabolism. The quantities excreted vary greatly. Table 12-1 gives approximations of the excretion of several nonprotein nitrogenous substances by healthy individuals on average diets.

TOTAL NITROGEN

Total nitrogen may be measured in urine using the methods outlined for NPN (first section of this chapter) except that a protein-free filtrate need not be prepared (unless a very high urine protein excretion is encountered). Urine requires a greater dilution because the nitrogen concentration is approximately 25 times the blood NPN level. Results for urine total nitrogen should be reported in grams per 24 hours.

AMMONIA

Since the determination of ammonia in urine is seldom performed in a clinical laboratory, the present discussion is confined to a general explanation of the process.

Ammonia may be determined with the aid of a sodium aluminum silicate called Permutit [5]. Permutit adsorbs ammonia at a neutral or acid pH and releases it at an alkaline pH. If Permutit powder is added to a measured amount of acidic urine, the ammonia in the urine will be adsorbed. Since the Permutit is heavy, it settles quickly, and the urine may be decanted. The Permutit is then washed. After the ammonia has been released by alkalinization, the ammonia concentration may be determined by nesslerization, by distillation and titration, or by the phenol-hypochlorite reaction. These processes are discussed earlier in this chapter. A very simple method for ammonia determination is the Conway microdiffusion method [1] described in this chapter in the section on Blood Ammonia.

Table 12-1. Excretion of Some NPN by Healthy People on Average Diets

Substance	g per 24 h
Total nitrogen	10 to 18
Urea nitrogen	7 to 16
Ammonia nitrogen	0.2 to 0.7
Amino acid nitrogen	0.2 to 0.7

Urea Nitrogen

Urea nitrogen may be determined in urine by the methods that were discussed for blood urea-nitrogen determinations. Since urine contains significant amounts of ammonia, this ammonia must be removed before the determination of urea nitrogen if a urease method is used. This may be accomplished with the aid of Permutit as follows:

Wash 2 g of Permutit in a 100 ml volumetric flask by adding water, shaking, allowing the Permutit to settle, and decanting. Repeat this operation at least once.

Into the volumetric flask containing the washed Permutit, measure 2 ml of urine plus approximately 10 ml of water. Agitate with a swirling motion for 5 minutes, then dilute with water to 100 ml. Mix well, and allow the Permutit to settle. An aliquot of the diluted urine then may be treated in the same way as blood or serum and the urea nitrogen concentration determined by any of the blood urea-nitrogen methods previously mentioned. Since the urine is diluted 2:100 or 1:50, the final result must be multiplied by 50 to obtain milligrams of urea nitrogen per 100 ml of urine. Then

$$\frac{mg}{dl} \times \frac{10}{1000} \times \frac{24\ h\ volume\ (ml)}{1000} = \frac{g}{24\ h}$$

Amino Acid Nitrogen

Amino acid nitrogen may be determined in urine with the same technics used for blood. (See the preceding section on Amino Acids.)

References

1. Conway, E. J.   *Microdiffusion Analysis and Volumetric Error* (4th ed.). New York: Macmillan, 1958.
2. Culley, W. J.   A rapid and simple thin-layer chromatographic method for amino acids in blood. *Clin. Chem.* 15:902, 1969.
3. Evans, R. T.   Manual and automated methods for measuring urea based on a modification of its reaction with diacetyl monoxime and thiosemicarbazide. *J. Clin. Pathol.* 21:527, 1968.
4. Fenton, J. C. B., and Williams, A. H.   Improved method for the estimation of plasma ammonia by ion exchange. *J. Clin. Pathol.* 21:14, 1968.
5. Folin, O., and Bell, R. D.   Applications of a new reagent for the separation of ammonia: I. The colorimetric determination of ammonia in urine. *J. Biol. Chem.* 29:329, 1917.
6. Folin, O., and Wu, H.   A system of blood analysis. *J. Biol. Chem.* 38:81, 1919.
7. Frame, E. G., Russell, J. A., and Wilhelmi, A. E.   The colorimetric estimation of amino nitrogen in blood. *J. Biol. Chem.* 149:255, 1943.
8. Gentzkow, C. J.   An accurate method for determination of blood urea nitrogen by direct nesslerization. *J. Biol. Chem.* 143:531, 1942.
9. Goodwin, J. F.   The colorimetric estimation of plasma amino nitrogen with DNFB. *Clin. Chem.* 14:1080, 1968.

10. Hamilton, P. B., and Van Slyke, D. D.   The gasometric determination of free amino acid in blood filtrates by the ninhydrin-carbon dioxide method. *J. Biol. Chem.* 150:231, 1943.
11. Hawk, P. B.   *Hawk's Physiological Chemistry* (14th ed.). New York: Blakiston, 1965.
12. Kjeldahl, J.   Neue Methode zur Bestimmung der Stickstoffs in organischen Koerpern. *Z. Anal. Chem.* 22:366, 1883.
13. Koch, F. S., and McMeekin, T. L.   A new direct nesslerization micro Kjeldahl method and a modification of the Nessler-Folin reagent for ammonia. *J. Am. Chem. Soc.* 46:2066, 1924.
14. Mateer, J. G., and Marshall, E. K., Jr.   The urease content of certain beans, with special reference to the jack bean. *J. Biol. Chem.* 25:297, 1916.
15. Meeker, E. W., and Wagner, E. C.   Titration of ammonia in the presence of boric acid. Macro and micro Kjeldahl analyses. *J. Ind. Eng. Chem. (Anal. Ed.).* 5:396, 1935.
16. Oreskes, I., Hirsch, C., and Kuffer, S.   Application of an enzymatic technique for measurement of ammonia in whole blood. *Clin. Chim. Acta* 26:185, 1969.
17. Peters, J. P., and Van Slyke, D. D.   Methods. In *Quantitative Clinical Chemistry.* Baltimore: Williams & Wilkins, 1932. Vol. II, P. 536.
18. Reif, A. E.   The ammonia content of blood and urine. *Anal. Biochem.* 1:351, 1960.
19. Seligson, D., and Katsugi, H.   The measurement of ammonia in whole blood, erythrocytes, and plasma. *J. Lab. Clin. Med.* 49:962, 1957.
20. Sobel, A. E., Hirschman, A., and Besman, L.   Estimation of ultramicroquantities of urea in Kjeldahl and amino acid nitrogen. *Anal. Chem.* 19:927, 1947.
21. Troughton, W. D., Brown, R. St. U., and Turner, N. A.   Separation of urinary amino acids by thin-layer high voltage electrophoresis and chromatography. *Am. J. Clin. Pathol.* 46:139, 1966.
22. Vickery, H. B.   The early years of the Kjeldahl method to determine nitrogen. *Yale J. Biol. Med.* 18:473, 1946.
23. Wolf, P. L., Williams, D., Tsudaka, T., and Acosta, L.   *Methods and Techniques in Clinical Chemistry.* New York: Wiley, 1972. P. 23.

13. Creatinine and Creatine

Creatinine and creatine are nonprotein nitrogenous compounds existing in the blood in relatively low concentrations in health. They are excreted in the urine. Increased concentration of creatinine or creatine in the blood is of diagnostic significance to the physician. Although both of these compounds are nonprotein nitrogenous substances, the specific methods for their quantitative determination do not depend upon this property.

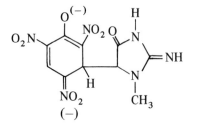

Creatine   Creatinine

Creatinine

Most creatinine methods are based on the Jaffe reaction [2, 4] a colori-

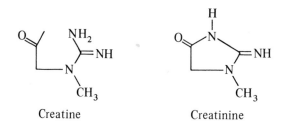

Creatinine $\xrightarrow[\text{NaOH}]{}$ red color

metric reaction between creatinine and alkaline picrate. This color may be due to formation of the complex illustrated below [2].

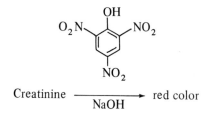

There is also a colorimetric creatinine method that employs dinitrobenzoate [1], but this method appears to offer no particular advantage over the standard picrate technics. The method of Folin [3] requires addition of picrate solution to a portion of protein-free filtrate. However,

170

because of the lack of specificity of the Jaffe reaction, this method gives falsely high values.

The method presented here employs Lloyd's reagent, which makes it one of the more specific and reliable modifications of the alkaline picrate method. Unfortunately, this modification is not adaptable to automation; so the automated picrate methods give high results.

CREATININE METHOD

METHOD
Hare [5], modified.

MATERIAL
Serum.

REAGENTS

Trichloroacetic acid, 20% (w/v). Dissolve 100 g of trichloroacetic acid (CCl_3COOH) in water and dilute to 500 ml.

Saturated oxalic acid. Add oxalic acid, $(COOH)_2 \cdot 2H_2O$, to a flask containing water, and shake until no more will dissolve. Oxalic acid crystals should always be visible on the bottom of the flask to ensure saturation.

Picric acid 0.04M (approximately). Weigh 10.0 g of picric acid, $C_6H_2OH(NO_2)_3$, crystals and dissolve in 1 liter of water.[1]

Sodium hydroxide 10% (w/v). Dissolve 20 g of sodium hydroxide (NaOH) in 200 ml of water.

Alkaline picrate. Mix:
    5 parts of picric acid
    1 part of 10% (w/v) sodium hydroxide
    30 parts of water
This reagent should be prepared just before use since it is unstable. Prepare enough solution to allow at least 10 ml for each Test plus 5.5 ml for a Blank.

Lloyd's reagent. A form of sodium aluminum silicate powder that is commercially available.

PROCEDURE
To a 16 × 125 mm test tube add 4 ml of water, 2 ml of serum, and 4 ml of 20% trichloroacetic acid.
Stopper, shake to mix, and allow to stand for 10 minutes.
Centrifuge for 5 minutes. Then filter through 7 cm filter paper.
Measure 5 ml of filtrate into a similar tube (Test).

---

[1] Since reagent-grade picric acid has about 10% of water added to it, this figure is approximately 10% higher than the calculated amount. The molarity need not be precise.

Measure 1 ml of creatinine working standard plus 4 ml of water into
    another tube (Standard).

Add 0.5 ml of saturated oxalic acid and approximately 40 mg of Lloyd's
    reagent on the tip of a small spatula.

Stopper the tubes, and shake intermittently for 10 minutes.[2] Then re-
    move the stoppers and save them.

Centrifuge for 10 min at approximately 2000 rpm.

With the aid of a Pasteur pipet, suction off the supernatant fluid and
    discard it, taking care not to lose any of the sediment.

Add 10 ml of alkaline picrate to the residue.

Replace the same stoppers and shake intermittently for 9 minutes.[2]

Centrifuge for 5 minutes. Then carefully decant into 19 mm cuvets.

Read the absorbances in a spectrophotometer 15 minutes from the time
    the picrate solution was added. Readings should be made at 490 nm,
    setting a Blank (consisting of at least 5.5 ml of the alkaline picrate
    solution) at zero.

CALCULATIONS

$$\text{absorbance of Test} \times \frac{\text{concentration of Standard (2 mg/dl)}}{\text{absorbance of Standard}}$$
$$= \text{creatinine (mg/dl)}$$

If a reading greater than 0.600 is obtained, the Test should be repeated
using 1 ml of filtrate plus 4 ml of water. The result must then be multiplied by 5.

STANDARDIZATION

Either creatinine or creatinine zinc chloride may be used to prepare a
standard.

Creatinine Stock Standard 100 mg/dl. Dissolve 0.100 g of pure creati-
    nine[3] in approximately 50 ml of 0.1N hydrochloric acid. Transfer to
    a 100 ml volumetric flask and dilute to 100 ml with 0.1N hydro-
    chloric acid. Store in a refrigerator.

Since the ratio of the molecular weights of creatinine zinc chloride to
    creatinine is $363/226$ or 1.61, preparation of the above stock standard
    requires 0.161 g of the former.

Creatinine Working Standard 2 mg/dl. Dilute 1 ml of the stock standard
    to 50 ml with 0.1N hydrochloric acid. When kept refrigerated, this
    standard is stable for at least one month.

This standard is treated in the same manner as the filtrate, but, since the

---

[2] The tubes may be shaken for the full period of time with a mechanical shaker.
[3] Available as a Standard Reference Material from the U.S. National Bureau of
Standards.

filtrate represents a 1:5 dilution of the sample, the amount of standard used should contain $\frac{1}{5}$ the amount of creatinine in the sample. Thus only 1 ml of this working standard, rather than 5 ml, is used to simulate a serum-creatinine level of 2 mg per dl.

NORMAL VALUES

With this method, the range of serum creatinine concentrations in healthy adults is 0.3 to 1.1 mg per dl. The levels are slightly higher in men than in women.

DISCUSSION

This method is capable of an accuracy of ±0.2 mg per dl. With some less specific methods, the normal range may extend up to 2 mg per dl.

In preparing the filtrate, it is important that the precipitant (trichloroacetic acid) be added to a mixture of serum and water. Addition of acid to the undiluted serum results in clumping of proteins, which may affect the results.

The addition of oxalic acid to a sample of serum filtrate results in the formation of creatinine oxalate. The creatinine oxalate is then adsorbed and held by the Lloyd's reagent in the acid solution.

After centrifugation, care must be exercised in removing the supernatant fluid so that none of the Lloyd's reagent is lost since this would represent a loss of creatinine.

Since alkaline picrate is not stable, it should be prepared just before use. When alkaline picrate is added to the Lloyd's reagent, the adsorbed creatinine is liberated and reacts with picrate to form a complex red-colored compound (Jaffe reaction) [4]. The quantity of red chromophore produced is directly proportional to the creatinine concentration.

Since the Jaffe reaction is not specific, the preliminary separation of creatinine from the other serum components, by use of oxalic acid and Lloyd's reagent, is necessary if the true creatinine level is to be determined.

Creatinine will decompose at room or refrigerator temperatures. Therefore, if serum or urine is to be stored for long periods of time before creatinine analysis is performed, the sample should be frozen to ensure valid results.

URINE CREATININE METHOD

The same method of analysis may be used to determine creatinine concentration in urine. There are two additional considerations in preparing a urine sample for creatinine analysis: (1) the concentration of creatinine in the urine is approximately 50 times the concentration in serum; and (2) normally the urine is virtually protein free.

A 1:50 dilution of urine should result in a creatinine concentration comparable to that in serum. Since preparation of a protein-free filtrate is unnecessary, the usual 1:5 dilution employed with serum may be combined with the 1:50 dilution. Hence a 1:250 dilution of urine should have a concentration of creatinine comparable to that in a serum filtrate. This dilution should contain 100 ml of 20% trichloroacetic acid in 250 ml.[4]

Then 5 ml of the dilution may be treated in the same manner as a serum filtrate. If the reading is too high or too low, another sample of urine should be diluted to bring the subsequent reading within a desirable range.

Since the creatinine calculations are based on a 1:5 dilution of a sample, correction is needed only for dilutions other than 1:5.

CALCULATIONS

$$\text{absorbance of Test} \times \frac{\text{concentration of Standard (2 mg/dl)}}{\text{absorbance of Standard}}$$

$$\times \frac{\text{urine dilution}^5}{5} = \text{creatinine (mg/dl)}$$

$$\text{mg/dl} \times 10 = \text{mg/liter}$$

$$\text{mg/liter} \times \frac{\text{24 h volume (ml)}}{1000} = \text{mg per 24 hours (mg/d)}$$

PHYSIOLOGICAL SIGNIFICANCE OF CREATININE
CONCENTRATIONS IN SERUM AND URINE

Creatinine is formed in the body by the dehydration of creatine. In healthy adults, the serum creatinine level and the total daily excretion of creatinine in the urine remain reasonably constant for each individual. The constancy of the daily creatinine excretion may be used as a verification of the completeness of a 24-hour urine collection. If a series of 24-hour urine collections from the same patient show much variation in total creatinine content, some collections probably have been incomplete. Healthy adults excrete approximately 700 to 2500 mg of creatinine per 24 hours in the urine. However, since creatinine production is related to muscle mass, it is more meaningful to express creatinine excretion in terms of body size. Normal values for men are 20 to 26 mg per kilogram of body weight per 24 hours; women, 14 to 22 mg per kilogram of body weight per 24 hours.

The serum creatinine level is useful in the diagnosis or evaluation of renal disease. In kidney dysfunction and uremia, creatinine is retained in the blood, and plasma creatinine levels may become grossly elevated. A slight elevation may indicate an early stage of renal disease.

The creatinine clearance test is described and discussed in Chapter 28.

---

[4] For practical purposes, 0.1 ml of urine and 10 ml of trichloroacetic acid may be diluted to 25 ml.
[5] If the urine is diluted 1:250 as suggested, this number is 250.

Creatine

Since creatinine is the anhydride of creatine, creatine may be converted to creatinine by boiling with acid. The creatinine thus formed then may be quantitated by the creatinine method.

SERUM CREATINE METHOD

PROCEDURE

To a 16 × 150 mm test tube, add 8 ml of water, 4 ml of serum, and 8 ml of 20% trichloroacetic acid. Mix well and allow to stand for 10 minutes.

Centrifuge for 10 minutes. Then add 5 ml of the clear supernate to a 16 × 125 mm screw-cap tube labeled A, and 5 ml to another such tube labeled B.

Mark tube A at the 5 ml level. Then add 0.2 ml of concd HCl to tube A only.

Heat tube A in a boiling-water bath for 1 h to convert the creatine into creatinine.

Cool tube A to room temperature, readjust the volume to 5 ml with water, and mix.

Proceed with both tubes (A and B) as for serum creatinine, considering them as serum *filtrates* which are next treated with 0.5 ml of saturated oxalic acid, etc., under Procedure in the preceding section on Creatinine. Prepare and process the Standard as described in that method.

CALCULATIONS

1. Calculate the creatinine concentration for both tubes A and B.
2. Subtract the value of B (mg per dl) from the value of A.
3. Multiply the difference by 1.16 to obtain creatine, mg per dl. (The figure $1.16 = {}^{131}/_{113}$, which is the ratio of the molecular weights of creatine and creatinine, converts a creatinine value into an equivalent amount of creatine.)

URINE CREATINE METHOD

PROCEDURE

To a 16 × 125 mm test tube calibrated at 5 ml add 1 ml of urine and 4 ml of 0.5N hydrochloric acid.

Place in a boiling-water bath for 1 h. Then cool to room temperature and readjust the volume to 5 ml with water.

Redilute a sample of this solution 1:50 (this makes a total dilution of 1:250), and proceed according to the urine creatinine method.

A regular urine creatinine determination should be performed simultaneously with the same urine, and the creatinine concentrations in both samples calculated in the same manner. The difference between the preformed creatinine level and the creatinine level after the boiling process represents the creatinine formed from creatinine present in the urine. This creatinine expression is converted to creatine by multiplying the figure by 1.16 (see the preceding #3 under Calculations). The creatine excretion should be reported as milligrams per 24 hours.

NORMAL VALUES OF SERUM AND URINE CREATINE

Normal serum creatine values are 0.2–0.7 mg per dl.

Healthy adults may excrete up to 200 mg of creatine per 24 hours in the urine. However, it is not unusual to find no creatine in urine from healthy individuals.

PHYSIOLOGICAL SIGNIFICANCE OF SERUM AND
URINE CREATINE CONCENTRATIONS

Creatine is a product of endogenous muscle breakdown. When this breakdown is accelerated, as in muscular dystrophy, large amounts of creatine may occur in the blood and be excreted in the urine. Urine excretions as high as 1400 mg per 24 hours have been reported.

References

1. Benedict, S. R., and Behre, J. A.   Some applications of a new color reaction for creatinine. *J. Biol. Chem.* 114:515, 1936.
2. Butler, A. R.   The Jaffe reaction. Identification of the coloured species. *Clin. Chim. Acta* 59:227, 1975.
3. Folin, O., and Wu, H.   A system of blood analysis. *J. Biol. Chem.* 38:81, 1919.
4. Greenwald, I.   The chemistry of Jaffe's reaction for creatinine: IV. A compound of creatinine, picric acid, and sodium hydroxide, *J. Biol. Chem.* 77:539, 1928.
5. Hare, R. S.   Endogenous creatinine in serum and urine. *Proc. Soc. Exp. Biol. Med.* 74:148, 1950.

14. Uric Acid

Like creatinine and creatine, uric acid is a nonprotein nitrogenous waste product which normally exists in relatively low concentrations in the blood and is excreted in the urine.

The reference method [4] for measuring uric acid employs the enzyme uricase to oxidize uric acid to form allantoin plus carbon dioxide.

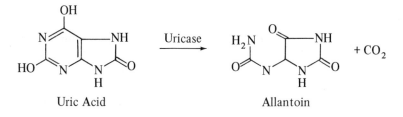

Uric Acid   Allantoin

Since uric acid absorbs light at 293 nm, but allantoin does not, the reaction is monitored by the decrease in absorbance at 293 nm. Because this is a time-consuming method, especially if appropriate controls are run, it is not used routinely.

Also uric acid can be oxidized to allantoin and carbon dioxide by phosphotungstic acid in alkaline solution. Conveniently, this is a colorimetric reaction, since the colorless phosphotungstic acid is reduced to an intensely blue substance called *tungsten blue*. Many different methods, all of which employ this same basic reaction, have been introduced [6]. They all share the common objective of trying to minimize interferences from the other reducing substances such as sulfhydryl compounds and ascorbic acid present in biological fluids. The four types of schemes which have been proposed are: (1) isolation of uric acid by precipitation with a metal ion [1] such as Ag^+; (2) isolation by an ion-exchange resin [7]; (3) precipitation of protein and interfering substances; and (4) colorimetric measurement before and after the addition of uricase, which hydrolyzes uric acid as discussed above. Cyanide has been used in some methods [2] to enhance the color development, but, because of its toxicity, instability (when dissolved), and high blank values, it is not recommended. Color enhancement can be achieved also by adding base.

Although the method presented here is not free of interferences, generally both positive and negative interferences are encountered in balancing amounts [3] so that the method is adequate for routine use.

Uric Acid Method

METHOD
Henry et al. [5].

MATERIAL
Serum.

REAGENTS

Sulfuric acid 0.667N. Dilute 10 ml of concd sulfuric acid to 500 ml. Determine the normality and dilute to 0.667N.

Sodium tungstate 10% (w/v). Dissolve 50 g of sodium tungstate ($Na_2WO_4 \cdot 2H_2O$) in water and dilute to 500 ml.

Sodium carbonate 10.3% (w/v). Dissolve 51.5 g of sodium carbonate (Na_2CO_3) in water and dilute to 500 ml.

Phosphotungstic acid reagent. To a 1500 ml Erlenmeyer flask marked at 1 liter add 40 g of sodium tungstate and 300 ml of water. Add 32 ml of 85% phosphoric acid and several glass beads. Attach a reflux condenser and boil gently for 2 hours.
Cool to room temperature and dilute to 1 liter. Add 32 g of lithium sulfate monohydrate ($Li_2SO_4 \cdot H_2O$) and dissolve. Store in a dark bottle in the refrigerator. This solution is stable indefinitely.

PROCEDURE

Into a 16 × 125 mm test tube measure 8 ml of water and 1 ml of serum.
Add 0.5 ml of 0.667N sulfuric acid and mix.
Add 0.5 ml of the sodium tungstate solution and shake to mix.
Filter through small (7 cm) filter paper.
Measure 3 ml of filtrate into a 19 mm cuvet (Test).
Into two other cuvets measure 3 ml of water (Blank) and 3 ml of working standard (Standard).
Add 1.5 ml of the sodium carbonate solution to each tube and mix.
Add 1 ml of the phosphotungstic acid reagent, invert to mix, and let stand for 30 minutes.
Read the absorbances of the Standard and Tests in a spectrophotometer at 700 nm, setting the Blank at zero.

CALCULATIONS

$$\text{absorbance of Test} \times \frac{\text{concentration of Standard (5 mg/dl)}}{\text{absorbance of Standard}}$$
$$= \text{uric acid (mg/dl)}$$

If a reading greater than 0.600 occurs, repeat the color development using 1 ml of filtrate and 2 ml of water. Then multiply the result by 3.

STANDARDIZATION

The preparation of a uric acid standard is somewhat more exacting than the usual technics for making a standard solution. If the directions are followed carefully, a stable and satisfactory standard results.

Uric Acid Stock Standard, 100 mg/dl. Dissolve 0.6 g of lithium carbonate (Li_2CO_3) in 150 ml of water in a 250 ml flask. Then filter.

Weigh 1.00 g of pure uric acid[1] on weighing paper and transfer in dry form to a 1 liter volumetric flask.

Warm the lithium carbonate solution to 60 C.

Warm the volumetric flask under running hot water, and pour the warm solution into the flask.

Shake the solution in order to dissolve the uric acid promptly. It should not take more than about 5 minutes to dissolve the uric acid, though the lithium carbonate solution itself may be slightly turbid.

After the uric acid is dissolved, cool the flask to room temperature, with shaking, under running cold water. Add 20 ml of 40% Formalin, and half fill the flask with water.

Add a few drops of methyl orange solution.

Then add from a pipet, slowly and with shaking, 25 ml of 1*N* sulfuric acid. The total acidity should be such that the methyl orange turns pink when there are about 2 or 3 ml of acid still left in the pipet.

Dilute to 1 liter. Mix well, and store in a tightly stoppered, dark glass bottle in the refrigerator.

This solution keeps indefinitely and contains 1 ml of uric acid per ml (100 mg per dl).

Working Standard. Dilute 1 ml of the stock solution to 200 ml with water. This dilute standard keeps for about 2 weeks in the refrigerator.

Since the stock standard is 100 mg per dl, the working standard has a uric acid concentration of $\frac{1}{200} \times 100 = 0.5$ mg per dl. Since the filtrate represents a 1:10 dilution of the serum, this standard is comparable to the *filtrate* from serum with a uric acid concentration of 5 mg per dl. When treated as a filtrate, 3 ml of this standard are equivalent to a 5 mg per dl serum concentration of uric acid.

Normal Values

The concentration of uric acid in the serum of healthy individuals ranges between 3 and 7 mg per dl. The levels are slightly lower (1 mg per dl) in women than in men.

Discussion

This method, with a precision of approximately ±0.5 mg per dl, is not the most specific one available for uric acid determination. However, the small increase in specificity offered by the more complex uricase methods does not merit the additional time and effort demanded.

Owing to the presence of interfering substances in red blood cells, uric

---

[1] Available as a Standard Reference Material from the U.S. National Bureau of Standards.

acid determinations should be performed using unhemolyzed serum or plasma. The tungstic acid precipitation technic is recommended because the uric acid comes through unaltered in the filtrate.

In the presence of sodium carbonate, uric acid reduces phosphotungstic acid to form a blue color, the intensity of which is directly proportional to the quantity of uric acid present. The intensity of the color produced with this method is not as great as that in the ureacyanide methods. However, under the conditions described here, the method affords adequate sensitivity without using the dangerous cyanide reagent.

Urine Uric Acid

The same method may be used to determine uric acid concentrations in urine. Since urine is relatively protein free, the preparation of a filtrate is unnecessary. The uric acid concentration in urine is approximately 10 times the serum uric acid level, so that a 1:10 dilution of the urine is necessary in order to bring the uric acid concentration into the serum range. The 1:10 dilution of serum resulting from the preparation of a protein-free filtrate must also be taken into account. The urine, therefore, is diluted 1:100 with water and treated in the same manner as a filtrate. The end result is multiplied by 100 to correct for the additional 1:10 dilution and to make the units milligrams per liter rather than mg per 100 ml. The results should be reported as milligrams per 24 hours (mg per day).

$$\text{mg/liter} \times \frac{24 \text{ h volume (ml)}}{1000} = \text{uric acid (mg/d)}$$

Healthy adults excrete in the urine approximately 600 to 800 mg of uric acid per 24 hours.

PHYSIOLOGICAL SIGNIFICANCE OF URIC ACID
CONCENTRATIONS IN SERUM AND URINE

Uric acid is the end product of purine metabolism in the human body. Its urinary excretion depends upon purine synthesis and metabolism as well as on renal function. In gout, the blood urate level is elevated. Often there is deposition of crystalline sodium urate around various joints in the form of so-called *tophi*. A high serum urate level usually confirms the diagnosis of gout. Since the level sometimes increases only during acute attacks, however, the gouty patient may not have a high concentration of serum urate all of the time.

Increases in serum uric acid level are often noted also whenever the blood urea nitrogen concentration is increased. However, the serum uric acid level is not considered to be a reliable or sensitive guide to kidney function. Other more suitable tests are therefore recommended for this purpose (see Chapter 28).

References

1. Benedict, S. R., and Behre, J. A.   The analysis of whole blood: III. Determination and distribution of uric acid. *J. Biol. Chem.* 92:161, 1931.
2. Folin, O.   Standardized methods for the determination of uric acid in unlaked blood and in urine. *J. Biol. Chem.* 101:111, 1933.
3. Hawk, P. B., Oser, B. L., and Summerson, W. H.   *Practical Physiological Chemistry* (13th ed.). New York: Blakiston, 1954. P. 559.
4. Henry, R. J., Cannon, D. C., and Winkelman, J.   *Clinical Chemistry Principles and Technics* (2nd ed.). New York: Harper & Row, 1974. P. 538.
5. Henry, R. J., Sobel, C., and Kim, J.   A modified carbonate-phosphotungstate method for the determination of uric acid and comparison with the spectrophotometric uricase method. *Am. J. Clin. Pathol.* 28:152, 1957.
6. Martinek, R. G.   Review of methods for determining uric acid in biologic fluids. *J. Am. Med. Technol.* 32:233, 1970.
7. Sambhi, M. P., and Grollman, A.   A simplified procedure for the routine determination of uric acid. *Clin. Chem.* 5:623, 1959.

15. Proteins

Proteins are polymers of alpha-amino acids occurring in both soluble and insoluble states in the body. Only tests for the soluble proteins presently are performed on a routine basis in the clinical chemistry laboratory because of the emphasis on fluid rather than solid-tissue samples. The three common fluids submitted for these analyses are serum, urine, and cerebrospinal fluid (CSF), all of which contain numerous types of proteins. This chapter deals with the serum proteins. The CSF proteins are discussed in Chapter 27, and the urinary proteins are covered in Chapter 28. However, occasional comments about the CSF and urinary proteins are made in this chapter to clearly establish the relationships among the proteins found in these three biological fluids.

Serum proteins can be analyzed in total, in groups, and individually. This is illustrated in Figure 15-1. The two major categories of serum proteins are albumin (an individual protein) and the globulins. These two categories are based on solubility, since the globulins are less soluble in general than is albumin. The globulins can be further separated into four groups or fractions called alpha-one (α_1), alpha-two (α_2), beta (β), and gamma (γ) globulin [14]. This separation is achieved by electrophoresis, a technic which is discussed in Chapter 7. Each of these categories is composed of a number of individual proteins, some of which are shown in Figure 15-1. Individual proteins usually are measured either based on their unique function (for example, enzymatic activity when the protein is an enzyme) or by means of immunochemical technics.

Each of the groups or individual proteins shown in Figure 15-1 may be of clinical interest in certain situations. Most laboratories measure total proteins and either the albumin or globulin fraction, or both. When only one of these fractions is measured, it may be subtracted from the total to give the other.

Another important point relative to Figure 15-1 is that many proteins are measured in a number of different ways in the clinical chemistry laboratory. For instance, immunoglobulin G (IgG) contributes to the values provided by four tests: total serum proteins, globulins, γ-globulins, and IgG. Only for the last test is it the only protein which is intended to be measured. The property of IgG that is used for its measurement in the test for total serum proteins is one which it shares with all the other serum proteins. A different property of IgG is used in the test for globulins, one which is shared by all the globulins but not by albumin. Yet another property of IgG is used in the test for γ-globulins, one which is not exhibited by the α_1, α_2, and β-globulins. Finally, the particular test for IgG is one which is based on still another property of IgG, a property which is totally unique to IgG and not shared by any other serum protein.

182

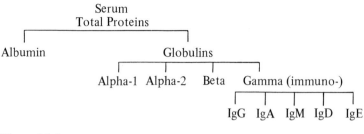

Figure 15-1
Classification of Serum Proteins

The actual properties of proteins that are used for these separations and measurements are discussed in subsequent sections.

Total Protein

KJELDAHL METHOD

All protein molecules contain a large number of nitrogen atoms. (Proteins are composed of amino acids, and each amino acid contains at least one nitrogen atom.) Since the approximate nitrogen composition of the proteins in body fluids is known, the concentration of protein in a given sample may be estimated from a measure of the protein nitrogen. The Kjeldahl method [21] is a nitrogen method used for this purpose and is considered a reference method for protein. However, the Kjeldahl method seldom is used in the clinical chemistry laboratory because it is complex and time consuming. With this method, the sample is heated in the presence of sulfuric acid, an oxidizing metal salt like copper sulfate, and another salt like potassium sulfate to raise the boiling point. The nitrogen in the sample is converted to ammonium ion, NH_4^+. The reaction mixture is cooled, and the nonvolatile NH_4^+ is converted to volatile NH_3 (ammonia) with aqueous NaOH. The mixture then is distilled for a short while with the tip of the condenser immersed in a standard solution of acid, which traps the distilled ammonia. The partly neutralized acid-ammonia solution is titrated back to its original acidity level with standard acid, using a colorimetric pH indicator to detect the endpoint. The amount of ammonia gives the amount of nitrogen in the original sample. After the digestion procedure, instead of distillation and titration, a colorimetric method may be used such as nesslerization or the Berthelot reaction (phenol-hypochlorite). These reactions are discussed in Chapter 12.

The Kjeldahl method gives the *total* nitrogen content of the sample. In order to calculate the *protein* nitrogen, the *nonprotein* nitrogen (NPN) must be determined and subtracted from the total. The average nitrogen content of serum protein is 15.3%. Therefore the protein nitrogen value is multiplied by $^{100}\!/_{15.3}$ or 6.54 to give the total serum protein concentration.

REFRACTIVE INDEX METHOD

Total protein concentration can be estimated by a refractive index reading made with a refractometer [10]. In this instrument, light is directed through the sample and the degree to which the light is bent (refracted) by the sample depends on the total amount of dissolved material in the sample. If all other constituents except proteins are present in their usual concentrations, changes in refractive index values are a measure of changes in protein concentration. However, because unusual amounts of any of the other substances (for example, glucose, urea, bilirubin, etc.) will affect the readings, the method is subject to interferences and should be used only for screening purposes with samples which are clear and of normal color.

BIURET METHOD

The biuret method is a rapid, simple, and accurate technic for protein determination. It depends upon the formation of a violet complex between cupric ions and protein. There are many biuret technics, the difference among them lying mainly in the preparation of the reagent used in the method. Although the essential ingredients of all biuret reagents are the same, ideal proportions of these ingredients should be used in order to produce a sensitive, stable solution capable of developing a stable color. Such a reagent is employed in the biuret method presented here.

TOTAL PROTEIN METHOD

METHOD
Biuret: Gornall, et al. [11].

MATERIAL
Serum.

REAGENTS

Biuret reagent. Dissolve 3 g of copper sulfate ($CuSO_4 \cdot 5H_2O$) and 12 g of potassium sodium tartrate ($KNaC_4H_4O_6 \cdot 4H_2O$) in approximately 1 liter of water. Transfer quantitatively to a 2 liter volumetric flask. Add enough carbonate-free sodium hydroxide solution to make a final base concentration of 3% (w/v). This requires 83 ml of 19N (50%) sodium hydroxide.[1] Shake the solution while diluting it to 2 liters to mix the base thoroughly. Mix well, and store in a polyethylene bottle. This reagent keeps indefinitely.

---

[1] See page 11 for preparation of 50% NaOH.

Sodium chloride 0.85% (w/v). Dissolve 8.5 g of sodium chloride (NaCl) in water and dilute to 1 liter.

PROCEDURE

To a 19 mm cuvet containing 1.9 ml of 0.85% saline, add 0.1 ml of serum (Test).

Set up a Standard in the same fashion, using standardized serum.

Add 2 ml of saline to another cuvet for a Blank.

Add 5 ml of biuret solution to each tube and invert to mix.

Allow to stand for 30 minutes.

Read absorbances at 550 nm, setting the Blank at zero.

CALCULATIONS

$$\text{absorbance of Test} \times \frac{\text{concentration of Standard}}{\text{absorbance of Standard}} = \text{protein (g/dl)}$$

STANDARDIZATION

Standard protein solutions are not as simple to prepare and do not keep as well as some of the standards for other methods. Because of the complexity of the human proteins, the best standard is human serum itself.

In order to standardize the biuret method, analyze a pool of human serum for protein concentration by either the Kjeldahl method or another reference method. Transfer 1 ml amounts of the serum into small tubes, stopper tightly, and store in a freezer. Each day, thaw the contents of one tube, mix thoroughly, and use this as a secondary protein standard, treating it in the same fashion as the tests. Protein concentration remains unchanged for at least 6 months in frozen serum.

Lyophilized human serum with assayed protein values is commercially available.

NORMAL VALUES

In healthy adults the serum total protein concentration ranges from 6.5 to 8.0 g per dl.

DISCUSSION

The biuret method is as simple and accurate a protein method as any currently available. It has a precision of approximately ±4%.

A compound that has in its molecular structure pairs of amide groups linked through nitrogen or carbon will show a positive biuret reaction. The name of the test is derived from the simplest of such compounds, biuret:

$CONH_2$
|
NH
|
$CONH_2$

When biuret is treated with an alkaline potassium copper tartrate solution, a violet-colored complex forms in which Cu^{2+} ions are surrounded by biuret molecules. Since protein molecules contain these linkages, they react with Cu^{2+} to give a violet color. The amount of this color is directly proportional to the protein concentration.

With turbid or highly colored samples, such as icteric sera, a serum Blank should be included to compensate for the turbidity or yellow color. This may be accomplished by measuring 0.1 ml of serum into 1.9 ml of saline solution in a cuvet, adding 5 ml of 3% NaOH, and reading the absorbance against a Blank of 2 ml of saline plus 5 ml of 3% NaOH. This absorbance is then subtracted from the reading of the biuret test before the final calculations are performed.

Protein methods should be performed using serum rather than plasma since the fibrinogen present in plasma is a protein and constitutes an interference when serum proteins are measured. Hemolyzed serum will yield inaccurate protein results and therefore should not be used.

The biuret method can be used to determine protein concentration in most transudates and exudates, but it is not directly applicable to urine or cerebrospinal fluid. Although generally the same kinds of protein are present in these fluids as in serum, the total protein concentration is much lower, making accurate analysis difficult. This problem is especially limiting in the case of CSF where the amount of fluid available for testing usually is quite small. The determination of protein in urine is made even more difficult by the presence of large amounts of interfering substances, especially inorganic ions, which precipitate when base is added.

For these reasons, the methods used routinely for total protein in CSF and urine are somewhat different from those employed for serum, although some overlaps occur. The Kjeldahl and biuret methods can be used, but extra steps must be taken to control the interferences. Turbidimetric and the Folin methods are widely used for measuring total protein in CSF, while turbidimetric and dye-binding methods are used for urine. These procedures are discussed in Chapters 27 and 28.

Groups of Proteins

The two main groups of proteins in body fluids are albumin and globulins. This classification is based on salt precipitation [24]. The globulins precipitate before albumin when salt (for example, ammonium sulfate, sodium sulfate, sodium sulfite) is added to serum. Therefore salt precipitation may be used to separate albumin and globulins prior to measurement. The concentration of the salt employed is an important consideration. Complete precipitation of globulins can be effected with 26.8% sodium sulfate [19] or 28% sodium sulfite [17]. After separation, one of the total protein methods (for example, Kjeldahl, biuret) may be used to determine the amount of protein in the precipitate (globulins) or in the supernatant (albumin). Once one of these components is determined, it may be subtracted from the total protein to give the other. Whereas albumin is regarded as a single protein, at least 20 to 30 major proteins comprise the globulins.

Albumin may be determined directly by dye binding since some of the dyes that undergo a change in color or color intensity upon protein binding tend to bind to albumin more readily than to the globulins. Three examples of such dyes are methyl orange [23], 2-(4'-hydroxyazobenzene)-benzoic acid (HABA) [2], and bromcresol green [9]. Bromcresol green has been found to be less subject to interferences and give the best correlations with other technics.

The globulins can be measured directly based on their content of the amino acid tryptophan since they contain much more tryptophan than does albumin. A useful reagent is an aqueous mixture of glyoxylic acid, cupric sulfate, sulfuric acid, and acetic acid [13]. This mixture gives a purple color with tryptophan in proteins, but not with free tryptophan.

Electrophoresis, discussed in Chapter 7, is widely used for separating protein fractions. In addition to separating albumin from the globulins, this technic further fractionates the globulins into four or five major groups: alpha-1 (α_1), alpha-2 (α_2), beta or beta-1 and beta-2 (β or β_1 and β_2)[2] and gamma (γ) globulins [4, 5].

Electrophoresis is of value not only for quantitating these proteins, but also for revealing the presence of an abnormal protein, that is, a protein associated only with certain diseases [18, 20]. (These abnormal proteins can further be identified by their immunochemical properties, as discussed in the section on Individual Proteins.)

The support material used for serum protein electrophoresis usually is paper, agar gel, or cellulose acetate. The exact details of the technic vary with the type of system employed; a typical method is described in the following paragraph.

Cellulose acetate strips are soaked in pH 8.8 tris-barbital-sodium barbital buffer. Excess buffer is removed by blotting with absorbent paper. Then 3–5 μl of each serum sample is applied. The strips are mounted in an electrophoresis chamber (illustrated in Figure 7-7) and subjected to 180 volts for 22 minutes. The strips are then immersed in a 0.1% solution of ponceau S dye in 5% trichloroacetic acid to precipitate and stain the proteins. The excess dye is rinsed out and the opaque strips are cleared (rendered transparent) by soaking in methanol-acetic acid (3:1) for 10 minutes. The location of each of the normal and abnormal protein fractions may be noted (see Figure 15-2).

Serum proteins migrate toward the positive electrode (anode) because

2 Depending on the system and buffer employed, one or two beta bands may occur.

these proteins are negatively charged in the pH range (8–9) normally used. Albumin migrates the farthest because it has the highest negative charge.

If the total protein concentration is known, the concentrations of the fractions may be quantitated in one of two ways:

1. The various spots are cut out and placed in solutions which will elute the dye from the strip. The absorbances of the solutions are then determined with a spectrophotometer.
2. The densities of the spots on the strip may be determined directly with a densitometer. In this method the strip is passed between a beam of monochromatic light and a detector. The spots absorb light in proportion to their densities, thereby changing the output of the detector. These variations are plotted by a recorder (Figure 15-2). The area under each of the resultant curves represents a percentage of the total protein.

Quantitation of the strips yields only the percentages of the various fractions. If these are to be converted to grams per 100 milliliters, the total protein concentration of the sample must be determined by some other method, for example, the biuret reaction.

The method presented here is a dye-binding technic suitable for determining albumin concentration in serum. If total protein is also determined, globulin can be calculated by difference.

ALBUMIN METHOD

METHOD
Doumas, et al. [8, 9].

MATERIAL
Serum.

REAGENTS
Succinate buffer, 0.1M, pH 4.0. Dissolve 11.8 g of succinic acid (HOOCCH$_2$CH$_2$COOH) in approximately 800 ml of water. Adjust the pH to 4.0 ± 0.05 with saturated (50%) sodium hydroxide. Add 0.1 g of sodium azide, then dilute to 1 liter with water. Store refrigerated.

Stock dye solution. Dissolve 210 mg of bromcresol green in 5 ml of 0.1N NaOH. Add 50 mg of sodium azide. Then dilute to 500 ml with water. Store refrigerated.

Working dye solution. Mix 125 ml of stock dye solution with 375 ml of succinate buffer. Add 2 ml of Brij-35, 30%. Adjust the pH to 4.2 ± 0.05 with saturated (50%) NaOH. Then store refrigerated.

Electrophoresis Pattern

Densitometer Scan

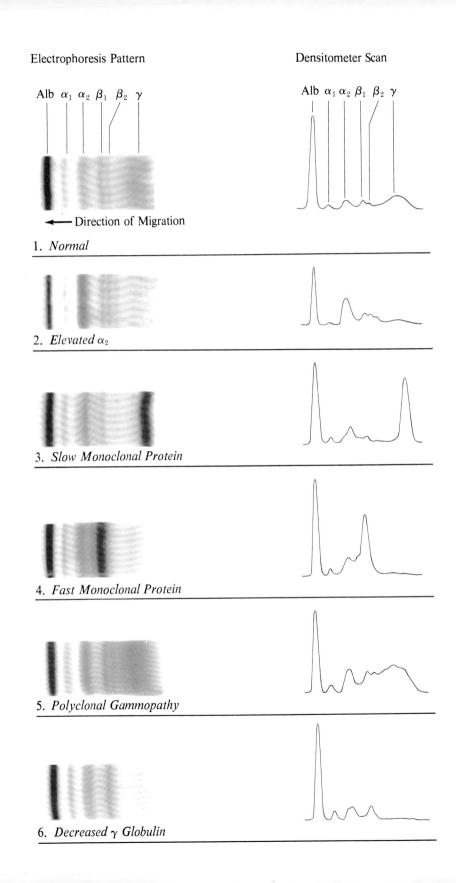

Alb α_1 α_2 β_1 β_2 γ

Alb α_1 α_2 β_1 β_2 γ

← Direction of Migration

1. *Normal*

2. *Elevated* α_2

3. *Slow Monoclonal Protein*

4. *Fast Monoclonal Protein*

5. *Polyclonal Gammopathy*

6. *Decreased* γ *Globulin*

Figure 15-2
Serum Protein Electrophoresis: Comments and Interpretations.
The electrophoresis patterns on the facing page were produced by the technic described on page 187.[a] The origin (point of application of the serum sample) is in the γ region, and the direction of migration is to the left.

1. *Normal.* All of these six bands, except for albumin, are composed of many individual proteins. For each band, however, a few proteins make up most of the intensity. Note that the β globulins have been separated into two bands.

2. *Elevated α_2.* This pattern is most commonly seen in nephrotic syndrome. Note the reduction in intensity of albumin and γ globulin bands.

3. *Slow Monoclonal Protein.* The intense, sharp band or spike in the γ region is an *M* protein, consistent with multiple myeloma, macroglobulinemia, or malignant lymphoma. Note the reduced intensity throughout the γ region and the increase in α_2 globulin.

4. *Fast Monoclonal Protein.* This *M* protein occurs in the β region. A general reduction in γ globulin accompanies this *M* peak, just as in the above case of a slow monoclonal protein.

5. *Polyclonal Gammopathy.* This general increase throughout the γ region is a very common abnormal pattern. It is seen in a wide variety of clinical disorders such as chronic liver disease, collagen disorders, and chronic infections.

6. *Decreased γ Globulin.* The most common causes of a general decrease in the γ globulins are congenital agammaglobulinemia, physiologic hypogammaglobulinemia (a transient condition in children after the first few months of life), and malignancy. One example of the latter is Bence Jones monoclonal gammopathy (light-chain disease).

[a] Courtesy of Dr. Stanley Elfbaum.

PROCEDURE

To 5 ml of working dye solution add 25 µl of serum. Mix, and allow to stand for 10 minutes. Read the absorbance at 630 nm in a 10 mm cuvet against dye solution set at zero.

CALCULATIONS

$$\text{absorbance of Test} \times \frac{\text{concentration of Standard}}{\text{absorbance of Standard}} = \text{albumin (g/dl)}$$

Total protein minus albumin = globulin.

STANDARDIZATION

Primary standard material for protein standardization is not available. Serum which has an assay value for total protein (for example, biuret assay) may be analyzed by electrophoresis and the albumin value calculated. This serum subsequently may be used as a secondary standard in the albumin method.

NORMAL VALUES

The range of serum albumin concentrations in healthy adults is approximately 3.8–5.0 g per dl, while the globulin concentrations are 2.5–3.5 g per dl.

DISCUSSION

A precision of approximately ±0.2 g per dl may be attained with this method.

Bromcresol green is a pH indicator which changes from yellow to blue in the pH range 3.8 to 5.4. At pH 4.2, albumin binds with bromcresol green to produce a related color change. This reaction is referred to as the *protein error of indicators*. Surfactants, such as Brij-35, decrease the color of the unreacted dye and also help prevent turbidity. These effects result in greater sensitivity and specificity.

The sensitivity of the method is excellent, as can be seen from the small amount of sample required. The method is fairly specific for albumin, although globulins react to a small degree. Bilirubin, hemoglobin, or lipids in moderately increased amounts do not affect the results.

Color development is virtually complete in 10 minutes and is stable for at least 30 minutes thereafter.

A standard should accompany each group of tests. Since the same amounts of different commercial albumin preparations may vary quantitatively in their reaction with bromcresol green, the best material to use is pooled human serum with an albumin concentration determined based on total protein electrophoresis analysis.

Results of this or other dye-binding methods do not always agree with values obtained by electrophoresis. This is not surprising because the two methods are based on entirely different principles and are measuring not a single substance but rather a group of very similar substances.

Results obtained by electrophoresis are considered more accurate than those of dye-binding or salting-out technics.

PHYSIOLOGICAL SIGNIFICANCE OF SERUM TOTAL PROTEIN, ALBUMIN, AND GLOBULIN CONCENTRATIONS

The serum protein concentration is dependent, to some degree, upon diet and intestinal absorption. Gross deficiencies in dietary protein intake and certain pathological conditions that impair intestinal absorption result in a low serum protein concentration. Although the total protein concentration will give the physician some information regarding the nutritional status of the patient, knowledge of the distribution of the protein fractions is of much more value [16].

Serum albumin values in hospitalized patients frequently are 0.5–1.0 g per dl lower than the usual normal values, even in the absence of disease.

There are five main types of protein disturbances that are of diagnostic value. These are summarized, together with their usual clinical implications, in Table 15-1.

Individual Proteins

Immunological technics are used to identify and quantitate the individual globulins in serum. Although a wide variety of such technics can be employed, they all depend on the availability of protein-specific antibodies, one for each individual serum protein that is to be measured. Immunodiffusion, immunoelectrophoresis, and competitive binding assays are the technics used most commonly [12]. The last is discussed in Chapter 8.

ANTIBODIES

When a specific protein is injected into an animal, a specific antibody against the injected protein (antigen) appears in the animal's blood. The animal is bled periodically and either its antibody-containing serum is used directly as antiserum (against the injected antigen) or the antibody

Table 15-1. Relationship of Serum Protein Concentrations to Disease

Total Protein	Albumin	Globulin	Usual Disorder
High	Normal	High	Multiple myeloma, sarcoidosis
High	High	High	Dehydration
Normal	Low	High	Hepatic damage, certain chronic infections
Low	Low	Normal	Renal disease, inadequate diet, intestinal malabsorption, nephrotic syndrome
Low	Low	Low	Third degree burns, water intoxication

fraction is isolated and used. When a mixture of antigens is injected (for example, a mixture of the normal serum proteins from a different animal), a mixture of antibodies results, one for each protein antigen. Antibodies themselves are proteins and usually possess two equivalent binding sites for antigen. Antibody and antigen fit together like a lock and key. Actually, since protein antigens provide a number of sites for antibody binding, the reaction of antibody with protein antigen is more like the combination of pairs of locks with rings of keys. Such a reaction begins with the invisible formation of soluble complexes between antibody and antigen, but a lattice eventually develops that expands into a visible precipitate. This is called a precipitin reaction and is illustrated in Figure 15-3. When the precipitate occurs in a gel, it is

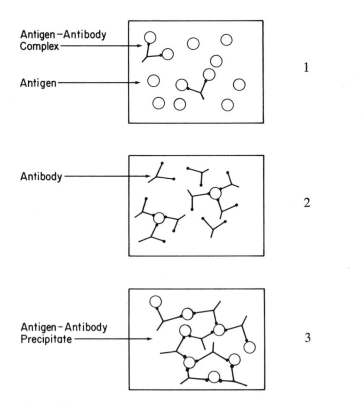

Figure 15-3
Concentration Dependence of the Precipitin Reaction. (1) *Antigen is in excess.* Soluble antigen-antibody complexes occur instead of an antigen-antibody lattice (precipitin reaction) because insufficient antibody is present to achieve crosslinking. (2) *Antibody is in excess.* Again no precipitate occurs, but now because antigen is so excessively coated with antibody that it cannot participate in crosslinking. (3) *The correct ratio* of antibody to antigen favors crosslinking (lattice formation). This lattice is insoluble, and its precipitation is called *the precipitin reaction.*

called a *precipitin line or band*. An excess of either antigen or antibody will not give a precipitin reaction because lattice formation is less likely. Small, soluble, individual antigen-antibody complexes occur instead. In each antigen-antibody reaction, there is a limited range of relative amounts of antigen to antibody which will give sufficient lattice formation to result in a precipitate. This range is called the *precipitin zone*. The precipitin reaction occurs both in solution and in gel media like wet agar or cellulose acetate and is used to identify and quantitate individual proteins in both immunodiffusion and immunoelectrophoresis analyses.

IMMUNODIFFUSION

Radial immunodiffusion (RID) currently is a widely used technic for quantitation of individual proteins in serum [1, 3, 7, 15]. The method is illustrated in Figure 15-4. An example is a specific antibody that is incorporated in an agar-gel which is spread on a flat surface. The sample is applied in a well (hole) in the gel and allowed to diffuse radially. An expanding ring of precipitate (precipitin ring) forms around the well as the protein antigen diffuses away from the well and encounters complementary antibody, that is, antibody which recognizes and therefore precipitates specifically this protein antigen. The precipitin ring expands because excess antigen continues to arrive (diffuse) from the well. (Antibody is unable to precipitate antigen when antigen occurs in excess, as illustrated in Figure 15-3.) However, eventually excess antigen is exhausted, and the circular precipitate sharpens somewhat and stops moving.

Both the rate of expansion of this precipitin ring and its final size are proportional to the amount of antigen in the well. The diameter of the ring is a convenient measure of its size both before and after it has stopped moving. In the former approach, a specific time is chosen before the ring stops expanding for measuring its diameter in order to quantitate the concentration of protein antigen contained in the sample being analyzed.

A series of standards containing known amounts of the antigen being measured accompany each set of samples, and a plot of antigen concentration versus precipitin ring diameter is constructed. The antigen concentrations in the tests are determined by referring their diameters to the standard curve. A different antibody in a different plate is used for each protein antigen to be measured in the sample. A practical application of this technic may be found in Chapter 27, where an RID method is presented for IgG.

IMMUNOELECTROPHORESIS

Immunoelectrophoresis [5, 6, 22] involves an electrophoretic separation followed by an immunodiffusion process, both in the same agar-gel spread. Although some semiquantitative information may be obtained by

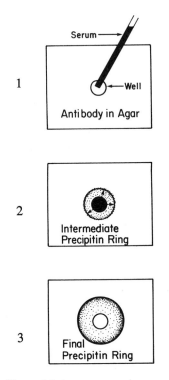

Figure 15-4
Radial Immunodiffusion Technic. (1) The well in the layer of antibody containing agar gel is filled with serum by means of a capillary tube. The antibody in the gel is complementary to (reactive with, specific for) just *one* of the proteins in the serum. (2) The serum diffuses into the surrounding gel. A ring of precipitate (antibody/protein-antigen lattice) grows around the well. Less precipitate occurs in the region immediately surrounding the well because the protein antigen is in excess in this region. (3) Eventually the ring stops growing and sharpens after the supply of protein antigen from the well is exhausted.

this technic, its primary value is in detecting the presence of abnormal proteins.

The method is illustrated in Figure 15-5. Serum is applied to an agar-gel plate and, under the influence of a controlled electrical current, the proteins migrate and separate. Subsequently, antiserum is applied in a trough parallel to the line of electrophoretic migration. An antiserum containing antibodies against only one of the proteins in the sample usually is used in each trough. The system then is incubated (allowed to stand) in a moist chamber, which prevents the agar gel from drying out.

During this period, the antibody and proteins diffuse in the gel in all directions, including toward one another. Upon contact, the antibody recognizes and precipitates only one of the proteins. The result is a

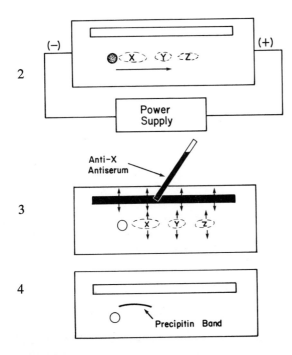

Figure 15-5
Immunoelectrophoresis Technic. (1) The well in the agar gel plate is filled with serum by means of a capillary tube. (2) Electrophoresis of the serum causes the proteins to migrate at different rates and become separated into components X, Y, and Z. (3) The trough is filled with an antiserum to one of the separated proteins. (In this case, antiserum against protein X is applied.) The antiserum diffuses into a gel and so do the separated protein bands. (4) A precipitin band (line of precipitate) forms only where anti-X antiserum (containing anti-X antibodies) meets protein X.

visible, arc-shaped precipitin line between this particular protein band and the antiserum trough. The general appearance of this line allows an approximation of the level (for example, absent, low, medium, high) of protein antigen. However, immunoelectrophoresis is considerably less quantitative than radial immunodiffusion and, therefore, is not recom-

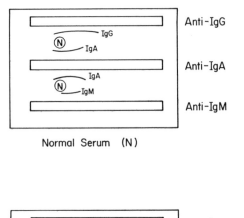

Normal Serum   (N)

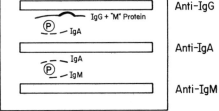

Abnormal Serum   (P)

Figure 15-6
Immunoelectrophoresis of a Normal and an Abnormal Serum. *Top.* Immuno-electrophoresis of a normal serum (the same serum was applied to each well) shows precipitin bands for IgG, IgA, and IgM when appropriate anti-sera are used. Although the precipitin bands differ somewhat in their length, intensity, and shape for these three proteins, the bands are located in the same general region. Application of a different, specific antiserum to each trough allows each of these three proteins to be visualized separately. *Bottom.* Immunoelectrophoresis of a serum from a patient with an IgM *M*-component shows an *M*-band as a heavy bow in the IgG precipitin line. IgA and IgM usually are decreased in concentration in the presence of IgG monoclonal gammopathies, as illustrated. (In analogous cases of IgA or IgM *M*-compo-nents, the heavy bow occurs instead as part of the IgA or IgM band, respec-tively.)

mended for this purpose. The primary value of immunoelectrophoresis is to identify abnormal proteins, particularly abnormal forms of the five classes of immunoglobulins (IgG, IgA, IgM, IgD, and IgE, see page 198) which are called *M-components.*[3] This is illustrated in Figure 15-6. Such an analysis is dictated whenever an extra band or an undefined

[3] The term *M-components* was adopted because these abnormal proteins are asso-ciated with myeloma, macroglobulinemia of Waldenstrom, and malignant lym-phoma, all of which begin with "M." The designation is unfortunate, however, because it frequently is confused with immunoglobulin-M (IgM).

elevation in a usual band is seen by conventional serum protein electrophoresis (page 189). In fact, such electrophoresis often is carried out prior to, or in conjunction with, an immunoelectrophoresis whenever a disease state is suspected in which an *M*-component (abnormal immunoglobulin or immunoglobulin fragment) is likely to occur in the blood. Such a diseased condition is called a monoclonal gammopathy.

Because monoclonal forms of IgG, IgA, and IgM occur much more often than monoclonal forms of IgD and IgE, initial immunoelectrophoretic analysis of serum usually is confined to only the first three. This can be achieved by simultaneously electrophoresing two wells of the serum on the same gel plate and then applying one of each of the three appropriate antisera to each of three troughs on the agar plate, as shown in Figure 15-6. As shown in the top part of this figure, such analysis of normal serum shows a single precipitin arc for IgG, IgA, and IgM. (The reason for using a monospecific antiserum for each of these proteins should be clear from the figure. Although the precipitin bands differ somewhat in their length, intensity, and shape, the bands would overlap considerably if a general antiserum containing antibodies to all three proteins were used.) The lower part of the figure illustrates what is seen when serum from a patient with an IgG *M*-component is analyzed. The *M*-component is seen as a heavy bow in an IgG precipitin line. IgA and IgM are shown to be decreased in concentration, which is the usual case in this situation. If the *M*-component had been of IgA or IgM origin, instead, the IgG precipitin line would have appeared normal, and the heavy bow would have occurred in the IgA or IgM band. If no monoclonal protein is revealed using anti-IgG, anti-IgA, and anti-IgM antibodies, the immunoelectrophoresis may be repeated using IgD and IgE antibodies to detect an *M*-component derived from IgD or IgE. Although normal serum contains amounts of IgD and IgE which are too low to be detected by this technic, the method will detect the increased amounts of these substances which occur in monoclonal gammopathies.

Cerebrospinal fluid and urine samples are subjected to radial immunodiffusion and immunoelectrophoresis essentially as described for serum, except that the samples are concentrated by a dialysis or filtration process prior to the analysis.

NORMAL VALUES

The normal concentrations of the major immunoglobins IgG, IgA, and IgM vary with age. Table 15-2 shows a listing of these ranges.

PHYSIOLOGICAL SIGNIFICANCE OF
INDIVIDUAL PROTEIN CONCENTRATIONS

Although many of the serum proteins can be analyzed by the immunological technics discussed in this chapter, these methods have been applied particularly to the analysis of the immunoglobulins.

Table 15-2. Normal Values for Immunoglobulins

Age	Concentration (mg/dl)		
	IgG range	IgA range	IgM range
Newborn	370–1300	0	<6–23
1–3 months	272–762	6–56	16–67
4–6	206–1125	8–93	10–83
7–12	279–1533	16–98	22–147
1 year	420–1290	16–96	30–188
2	490–1390	24–164	36–200
3	520–1450	32–196	36–202
4	550–1490	36–232	36–204
5	590–1520	42–260	36–206
6	620–1550	46–284	36–208
7	660–1570	46–308	36–210
8	680–1590	46–332	36–214
9	700–1600	46–350	36–220
10	720–1600	48–368	36–222
11	740–1600	50–394	36–224
12	750–1600	56–404	36–224
13	760–1600	62–424	36–226
14	760–1600	66–444	36–228
15	770–1600	70–456	36–228
16	770–1600	76–464	36–228
17	760–1600	78–472	36–228
18	740–1590	80–480	36–228
19	720–1560	74–486	36–222
20	721–1550	60–494	36–216
21	690–1540	44–506	36–210
Adult	710–1540	60–490	37–204

REFERENCES:
1–11 months: Stiehm, E. R. & Fudenberg, H. H.   *Pediatrics* 37:715, 1966.
1 year to adult: Allensmith, M., et al.   *J. Pediatr.* 72:276, 1968.

The immunoglobulins (or gamma globulins, a name based on their electrophoretic mobility) are plasma proteins which function as antibodies. Also they elicit antibody formation (anti-antibodies) themselves, just as most any other large proteins do when injected into an animal, as long as they belong to a species different from that of the injected animal.

Five basic types or classes of immunoglobulins occur: IgG, IgA, IgM, IgD, and IgE. They all have similar structures composed of two light (short) amino acid chains and two heavy (long) amino acid chains, except IgM, which has five pairs of each. Two different types of light chains, designated *kappa* and *lambda,* occur in all of the immunoglobulin classes.

The three types of immunoglobulin abnormalities are immune-deficiency diseases, polyclonal gammopathies, and monoclonal gammopathies.

Immune-deficiency diseases involve a decrease in the concentration of one or more of the immunoglobulin classes. A typical symptom is increased susceptibility to infection since fewer antibodies are available to provide immunity against infectious agents like bacteria and viruses.

Polyclonal gammopathies involve a generalized increase in most or all of the immunoglobulins largely because some disease process involves, or has triggered, an antigenic challenge in the body, so that excess antibodies are being produced. Infection, liver diseases, collagen diseases, sarcoidosis, and malignancies such as Hodgkin's disease and monocytic leukemia all are classified as polyclonal gammopathies. A broad, diffuse increase in the gamma globulin concentration occurs in the electrophoresis pattern. This type of general increase in one of the normal immunoglobulins should not be confused with the presence of a sharp, abnormal band in one of the normal immunoglobulin regions, discussed in the next paragraph. (Also see Figure 15-2.)

Monoclonal gammopathies usually are cancers of the immune system. (Only rarely are they benign.) Examples of these diseases are multiple myeloma, macroglobulinemia, and heavy-chain disease. Each of these disorders involves the production of an excessive amount of abnormal protein(s). In the case of multiple myeloma, the abnormal protein is either an IgG, IgA, IgD, IgE, or free light-chain (Bence Jones protein). Frequently one of the cited immunoglobulins *plus* a Bence Jones protein is produced. For macroglobulinemia, the protein is an IgM, and for heavy-chain disease, the protein is the free heavy-chain fragment derived either from IgG, IgA, or IgM. These disorders also can be accompanied by a Bence Jones protein.

In these disorders an intense, sharp band of abnormal protein occurs in the electrophoresis pattern (see Figure 15-1). Such a band gives rise to a sharp peak in the densitometer curve and is referred to as a *spike* or *M-component*. In this group of diseases, one of the specific types of plasma cells (plasma cells produce antibodies) proliferates into a clone (a large number of cells of the same exact type). This clone of plasma cells produces the particular homogeneous antibody (all of the antibody is exactly the same), which constitutes the *M*-component. The specific class of immunoglobulins involved is determined by immunoelectrophoresis using antibodies specific to each of the major immunoglobulin classes.

Monoclonal gammopathies frequently result in an excess of Bence Jones protein (homogeneous immunoglobulin light chains) in the blood. Because of their small size, they readily pass through the filtration zone of the kidney into the urine. They are most reliably detected by the immunological technics described in this chapter. However, their unusual ability to precipitate when the urine is heated to 60 C, redissolve at 100 C, and reprecipitate upon cooling to 60 C has been used as the basis of a screening test. The problem with this heat test is that the results sometimes are not reliable, particularly in the presence of other proteins, like albumin, which tend to mask the changes in the Bence Jones precipitate.

References

1. Arvan, D. A.   Antigen-antibody crossed electrophoresis. *Lab. Management* 7:21 (Aug.), 1969.
2. Arvan, D. A., and Ritz, A.   Measurement of serum albumin by the HABA-dye technique: A study of the effect of free and conjugated bilirubin, of bile acids and of certain drugs. *Clin. Chim. Acta* 26:505, 1969.
3. Berne, B. H.   Differing methodology and equations used in quantitating immunoglobulins by radial immunodiffusion—A comparative evaluation of reported and commercial techniques. *Clin. Chem.* 20:61, 1974.
4. Briere, R. O., and Mull, J. D.   Electrophoresis of serum protein with cellulose acetate. *Am. J. Clin. Pathol.* 42:547, 1964.
5. Cawley, L. P.   *Electrophoresis and Immunoelectrophoresis.* Boston: Little, Brown, 1969.
6. Clarke, M. H. G., and Freeman, T.   Quantitative immunoelectrophoresis of human serum proteins. *Clin. Sci.* 35:403, 1968.
7. Crowle, A. J.   *Immunodiffusion* (2nd ed.). New York: Academic, 1973.
8. Doumas, B. T., and Biggs, H. G.   Determination of Serum Albumin. In Cooper, G. R. (Ed.), *Standard Methods of Clinical Chemistry.* New York: Academic, 1972. Vol. 7.
9. Doumas, B. T., Watson, W. A., and Briggs, H. G.   Albumin standards and the measurement of serum albumin with bromcresol green. *Clin. Chim. Acta* 31:87, 1971.
10. Drickman, A., and McKeon, F. A., Jr.   Determination of total serum protein by means of the refractive index of serum. *Am. J. Clin. Pathol.* 38:392, 1962.
11. Gornall, A. G., Bardawill, C. J., and David, M. M.   Determination of serum proteins by means of the biuret reaction. *J. Biol. Chem.* 177:751, 1949.
12. Grant, G. H., and Butt, W. R.   Immunochemical Methods in Clinical Chemistry. In Sobotka, H., and Stewart, C. P. (Eds.), *Advances in Clinical Chemistry.* New York: Academic, 1970. Vol. 13.
13. Henry, R. J., Cannon, D. C., and Winkelman, J. W.   *Clinical Chemistry Principles and Technics* (2nd ed.). New York: Harper & Row, 1974. P. 453.
14. Hobbs, J. R.   Immunoglobulins in Clinical Chemistry. In Sobotka, H., and Stewart, C. P. (Eds.), *Advances in Clinical Chemistry.* New York: Academic, 1971. Vol. 14.
15. Kalff, M. W.   Quantitative determination of serum immunoglobulin levels by single radial immunodiffusion. *Clin. Biochem.* 3:91, 1970.
16. Kawai, T.   *Clinical Aspects of the Plasma Proteins.* Philadelphia: Lippincott, 1973.

17. Kingsley, G. R.   A rapid method for separation of serum albumin and globulin. *J. Biol. Chem.* 133:731, 1940.
18. Laurell, C. B.   Electrophoresis, specific protein assays, or both in measurement of plasma proteins? *Clin. Chem.* 1:99, 1973.
19. Milne, J.   Serum protein fractionation comparison of sodium sulfate precipitation and electrophoresis. *J. Biol. Chem.* 169:595, 1947.
20. Nerenberg, S. T.   *Electrophoretic Screening Procedures.* Philadelphia: Lea & Febiger, 1973.
21. Peters, J. P., and Van Slyke, D. D.   Methods. In *Quantitative Clinical Chemistry.* Baltimore: Williams & Wilkins, 1932. Vol. II, pp. 516, 691.
22. Verbruggen, R.   Quantitative immunoelectrophoretic methods: A literature survey. *Clin. Chem.* 21:5, 1975.
23. Watson, D., and Nankiville, D. D.   Determination of plasma albumin by dye-binding and other methods. *Clin. Chim. Acta* 9:359, 1964.
24. Wolfson, W. Q., Cohn, C., Calvary, E., and Ichiba, F.   Studies in serum proteins: V. A rapid procedure for the estimation of total protein, true albumin, total globulin, alpha globulin, beta globulin, and gamma globulin in 1.0 ml of serum. *Am. J. Clin. Pathol. (Tech. Sec.)* 18:723, 1948.

16. Calcium

Calcium is present in serum in three basic forms:

1. Protein-bound (40%)
2. Complexed (47%)
3. Ionized (13%)

The percentages given are approximate values. Protein-bound calcium is distinguished by its ability to resist transport across a dialysis or ultra-filtration membrane. Complexed calcium consists of complexes of calcium with such anions as citrate, phosphate, bicarbonate, and sulfate. It is distinguished by its ability to resist binding to an appropriate ion exchanger. Ionized calcium is the form of calcium in serum which is physiologically active. The method of choice for its measurement is the calcium ion-selective electrode [6]. This electrode contains a liquid ion exchanger, usually a calcium organophosphate, and a calcium chloride solution. These constituents are in contact with each other along the inside surface of a special membrane and an electrical potential is maintained. When the electrode is immersed in a solution containing calcium ions, the calcium diffuses across the membrane and changes the electrical potential. Since the potential changes only with calcium-ion concentration, the method is specific for ionic calcium. The general principles of electrode measurements are discussed in conjunction with pH electrodes on page 121.

The sum of the three calcium forms is total calcium, which is best measured by atomic absorption spectrophotometry. In the atomic absorption method either a very hot flame or a lanthanum diluent is used to overcome the tendency of protein-bound and complexed calcium to resist detection. The hotter flame breaks these forms of calcium down so that a more complete calcium signal results. The lanthanum diluent (for example, 0.1% or 0.5% $LaCl_2$ in dilute HCl) liberates the calcium which is bound by proteins and chelating anions by furnishing lanthanum ions for these substances to bind instead. A general discussion of atomic absorption is given in Chapter 5.

Various calcium methods were developed for determination of total calcium in serum prior to the perfection of the atomic absorption spectrophotometer. Although most of them do not offer the overall accuracy, precision, and convenience of atomic absorption, these methods still are used in laboratories where an atomic absorption instrument is not available. The more widely used methods are colorimetric or titrimetric and fall into one of two categories: (1) reaction of calcium with an anion which can be quantitated, or (2) reaction of calcium with ethylenediaminetetraacetate (EDTA) in the presence of a colorimetric indicator. To illustrate the first category, calcium can be precipitated with the oxalate anion, and the oxalate in the calcium oxalate precipitate can be titrated with standard permanganate [4]. Permanganate serves as its own indicator since the titration mixture does not exhibit the pink color (endpoint) of permanganate until all of the oxalate has been oxidized and a slight excess of permanganate has been titrated into the solution.

A redox indicator like ferroin (ferrous o-phenanthroline) is used in the ceric-ion titration. As soon as all of the oxalate has been oxidized by ceric-ion titrant, ceric ion added subsequently oxidizes the orange ferroin to an almost colorless ferric state, denoting the endpoint.

The red-purple chloranilate anion, which precipitates calcium, is an example of an anion which can be quantitated colorimetrically in a calcium method [5]. The calcium chloranilate precipitate can be re-dissolved and the chloranilate quantitated based on its absorbance.

In the second category, anionic dyes such as Eriochrome Black-T, ammonium purpurate, calcein, and Cal Red are used. All of these dyes form colored or fluorescent complexes with calcium ions. When the colorless chelating agent, ethylenediaminetetraacetate (EDTA), is added, the calcium is bound, instead, by this agent, and a color or fluorescence change results. Methods based on this principle [1, 3, 8] can be performed rapidly, but in some cases the endpoint is difficult to determine. However, using the proper indicator and optimal chemical conditions, these methods give satisfactory results [7] and are much simpler to perform than the precipitation methods.

The method described below employs the complexing dye o-cresol-phthalein complexone. Readings are made in a spectrophotometer, which avoids the problem of determining a visual endpoint.

Calcium Method

METHOD
Baginski et al. [2].

MATERIAL
Serum or urine.

REAGENTS
Cresolphthalein complexone (CPC). To a small plastic beaker add 40 mg of o-cresolphthalein complexone (phthalein purple) and 1 ml of concd HCl. Swirl gently until the powder dissolves. (A few drops of water may be used to effect complete solution if necessary.) Rinse the dissolved dye into a 1 liter volumetric flask with 100 ml of dimethyl sulfoxide [$(CH_3)_2SO$; grade 1]. Dissolve 2.5 g of 8-hydroxyquinoline (C_9H_7NO) in this solution, then dilute to 1 liter with water. Store in a plastic bottle.

Diethylamine buffer. Dissolve 0.5 g of potassium cyanide (KCN) in approximately 500 ml of water in a 1 liter volumetric flask. Add 40 ml of diethylamine, $(C_2H_5)_2NH$, and dilute to 1 liter with water. Store in a plastic bottle.

EGTA reagent. To 100 ml of water add 0.5 g of ethylene glycol bis-(2-aminoethyl ether) tetraacetic acid (EGTA). Place on a magnetic stirrer and, while stirring, add 50% (w/w) NaOH (approximately 19N) dropwise until solution is complete. At this point the pH should be approximately 9.

PROCEDURE

Note: It is advisable to set up the Test and Standard in duplicate.

Measure 25 μl of serum into the bottom of a 12 × 75 mm polystyrene tube (Test).

Measure 25 μl of standard into a similar tube (Standard).

Add 1.25 ml of CPC reagent to each, and to a third (empty) tube (Blank).

Mix all tubes. Then add 1.25 ml of buffer to each tube.

Cover with Parafilm and invert to mix.

Read the absorbances of the Test and Standard at 575 nm in a spectrophotometer against the Blank set at zero. For these readings use 2 *clean* 12 mm cuvets. Pour the Blank into one and pour each Test into the other in sequence.

If a Test is turbid or lipemic add 25 μl of EGTA reagent to it and to the Blank after reading the absorbance. Set the Blank at zero again and reread the Test against it, thereby obtaining an EGTA reading.

CALCULATIONS

Serum.

$$\frac{\text{concentration of Standard (10 mg/dl)}}{\text{absorbance of Standard}} \times \text{absorbance of Test}$$

$$= \text{calcium (mg/dl)}$$

If an EGTA reading was made, subtract it from the absorbance obtained prior to the EGTA addition, and use only the resultant absorbance in the above formula.

Urine. Calculate as above, then:

$$\text{mg/dl} \times 10 \times \frac{24 \text{ h volume (ml)}}{1000} = \text{calcium (mg/d)}$$

STANDARDIZATION

Stock standard, 100 mg/dl. Dissolve 0.2498 g of pure calcium carbonate[1] in approximately 50 ml of water containing 1 ml of concd HCl. Dilute to 100 ml with water.

---

[1] Available as a Standard Reference Material from the U.S. National Bureau of Standards.

Working standard, 10 mg/dl. Dilute 10 ml of stock standard to 100 ml
with water. If desired, a series of standards may be prepared by using
different amounts of stock standard diluted to 100 ml.

NORMAL VALUES

The calcium concentration in the serum of healthy adults is between 8.6
and 11.0 mg per dl (4.3 and 5.5 meq per liter).

DISCUSSION

Cresolphthalein complexone (CPC) forms a complex with calcium
which is purple in an alkaline medium. In the method described above,
diethylamine buffer is used to produce such a medium of approximately
pH 9. Cyanide is included to help stabilize the diethylamine, probably
by complexing metal ions which might otherwise catalyze the oxidation
of diethylamine. The dimethyl sulfoxide in the CPC reagent helps increase the solubility of the reagents and stabilizes the system. The
8-hydroxyquinoline complexes magnesium and prevents it from reacting
with the CPC.

The extreme sensitivity of this method is a great asset; however, it also
requires great care in preventing contamination. Since glass can be a
source of calcium, reagents should be stored in plastic bottles, and
disposable plastic tubes should be used for the tests. Very pure water
must be used in preparation of the reagents. Use of a pipet with disposable plastic tips is recommended. If a flow-through cuvet is not used,
each sample should be poured into the same cuvet for the absorbance
reading and returned to its plastic tube immediately thereafter.

The integrity of the reagents may be monitored by reading the absorbance of the reagent Blank against water each time before setting the
absorbance of the reagent Blank itself at zero. The absorbance of this
Blank should be reasonably constant. With the method described here,
a blank absorbance of approximately 0.2 might be expected.

Bilirubin and hemoglobin in moderate amounts do not interfere with
this method. However, turbidity due to lipids can cause spurious results.
Addition of EGTA reagent to a test performed with lipemic serum
chelates the calcium, thereby destroying the colored complex. Any remaining absorbance, over and above that of a similarly treated reagent
blank, is due to the turbidity and can be subtracted out.

Urine Calcium

The method described above may be used for determination of calcium
in urine with no special modification. However, the validity of the
specimen itself must be assured. If the urine is not acid enough, calcium
salts may precipitate and low analytical values will be obtained. For this
reason, urine should be acidified before a sample is measured for calcium. Measure the volume of a 24-hour urine, mix well, and pour
approximately 20 ml into a flask. Using 1N HCl and pH paper, adjust

the pH to approximately 5. The calcium concentration then may be determined as described for serum.

The concentration of calcium in urine is greatly dependent upon diet, and pathological variations can be determined only through controlled study. With average normal diets, healthy adults excrete approximately 100 to 300 mg of calcium per 24 hours in the urine. After 3 days on a low-calcium diet, a person should excrete in the urine less than 100 mg of calcium per 24 hours.

Physiological Significance of Serum and Urine Calcium Concentrations

Calcium exists in the blood in three forms: ionized, complexed, and bound to protein (mainly albumin). When calcium determinations are performed, the total calcium concentration is determined regardless of the amount of calcium present in each form. A depressed concentration of total calcium can be due to hypoproteinemia (low protein concentration), but the concentration of physiologically active (ionized) calcium in such a case may be normal. For this reason, a protein determination should accompany each calcium analysis so that the calcium value can be interpreted properly.

Depressed serum calcium levels usually accompany hypoparathyroidism, some bone diseases, certain kidney diseases, and low serum protein levels. A very low level of ionized calcium results in a neuromuscular disorder called tetany, characterized by uncontrollable muscular cramps and tremors. In cases of suspected tetany it is of great value to the physician if a calcium level can be determined rapidly so that therapy can be instituted immediately. Elevated serum calcium levels occur in hyperparathyroidism, certain types of bone disease, vitamin-D poisoning, and sarcoidosis.

Ionic calcium and phosphorus tend to maintain an equilibrium in the blood. Changes in the calcium-ion level are often reflected reciprocally in the phosphorus level. This is because a high level of one causes precipitation of most of the other out of the blood as bone, which is largely calcium phosphate. Therefore, in the cases cited above, when low serum calcium levels occur they are often accompanied by elevations in the phosphorus level. The converse holds for situations involving hypercalcemia or elevated serum calcium concentration.

Normally, the urine calcium excretion parallels the serum calcium level. Large amounts of calcium are excreted in the urine in hyperparathyroidism, metabolic acidosis, renal tubular insufficiency, multiple myeloma, bone malignancies, and following excessive calcium intake (usually in the form of milk). Low urine calcium excretion usually accompanies hypoparathyroidism, sprue and steatorrhea, and inadequate calcium intake.

References

1. Bachra, B. N., Daver, A., and Sobel, A. E.   The complexometric titration of micro and ultramicro quantities of calcium in blood serum, urine, and inorganic salt solutions. *Clin. Chem.* 4:107, 1958.
2. Baginski, E. S., Marie, S. S., Clark, W. L., and Zak, B.   Direct micro-determination of serum calcium. *Clin. Chim. Acta* 46:49, 1973.
3. Buckley, E. S., Gibson, J. G., and Bortolotti, T. R.   Simplified titrimetric techniques for the assay of calcium and magnesium in plasma. *J. Lab. Clin. Med.* 38:751, 1951.
4. Clark, E. P., and Collip, J. B.   A study of the Tisdall method for the determination of blood serum calcium with a suggested modification. *J. Biol. Chem.* 63:461, 1925.
5. Ferro, P. V., and Ham, A. B.   A simple spectrophotometric method for the determination of calcium. *Am. J. Clin. Pathol.* 28:689, 1957.
6. Ladenson, J. H., and Bowers, G. N., Jr.   Free calcium in serum: I. Determination with the ion-specific electrode, and factors affecting the results. *Clin. Chem.* 19:565, 1973.
7. Sadek, F. S., and Reilley, C. N.   A survey on the application of visual indicators for the chelometric calcium determination in serum. *J. Lab. Clin. Med.* 54:621, 1959.
8. Sobel, A. E., and Hanok, A.   Rapid method for determination of ultra-micro quantities of calcium and magnesium. *Proc. Soc. Exp. Biol. Med.* 77:737, 1951.

17. Phosphorus

Most of the phosphorus in the body is present in the bones as calcium phosphate, $Ca_3(PO_4)_2$. The phosphorus which is present in the blood occurs also as phosphate, either free (inorganic phosphate, PO_4^{3-}) or in combination with an organic group (organic phosphate, $ROPO_3^{2-}$). The latter is confined essentially to erythrocytes. So phosphate determinations on protein-free filtrates of serum samples contain essentially only inorganic phosphate.

The molybdate ion, $Mo_7O_{24}^{6-}$, is used in many phosphate methods. The molybdenum in this ion turns blue when it is reduced. Interestingly, molybdate reacts with phosphate in the presence of acid to form a product (phosphomolybdate) in which the molybdenum can be reduced to its blue oxidation state much more easily than in the case of molybdate. Phosphate therefore can be quantitated by mixing it with molybdate, then adding a mild reducing agent which will reduce the phosphomolybdate, but not the molybdate, to form *molybdenum blue*. Among the reducing agents which have been employed are aromatic amines like 1,2,4-aminonaphtholsulfonic acid [2], *p*-methylaminophenol sulfate (Elon) [3], ferrous sulfate [7], stannous chloride [1], and ascorbic acid [6].

Phosphate can be measured also by its reaction with the dye malachite green [5]. This reaction, which is very sensitive, can be quantitated, based on the change which occurs in absorbance of the dye when it reacts with phosphate.

A molybdate method for phosphorus is presented here. The method actually measures phosphate, but it is customary to express the results as phosphorus and to call the test *phosphorus* or *inorganic phosphorus*.

Phosphorus Method

METHOD
Fiske and Subbarow [2], modified.

MATERIAL
Serum.

REAGENTS

Trichloroacetic acid 10% (w/v). Dissolve 100 g of trichloroacetic acid (CCl_3COOH) crystals in water and dilute to 1 liter.

Molybdate reagent. To 200 ml of water add 83 ml of sulfuric acid. Dissolve 25.0 g of ammonium molybdate, $(NH_4)_6Mo_7O_{24} \cdot 4H_2O$, in this solution. Pour into a one-liter volumetric flask and dilute to 1 liter with water.

Sulfonic acid reagent. Weigh out:
0.125 g of 1,2,4-aminonaphtholsulfonic acid

7.28 g of sodium bisulfite ($NaHSO_3$)

0.25 g of sodium sulfite (Na_2SO_3)

Dissolve these salts in 50 ml of water and filter if necessary. Store in a well-stoppered dark bottle.

PROCEDURE

To a 16×125 mm test tube add 0.5 ml of serum and 9.5 ml of 10% trichloroacetic acid (Test).

Shake well to mix, centrifuge for 5 minutes, then filter through 7 cm filter paper.

Measure 5 ml of filtrate into a 19 mm cuvet.

Measure 5 ml of filtered 10% trichloroacetic acid and 5 ml of working standard into two cuvets to serve as Blank and Standard respectively. Treat these the same as the sample filtrate.

Add 0.5 ml of molybdate reagent to each tube and mix.

Add 0.2 ml of sulfonic acid reagent.

Stopper, mix, and let stand 10 minutes.

Read the absorbances of the Test and Standard in a spectrophotometer at 660 nm, setting the Blank at zero.

CALCULATIONS

$$\text{absorbance of Test} \times \frac{\text{concentration of Standard (4 mg/dl)}}{\text{absorbance of Standard}} = P \text{ (mg/dl)}$$

If a value greater than 10 mg/dl is obtained, the determination should be repeated using 2 ml of filtrate plus 3 ml of 10% trichloroacetic acid. The result then must be multiplied by $\frac{5}{2}$ or 2.5.

STANDARDIZATION

A phosphorus standard may be prepared from one of the following salts:

1. Dibasic sodium phosphate (Na_2HPO_4, mol wt $= 142$).
2. Monobasic sodium phosphate or sodium acid phosphate ($NaH_2PO_4 \cdot H_2O$, mol wt $= 138$).
3. Dibasic potassium phosphate (K_2HPO_4, mol wt $= 174$).
4. Monobasic potassium phosphate or potassium acid phosphate (KH_2PO_4, mol wt $= 136$).

If the standard is to contain 200 mg of phosphorus per liter and if it is prepared from dibasic sodium phosphate, the calculations are as follows:

1. Molecular weight of dibasic sodium phosphate $= 142$.
2. One mole of the salt contains 1 mole or 31 g of phosphorus.
3. In order to get 200 mg of phosphorus, $\frac{142}{31} \times 200 = 916$ mg of dibasic sodium phosphate are required.

The amount of each of the salts listed above required for 200 mg of phosphorus is: 0.9160 g, 0.8896 g, 1.1224 g, and 0.8776 g respectively.[1]

Phosphorus Stock Standard 20 mg/dl. Weigh out the stated amount of one of the salts listed above. Dissolve in approximately 500 ml of water and transfer to a one-liter volumetric flask. Add 8 ml of concentrated HCl and dilute to 1 liter with water. Store in a refrigerator.

Working Standard 0.2 mg/dl. Dilute 1 ml of the stock standard to 100 ml with 10% trichloroacetic acid. Store in a refrigerator.

The phosphorus test is performed with a protein-free filtrate, so the standard must have a phosphorus concentration corresponding to that in the serum filtrate. The filtrate made from serum having a phosphorus concentration of 4 mg per dl will contain $\frac{1}{20} \times 4$ or 0.2 mg of phosphorus per 100 ml. Therefore, when 5 ml of the working standard described above are treated as a filtrate, the amount of phosphorus provided is equivalent to a serum phosphorus concentration of 4 mg/dl.

NORMAL VALUES
The range of concentrations of inorganic phosphorus in the serum of healthy adults is 2.7 to 4.5 mg per dl and varies with age.

DISCUSSION
The Fiske-Subbarow method is capable of precision of ±5%.

Trichloroacetic acid that has been passed through filter paper may develop a small degree of blue coloration when treated with the phosphorus reagents. In order to compensate for any possible error from this source, the Blank is passed through the filter paper before it is used.

When a molybdic acid reagent is added to a sample of serum filtrate, phosphomolybdic acid is formed from the phosphorus present. In order to produce a color, the phosphomolybdic acid must be reduced, but most reducing agents will also reduce the excess molybdic acid present. The 1,2,4-aminonaphtholsulfonic acid reagent is a mild reducing agent capable of reducing phosphomolybdic acid without acting on molybdic acid. Therefore the intensity of color produced is dependent only upon the amount of phosphomolybdic acid present and consequently upon the amount of phosphorus in the filtrate.

The phosphorus tests should not be left standing longer than the prescribed 10 minutes before reading since reoxidation with subsequent fading of the blue color may occur.

An elevated blood urea nitrogen level is usually accompanied by an elevated serum phosphorus level. Therefore, if a patient's blood urea nitrogen is greater than 75 mg/dl, it is quite probable that the phos-

[1] If hydrates of the salts are used this must be taken into account in calculating the amount of salt required.

phorus level will be elevated. So a repeat test may be avoided by using a smaller amount of filtrate for the phosphorus analysis.

As described here, this method determines inorganic phosphorus. The method can be used to determine also the phosphorus bound to organic compounds such as proteins and lipids, but extraction with an organic solvent and digestion of the sample is a necessary prerequisite. (See Phospholipid Method in Chapter 21.) After the organic phosphorus has been digested, the analysis is carried out in the same manner as for inorganic phosphorus.

Urine Phosphorus

Phosphorus concentration in urine may be determined by the same method as that presented for serum phosphorus determinations. However, since phosphate may precipitate in alkaline urine, the sample must first be acidified. Measure the volume of a 24-hour urine sample and mix well.[2] Adjust the pH of a portion of the urine to approximately 5 with $1N$ HCl. Dilute 1 ml of this to 10 ml with water. Mix. Then treat 0.5 ml of this dilution the same as serum. (See Procedure in preceding section, Phosphorus Method.) This extra 1:10 dilution usually brings the absorbance reading into the range of the serum values. The results must be multiplied by 100 to correct for the extra dilution and to get milligrams of phosphorus per liter of urine.

The determination should always be performed with a 24-hour specimen and the result (mg per liter) multiplied by

$$\frac{\text{24-h volume (ml)}}{1000}$$

and reported as milligrams per 24 hours (mg per day).

NORMAL VALUES

The excretion of phosphorus varies widely with diet. It is difficult to establish a definite normal range. However, healthy adults with average diets usually excrete 900 to 1300 mg of phosphorus in the urine per 24 hours.

Physiological Significance of Phosphorus Concentrations

Large quantities of phosphate are excreted by the kidneys. When the kidneys are not functioning properly, the phosphorus concentration in the blood increases. Since the blood nitrogen values are an indication of kidney function, it may be expected that elevations in blood urea nitrogen will be accompanied by elevated phosphorus levels.

As discussed in the section on the physiological significance of calcium

---

[2] This is particularly important if the urine is turbid since the turbidity may be due to precipitated phosphate salts.

(Chapter 16), phosphorus and calcium are intimately related, because together they form the inorganic matrix of bone. When the concentration of one is elevated in the blood, the concentration of the other tends to be depressed, because an excess of one will precipitate the normal amount of the other in bone.

References

1. Berenblum, I., and Chain, E.   An improved method for the colorimetric determination of phosphate. *Biochem. J.* 32:295, 1938.
2. Fiske, C. H., and Subbarow, Y.   The colorimetric determination of phosphorus. *J. Biol. Chem.* 66:375, 1925.
3. Goodwin, J. F., Thibert, R., McCann, D., and Boyle, A. J.   Estimation of serum phospholipid and total phosphorus using chloric acid. *Anal. Chem.* 30:1097, 1958.
4. Greenberg, B. G., Winters, R. W., and Graham, J. B.   The normal range of serum inorganic phosphorus and its utility as a discriminant in the diagnosis of congenital hypophosphatemia. *J. Clin. Endocrinol.* 20:364, 1960.
5. Itaya, K., and Ui, M.   A new micromethod for the colorimetric determination of inorganic phosphate. *Clin. Chim. Acta* 14:361, 1966.
6. Lowry, O. H., and Lopez, J. A.   The determination of inorganic phosphate in the presence of labile phosphate esters. *J. Biol. Chem.* 162:421, 1946.
7. Taussky, H. H., and Shorr, E.   A microcolorimetric method for the determination of inorganic phosphorus. *J. Biol. Chem.* 202:675, 1953.

18. Magnesium

Magnesium is one of many metals present in the serum in low concentration. Some of its chemistry and physiology is analogous to that of calcium. For instance, like calcium, it is divalent and exists in the serum in both protein-bound and ionized forms. However, unlike calcium, it is known primarily for its high intracellular concentration (second only to potassium) and its role as an activator of many enzymes.

The method of choice for magnesium determination is atomic absorption spectrophotometry (see Chapter 5). The method is simple to perform, is sensitive, and can easily be freed of interferences. Under appropriate conditions, calcium and magnesium may be measured simultaneously with a flame atomic absorption technic. Emission flame photometry also has been used to measure these metals, but not as successfully as with atomic absorption.

For laboratories which lack atomic absorption instrumentation, various titrimetric, colorimetric, and fluorescent methods are available for measurement of magnesium. For instance, magnesium can be precipitated as magnesium ammonium phosphate or as magnesium-8-hydroxyquinoline with the phosphate or 8-hydroxyquinoline subsequently determined by a variety of methods. The sulfonate of 8-hydroxy-5-quinoline [4] is an example of a reagent which forms a fluorescent complex with magnesium, and Magon [1] and titan yellow [3] are each dyes which form colored complexes with magnesium.

The method presented here is one of the variations of the titan-yellow method.

Magnesium Method

METHOD
Heagy [2]; Orange and Rhein [3], modified.

MATERIAL
Serum.

REAGENTS

Trichloroacetic acid, 10%. Dissolve 100 g of trichloroacetic acid crystals in water and dilute to 1 liter. Reagent made from redistilled trichloroacetic acid crystals yields more satisfactory results. The trichloroacetic acid solution made for iron determination (see page 329) may be diluted for use in this method.

Polyvinyl alcohol, 0.1%. Place 1.0 g of polyvinyl alcohol in 1 liter of water in a beaker. Heat and stir until the solution clears. Store at room temperature.

Titan yellow (clayton yellow), 0.025%. Dissolve 0.250 g of titan yellow in 1 liter of water.

213

Sodium hydroxide, 4N. Dissolve 160 g of sodium hydroxide in water and dilute to 1 liter. Store in a plastic bottle.

PROCEDURE

To a 16 × 100 mm test tube add 2 ml of water, 1 ml of serum, and 3 ml of trichloroacetic acid (Test).

Stopper the tube, shake well, and allow to stand for 5 minutes.

Centrifuge for 10 minutes. Then carefully pipet 3 ml of the supernatant into a 19 mm cuvet.

Into another cuvet measure 1.5 ml of water and 1.5 ml of trichloroacetic acid (Blank).

Into a third cuvet measure 1.0 ml of working standard, 0.5 ml of water, and 1.5 ml of trichloroacetic acid (Standard).

To all cuvets add 0.5 ml of polyvinyl alcohol and mix.

Add 1 ml of titan yellow and mix.

Add 1 ml of sodium hydroxide, stopper, and mix well.

Read the absorbances of the Test and Standard in a spectrophotometer at 540 nm, setting the Blank at zero.

CALCULATIONS

$$\text{absorbance of Test} \times \frac{\text{concentration of Standard (2.0 meq/liter)}}{\text{absorbance of Standard}}$$
$$= \text{Mg (meq/liter)}$$

The magnesium (Mg) results may be converted to milligrams per 100 ml as follows:

$$\text{meq/liter} \times 1.22 = \text{mg/dl}$$

STANDARDIZATION

Magnesium Stock Standard 100 meq/liter. Weigh 1.216 g of reagent-grade magnesium metal in a 100 ml beaker. Add approximately 50 ml of water. Begin stirring. Then add 1 ml of concd hydrochloric acid.[1] When the effervescence subsides, add another 1 ml of acid, repeating this procedure until a total of 9 ml of hydrochloric acid has been added. Continue stirring until the solution is clear. Transfer the solution quantitatively to a 1 liter volumetric flask and dilute to 1 liter with water.

Magnesium Working Standard. Dilute the stock standard 1:100 with water. The actual concentration of this standard is 1.0 meq per liter. However, in the method described above the sample is diluted 1:6. Then 3 ml of this dilution are used, which is equivalent to using 0.5

---

[1] Since hydrogen gas is released, this procedure should not be carried out near an open flame.

ml of sample. Therefore 1 ml of this standard, when treated as a *filtrate,* is equivalent to a serum magnesium concentration of 2.0 meq per liter.

NORMAL VALUES

Magnesium levels in the serum of healthy adults range from 1.5 to 2.3 meq per liter (1.7 to 2.8 mg per dl).

DISCUSSION

The precision of this method is approximately ±0.2 meq per liter.

The numerous minor modifications of the titan-yellow method that have been published testify to the fact that it is not one of the more stable and reliable technics in clinical chemistry. The optimal conditions appear to vary from one laboratory to another. Once the method is properly installed, however, it yields satisfactory results and requires minimal technical effort.

The deproteinization technic described above usually yields clear solutions. However, occasionally a supernate will be somewhat cloudy, but this small amount of protein dissolves when the sodium hydroxide is added and does not affect the results.

The polyvinyl alcohol is added to stabilize the dye lake. In some modifications of this method, gum ghatti is used for this purpose.

When titan yellow is added to the filtrate, a complex is formed with magnesium. Addition of base dissolves the dye lake and intensifies the color. One of the difficulties inherent in this method is the intense color of the Blank. With some spectrophotometers difficulty may be experienced in setting the Blank at zero. In these cases, switching to a wavelength of 560 nm often will solve the problem: Although the linearity of the reaction is the same as at 540 nm, the sensitivity is somewhat less.

Some reports indicate that the color formation is influenced by calcium or protein (or both), but these observations are not universally accepted.

When problems occur with the titan-yellow magnesium method, usually a reagent is at fault. The titan yellow, polyvinyl alcohol, and trichloroacetic acid solutions probably should be suspected in that order.

Physiological Significance of Magnesium Concentrations

Magnesium is essential to many enzymatic processes and also has a profound effect on muscle activity. High serum magnesium concentrations produce central-nervous-system depression, and low concentrations result in tetany and convulsions. Abnormally high serum magnesium levels usually are found only in severe renal disease or magnesium poisoning. Low levels have been reported for a variety of conditions including diarrhea, alcoholic cirrhosis, diabetic coma, and hyperparathyroidism. However, at the present time the clinical usefulness of serum magnesium levels is relatively limited.

References

1. Burcas, P. J., Boyle, A. J., and Mosher, R. E.   Spectrophotometric determination of magnesium in blood serum using magon. *Clin. Chem.* 10: 1028, 1964.
2. Heagy, F. C.   The use of polyvinyl alcohol in the colorimetric determination of magnesium in plasma or serum by means of titan yellow. *Can. J. Res.* 26:295, 1948.
3. Orange, M., and Rhein, H.   Microestimation of magnesium in body fluids. *J. Biol. Chem.* 189:379, 1951.
4. Schacter, D.   Fluorometric estimation of magnesium with 8-hydroxy-5-quinolinesulfonate. *J. Lab. Clin. Med.* 58:495, 1962.

19. Liver Function Tests

The liver is a complex organ which performs numerous metabolic, excretory, storage, and detoxifying functions. It is instrumental in the metabolism of carbohydrate, fat, and protein, the excretion of bilirubin, the alteration of many toxic substances, and the storage of essential nutrients such as glycogen. Damage to the liver or biliary tract may affect any or all of these functions.

Laboratory data are helpful not only in detecting the existence of liver disease but also in establishing the site, type, and extent of the damage. Due to the multiplicity and diversity of liver functions, there is no single laboratory test that yields all the necessary information [21]. To obtain a complete clinical picture of liver function, it is often necessary for the physician to order several carefully selected liver function tests. The number and types of tests requested vary greatly. Each laboratory should be prepared to perform those liver function tests deemed valuable by the medical staff. A few of the more commonly used liver function tests are described in this chapter. Other tests for liver function may be found in Chapter 15 (Proteins), Chapter 20 (Enzymes), and Chapter 21 (Lipids, the section on cholesterol esters).

Bilirubin

The chemical methods for quantitative estimation of bilirubin in serum are based on diazotization of the bilirubin and measurement of the azo dye. Most methods measure the red dye in an acid medium, but one popular method [9] employs caffeine and sodium benzoate to enhance the reaction and measures a blue form of the azo dye in an alkaline solution. The method presented here is one of many modifications of the Malloy-Evelyn technic.

BILIRUBIN METHOD

METHOD
Malloy and Evelyn [7], Kingsley et al. [5].

MATERIAL
Serum.

REAGENTS
Methyl alcohol. Absolute, reagent grade.

Diazo blank solution. Dilute 15 ml of concd HCl to 1 liter with water.

Diazo reagent. Make up fresh as needed from two stock solutions which are prepared as follows:

Solution A: To 985 ml of water add 15 ml of concd HCl. Add 1 g of sulfanilic acid ($NH_2C_6H_4SO_3H \cdot 2H_2O$) and stir until dissolved.

Solution B: Dissolve 0.5 g of sodium nitrite ($NaNO_2$) and dilute to 100 ml with water. This solution remains stable for several months if kept refrigerated.

To prepare the diazo reagent, add 0.3 ml of solution B to 10 ml of solution A, and mix. This reagent is unstable. A fresh solution should be prepared for each set of determinations.

PROCEDURE

To each of two 19 mm cuvets add 4 ml of water and 0.5 ml of serum.

To each of a second pair of 19 mm cuvets add 4 ml of water and 0.5 ml of standard.

To one tube of each pair add 1 ml of diazo blank solution. To the other add 1 ml of diazo reagent (Test and Test Blank, Standard and Standard Blank).

Mix each Test and Test Blank by swirling gently and read the absorbances in exactly 1 minute at 540 nm against a Water Blank. This is the *direct* bilirubin. The Standard tubes need not be read at this point.

To all tubes add 5.5 ml of methyl alcohol. Mix by inverting several times. Read the absorbances (including the Standard tubes) after 20 minutes. This is the *total* bilirubin.

CALCULATIONS

Total bilirubin. Subtract the absorbance of the Test Blank from the Test, and of the Standard Blank from the Standard to obtain net absorbance readings. Then:

$$\text{net Test absorbance} \times \frac{\text{concentration of Standard}}{\text{net Standard absorbance}}$$
$$= \text{total bilirubin (mg/dl)}$$

The calculations for *direct* bilirubin are the same except that the final result is divided by 2 because the direct readings are made in a volume of 5.5 ml, while the totals are read in 11 ml.

STANDARDIZATION

Lyophilized serum containing known amounts of bilirubin is commercially available.[1] Although this material is actually sold as control serum, it can be used as a secondary standard. It should be reconstituted in small amounts and stored refrigerated (for not more than 1 week) when not in use.

Bilirubin standard, 20 mg/dl

Phosphate buffer pH 7.4. Dissolve 11.92 g of anhydrous dibasic sodium phosphate (Na_2HPO_4) and 2.8 g of anhydrous monobasic potassium phosphate (KH_2PO_4) in water and dilute to 1 liter.

[1] Versatol Pediatric; General Diagnostics, Morris Plains, N.J.

Sodium hydroxide 5N. Make an appropriate dilution of concd carbonate-free NaOH.

Alkaline solvent pH 11.3. To a 100 ml volumetric flask add 1 ml of 5N NaOH and dilute to 100 ml with phosphate buffer (pH 7.4).

Carefully weigh 20 mg of pure bilirubin[2] in a 30 ml beaker. Add 4 ml of the alkaline solvent (pH 11.3) and stir with a glass rod until the bilirubin is dissolved. Two to four minutes after adding the solvent add 10 ml of a 25 g per 100 ml human albumin solution. Transfer the solution to a 100 ml volumetric flask, rinse the beaker several times with phosphate buffer (pH 7.4), and dilute to 100 ml with this buffer. The pH of the final solution should be 7.4 ± 0.1.

Prepare another solution in the same manner but omit the bilirubin. This solution serves as a blank.

Appropriate dilutions of this Standard (and Blank) may then be made and used in the same manner as serum for standardization of the chemical procedure. For example, dilute the Standard 1:10 with water to give a bilirubin concentration of 2.0 mg per 100 ml. Dilute the Blank solution in the same manner and analyze both dilutions according to the procedure described above. Subtract the absorbance (if any) of the Blank solution from that of the Standard to obtain the net absorbance of the standard.

NORMAL VALUES

Serum bilirubin levels in healthy adults are less than 0.4 mg per dl direct and less than 1.0 mg per dl total.

DISCUSSION

The Malloy-Evelyn method has a precision of approximately ±10% in the low range and approximately ±5% with levels over 5 mg per dl.

Elevations in bilirubin concentration are reflected in the appearance of the serum, but a highly colored serum does not always indicate an elevated bilirubin level since hemoglobin or carotene may impart a similar color. Discernible hemolysis may cause negative errors of 5% to 15% in most diazo methods [8].

The *icterus index test* may be used as a screening test for bilirubin. This test consists merely of comparing the color of the serum with that of prepared icterus index standards. These standards have assigned values and an icterus index value of 5 has been chosen as a conservatively maximum normal color. Since bilirubin imparts its characteristic golden-yellow color to serum, the amount of yellow color is a rough indication of the quantity of bilirubin present. If the serum has an icterus value of less than 5, the total bilirubin concentration is not elevated.

[2] Pure crystalline bilirubin is available as a Standard Reference Material from the U.S. National Bureau of Standards.

If the icterus index value is over 5 or if it cannot be determined accurately because of hemolysis or lipemia, the bilirubin test should be performed.

An icterus index standard of 5 units may be prepared as follows:

Dissolve 0.100 g of potassium dichromate ($K_2Cr_2O_7$) in approximately 100 ml of water in a 200 ml volumetric flask. Add 4 drops of concentrated sulfuric acid, and dilute to 200 ml with water.

When comparisons are being made in the icterus index method, the samples and standard should be in test tubes of equal diameter.

In the bilirubin method it is important that the reagents and sample be added in the exact proportions and sequence listed or a precipitate may form that will invalidate the results. Effective mixing after each step is essential. Before readings are made, the tubes should be checked for the presence of bubbles. Bubbles may be dispersed by gentle tapping of the tube or by using a stirring rod.

When solution A (sulfanilic acid plus hydrochloric acid) and solution B (sodium nitrite) are mixed, nitrous acid is formed. Nitrous acid is very unstable and should be freshly prepared for each group of analyses.

The sulfanilic acid and bilirubin, in the presence of nitrous acid, undergo a diazo reaction resulting in the pink-colored compound azobilirubin. The amount of azobilirubin formed, and thus the amount of pink color, is directly proportional to the serum bilirubin concentration. This is the van den Bergh reaction [16].

The amount of time required for full color development in the total reaction is a matter of some uncertainty. Intervals ranging from 10 to 30 minutes have been suggested, but it appears that 20 minutes is suitable.

Serum bilirubin is not very stable. Determination should be made, therefore, as soon as possible. If the sample must be stored, it should be kept out of the light and in a refrigerator. Even under these conditions, deterioration still takes place, but at a slower rate. In the frozen state bilirubin in *serum* is stable for several months; however, this does not appear to be true of most prepared bilirubin standards.

It is generally accepted that bilirubin exists in the serum in two forms: direct and indirect. The direct-reacting type is thought to be bilirubin conjugated with glucuronic acid, rendering it water soluble so that it undergoes a direct diazotization reaction in the (aqueous) serum. The indirect bilirubin is bound to protein and must be extracted from this protein and rendered soluble with alcohol before it will undergo diazotization. Hence the method presented here measures the direct bilirubin in water, then the *total* bilirubin after the alcohol is added. The indirect bilirubin may be calculated by subtracting the direct from the total.

PHYSIOLOGICAL SIGNIFICANCE OF BILIRUBIN CONCENTRATIONS

Bilirubin is a product of red-cell destruction and is excreted by the liver. An elevated serum bilirubin level may indicate impairment of the excretory function of the liver, excessive hemolysis (destruction of red cells), or obstruction of the biliary tract. Regardless of the cause, gross elevations of serum bilirubin concentration are evidenced by a yellow skin color, the condition known as jaundice.

The physiological significance of the direct bilirubin concentration is not generally agreed upon by clinicians. Free bilirubin normally is conjugated (esterified) in the liver and excreted in the bile, with a small amount of the conjugate being absorbed into the blood. An increased concentration of this direct bilirubin in the blood indicates an hepatic or a biliary disorder, while a normal concentration of direct bilirubin with an increased total concentration suggests a hemolytic problem in which the liver cannot accommodate the large amounts of free bilirubin presented to it. It has also been suggested, however, that the direct bilirubin concentration has no clinical significance [3].

MICRO BILIRUBIN METHOD

Measurement of bilirubin concentration in the blood of newborn babies frequently is of clinical importance. Because of the limited amount of blood available from infants, a micro procedure must be used. Most of the micro bilirubin methods proposed are modifications of the Malloy and Evelyn procedure described above.

If a spectrophotometer of good quality is available, the simple and accurate direct absorption method [20] may be used. With this method, the absorption of a diluted sample is measured at two wavelengths (575 and 455 nm), one of which (455 nm) is optimum for bilirubin absorption. Bilirubin exhibits virtually no absorption at 575 nm. Because fetal hemoglobin absorbs with equal intensity at both wavelengths, its interference must be subtracted since blood samples from newborn babies frequently are hemolyzed. This ability to correct for hemoglobin is an important advantage of the spectrophotometric method, because hemoglobin can cause spuriously low values with the chemical methods for bilirubin [8].

The method described below is a micro adaptation of the Malloy-Evelyn technic.

SAMPLE COLLECTION

Blood should be collected from a free-flowing heel puncture in 80 mm lengths of 4 mm glass tubing. Three such full tubes usually will yield enough serum for the test. Stopper both ends of the tubing with rubber vaccine stoppers. Allow the blood to clot. Then centrifuge at high speed for 10 minutes. After centrifugation, score the tubes with a file just above the cells and break the tubes. The serum then may be pipetted directly from the tubes with a micropipet.

REAGENTS
See page 217.

PROCEDURE

Into each of two 19 mm cuvets add 4.4 ml of water and 0.1 ml of serum.

Into each of a second pair of 19 mm cuvets add 4.4 ml of water and 0.1 ml of standard.

Proceed the same as the macro method on page 218, starting at the step involving addition of the diazo solution.

CALCULATIONS

The calculations are the same as for the macro procedure (page 218) except that the results must be multiplied by 5 because 0.1 ml of serum was used instead of 0.5 ml.

DISCUSSION

This method is not very precise with low concentrations of bilirubin because of the low absorbances obtained. However, since serum bilirubin values in the newborn infant of less than 5 mg per dl are seldom of clinical importance, this method usually is adequate.

Elevated serum bilirubin levels occur in many infants during the first few days of life. This generally is the result of a temporary deficiency of the enzyme, glucuronyl transferase, which converts unconjugated (indirect) bilirubin into its water-soluble (conjugated, direct) form. (This enzyme, as implied by its name, transfers glucuronate groups onto bilirubin. This renders the bilirubin water soluble because the glucuronate groups are charged.) Unlike unconjugated bilirubin, which is water insoluble and tends to accumulate in the body because it sticks to albumin, conjugated bilirubin is rapidly excreted, which is the intended metabolic route for bilirubin. Thus a temporary deficiency of glucuronyl transferase in the newborn results in increased serum levels of unconjugated bilirubin. Blood levels of bilirubin are monitored carefully to ensure that toxic concentrations (over 20 mg per dl) are not reached. If the albumin binding capacity for bilirubin is exceeded, the unconjugated bilirubin may enter the brain and cause serious damage (kernicterus). Phototherapy or exchange transfusion [14] are the usual courses of treatment for hyperbilirubinemia.

Frequently the precision of the micro method in a given laboratory is excellent but the accuracy is poor. This is usually due to improper standardization.

Bromsulfophthalein (BSP)

The organic dye disodium phenoltetrabromphthalein sulfonate (bromsulfophthalein, bromsulphalein, BSP) is used to test the excretory function of the liver. The laboratory methods involve measurement of the dye color at alkaline pH.

BROMSULFOPHTHALEIN (BSP) TEST

METHOD

Seligson et al. [12].

MATERIAL
Serum.

REAGENTS

Alkaline buffer. Weigh out:
   6.4 g of anhydrous dibasic sodium phosphate (Na_2HPO_4)
   1.77 g of tribasic sodium phosphate, dodecahydrate ($Na_3PO_4 \cdot 12H_2O$)
   3.20 g of sodium para-toluene sulfonate (para-toluene sulfonic acid, sodium salt, $CH_3C_6H_4SO_3Na$)
Dissolve these salts in water and dilute to 500 ml. With the aid of a pH meter adjust the pH of the solution to 10.7 (± 0.1) using $1N$ HCl or $1N$ NaOH as necessary. Store in a plastic container.

Acid reagent. Dissolve 98.7 g of monobasic sodium phosphate, monohydrate (sodium acid phosphate, $NaH_2PO_4 \cdot H_2O$) in approximately 175 ml of water and dilute to 250 ml. The pH should be approximately 3.6.

PROCEDURE

Five milligrams of BSP dye per kilogram of body weight are given intravenously (see page 224). One blood sample (10 ml, clotted) is obtained 45 minutes after the injection.

To a 19 mm cuvet add 1 ml of serum and 5 ml of alkaline buffer.

Treat 1 ml of working standard in the same fashion.

Mix by inverting and read the absorbance (A_1) against a water Blank at 575 nm.

Add 0.1 ml of acid reagent, mix, and read the absorbance again (A_2).

CALCULATIONS

$$(A_1 - A_2) \times \frac{\text{concentration of Standard (10\%)}}{\text{absorbance of Standard}} = \% \text{ dye}$$

If a dye concentration greater than 40% occurs, repeat the determination using 0.5 ml of serum and 5.5 ml of alkaline buffer. Calculate as above, then multiply the result by 2.

STANDARDIZATION

The BSP standard is prepared on the assumption that the body contains 50 ml of plasma per kilogram of weight. Therefore injection of 5 mg of dye per kilogram of body weight makes a 5 mg per 50 ml or a 10 mg per 100 ml solution of dye in the plasma. Hence a solution containing 10 mg of BSP per 100 ml represents a 100% dye concentration.

BSP Stock Standard, 5000 mg per 100 ml. The stock BSP solution may be obtained commercially in ampules, each milliliter containing 50 mg of BSP.

BSP Working Standard, 1 mg per 100 ml. Dilute 1 ml of the stock dye
solution to 50 ml with water. Then redilute this solution 1:100. This
working standard is equivalent to 10% dye in serum.

DISCUSSION

The BSP dye should be injected intravenously by a physician. Care
should be taken to avoid injection of dye into the subcutaneous tissue.

To calculate the volume of dye solution to be injected:

$$\frac{W \times 5}{2.2 \times C} = \text{ml of dye solution}$$

where W is the patient's weight in pounds and C is the concentration of
the dye solution in milligrams per milliliter (usually 50 mg per ml).
If C equals 50 mg per ml, the formula reduces to:

$$\frac{W}{22} = \text{ml of dye solution to be injected}$$

Only one blood sample is necessary. It should be collected with a
clean, dry syringe exactly 45 minutes after the injection of the dye.

Normally, less than 5% of the dye should remain in the blood at the
end of 45 minutes. If preferred, the sample can be drawn at 30 minutes,
in which case there may normally be up to 10% of the dye remaining in
the blood.

Earlier directions for the BSP test specified that 2 mg of dye per
kilogram of body weight be administered. The standards were made up
accordingly. Current methods employ 5 mg of dye per kilogram of body
weight. The standards described above are prepared on the basis of the
5 mg dosage.

If a BSP test is to be repeated, an interval of at least 24 hours should
elapse between the two tests in order to ensure complete clearance of the
first dosage of dye from the blood.

Phenolsulfonphthalein (PSP), a dye used in a kidney-function test,
will interfere with the BSP test. Since each of these dyes assumes a
characteristic color in base, care must be taken that no PSP dye is
present in the blood when a BSP test is performed.

A grossly hemolyzed blood sample may yield erroneous results because of the brownish color assumed by the hemoglobin upon the addition of base. It is important, therefore, that care be exercised in the
collection and handling of the blood for the BSP test. The use of the
Blank reading and the 575 nm wavelength help to minimize interference
from bilirubin and hemoglobin in the serum.

There are simpler methods for the determination of BSP dye which
merely employ excesses of NaOH and HCl for the readings. However,
excessive changes in the pH bring about increased errors from interfering

color (hemoglobin) and turbidity (lipids). The conditions of the method presented above give maximum dissociation of BSP dye with minimum pH change, thereby minimizing errors from interfering substances.

BSP dye interferes with colorimetric tests employing alkaline solutions. For this reason, blood for other chemical analyses should be obtained before injection of the dye.

There are some errors that are committed often enough in the performance of the BSP test to warrant a word of caution here.

1. The amount of dye to be administered must be calculated and measured accurately.
2. All of the solution must go into the vein.
3. The 45-minute interval should be carefully observed.
4. The syringe or needle used for measuring or injecting the dye should not be used to draw the blood sample.

PHYSIOLOGICAL SIGNIFICANCE OF THE BSP TEST

Normally the liver removes BSP dye from the blood stream and excretes it in the bile. The BSP test, therefore, is a valuable aid in detecting parenchymal liver damage from any cause. Retention of this dye may occur in infectious hepatitis or portal cirrhosis.

The performance of BSP tests for patients with obvious jaundice due to disorders of the liver or biliary tract is of questionable value because retention of bilirubin indicates that BSP will also be retained.

Bile Pigments in Urine

Certain bile pigments, such as bilirubin and biliverdin, may appear in excess in the urine of patients with liver disease. The qualitative tests for bilirubin in urine are simple:

If present in sufficient quantity, bilirubin in urine produces a deep amber color, or a yellow foam when shaken vigorously. These visual tests for bilirubin, however, usually are unreliable because of the various pigments and dyes that may be present in urine. Therefore, a chemical test is recommended. Screening tests are available commercially (for example, Ictotest[3]) which detect the presence of bilirubin by its reaction with diazo reagents to form a purple azo dye. The reagent is essentially a buffered preparation of p-nitrobenzenediazonium p-toluene-sulfonate.

Urobilinogen

Urobilinogen is formed by the action of intestinal bacteria upon bilirubin. Some urobilinogen is reabsorbed into the blood, returned to the liver, and ultimately excreted via the kidneys. Therefore small amounts of urobilinogen are normally present in the urine, but increased urinary

---

[3] Ames Co., Elkhart, Ind.

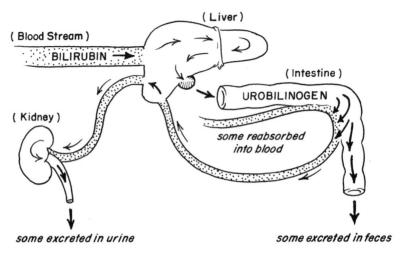

Figure 19-1
Formation and Fate of Urobilinogen

excretion of urobilinogen occurs in many types of liver disease and in hemolytic anemias. The consistent absence of urobilinogen from the urine may indicate obstruction of the biliary tract since the obstruction would prevent excretion of bilirubin into the intestine (Figure 19-1).

Urobilinogen reacts with Ehrlich's aldehyde reagent (*p*-dimethylaminobenzaldehyde) to give a red-colored compound. The simplest procedure for quantitating urine urobilinogen is one involving serial dilution of the urine, addition of Ehrlich's reagent, and reporting the results in terms of the last dilution in which a pink color is visible [17]. However, the presence of interfering substances frequently makes it difficult to judge the presence of a pink color. At best the quantitation is only a rough estimate.

A precise quantitative test involves the reduction of urobilin to urobilinogen with ferrous sulfate and extraction with petroleum ether prior to performing the Ehrlich reaction [10]. This procedure is cumbersome for routine use.

The method described below is a suitable compromise between the two methods mentioned above.

UROBILINOGEN METHOD
METHOD
Watson et al. [19].

MATERIAL
Urine (freshly voided).

REAGENTS

Sodium acetate saturated. Crystals of sodium acetate ($NaC_2H_3O_2 \cdot 3H_2O$)
  should always be visible at the bottom of the flask to ensure saturation.

Ehrlich's reagent modified. Weigh out 0.7 g of para-dimethylamino-
  benzaldehyde, $(CH_3)_2NC_6H_4CHO$, and dissolve in 150 ml of concentrated HCl. Add 100 ml of water, mix, and store in a dark bottle.

Barium chloride, 10%. Dissolve 10 g of barium chloride ($BaCl_2 \cdot 2H_2O$)
  in water and dilute to 100 ml.

PROCEDURE

If bile is present it should be removed before the test is performed.
  Removal of bile: To 5 ml of urine add 5 ml of 10% $BaCl_2$. Mix
  well and filter.
To each of two 50 ml Erlenmeyer flasks add 2.5 ml of bile-free urine.
To one portion (Blank) add 5 ml of sodium acetate solution and mix
  well.
Then add 2.5 ml of modified Ehrlich's reagent to the same flask *slowly*
  and with constant mixing.
Pour into a 19 mm cuvet and set at zero absorbance at 565 nm in a
  spectrophotometer.
To the other portion of urine add 2.5 ml of modified Ehrlich's reagent
  and mix well (Test).
Add 5 ml of sodium acetate solution and mix again. Pour into a cuvet
  and read the absorbance of the Test immediately against the Blank.

CALCULATIONS

Use the standard curve or the following formula:

Absorbance of Test × constant (see Standardization)

$$= \text{urobilinogen (mg/dl)}$$

If the specimen was collected over a timed interval, measure the volume
and multiply the result by

$$\frac{\text{volume (ml)}}{100}$$

to get milligrams of urobilinogen for the time interval. If bile was removed from the sample, the result should be multiplied by 2 to compensate for the dilution with $BaCl_2$.

STANDARDIZATION

The usual standardization procedure [10] for the urobilinogen methods
employs a solution of dyes that simulates the color of urobilinogen-
aldehyde.

Stock Standard. Weigh 0.100 g of pontacyl carmine 2B.[4] Dissolve in 0.5% (v/v) acetic acid and dilute to 500 ml with the acetic acid solution.

Weigh 0.09 g of pontacyl violet 6R, 150%.[4] Dissolve in 0.5% (v/v) acetic acid and transfer to a 1 liter volumetric flask. Add 25 ml of the pontacyl carmine solution and dilute to 1 liter with the acetic acid solution.

Working Standard. Measure 10.2 ml of the stock standard into a 100 ml volumetric flask and dilute to volume with 0.5% acetic acid. This solution represents a urobilinogen concentration of 1.2 mg per dl.

A series of standards may be prepared in cuvets as follows:

ml of Working Standard	ml of 0.5% Acetic Acid	Urobilinogen (mg/dl)
10	0	1.20
6	4	0.72
4	6	0.48
2	8	0.24
1	9	0.12

Absorbance readings should be made at 565 nm against a water Blank. The curve may be plotted or a constant calculated by dividing each concentration by its respective reading and averaging those figures.

NORMAL VALUES

The optimum excretion of urobilinogen occurs during the afternoon hours. Urine should be collected at this time whenever possible. Normal excretion is 0.1 to 1.2 mg per 2 hours.

DISCUSSION

Urine for the urobilinogen test should be fresh and should not be exposed to light for long periods of time, since the colorless urobilinogen is easily oxidized to the pigment urobilin which does not react with Ehrlich's reagent.

In the method described above, the urobilinogen reacts with Ehrlich's reagent to give a red-colored condensation product. Sodium acetate is added to buffer the hydrochloric acid and to enhance the color of the product. The urine sample used as the Blank is treated *first* with sodium acetate so that when Ehrlich's reagent is added the hydrochloric acid is buffered and no reaction with urobilinogen takes place.

---

[4] This dye is available from E. I. du Pont de Nemours and Company, Wilmington, Delaware, and some laboratory chemical retailers.

It is important that bile be removed prior to performing the urobilinogen test, because bile interferes with formation of the proper color.

The method described here is not as accurate as the reduction and extraction method because urobilin is not first converted to urobilinogen and the urobilinogen is not separated from other interfering substances. However, the test is considered accurate enough for the usual clinical interpretations.

Porphyrins

The term porphyrins refers to a group of substances which share in common a tetrapyrrole ring system called *porphin*. In addition to the

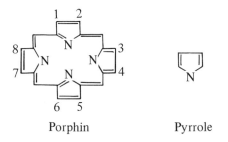

Porphin   Pyrrole

tetrapyrrole ring, all porphyrins have groups like acetyl (CH_2COOH) or methyl (CH_3) attached to various of the numbered positions on porphin. Each porphyrin is identified according to which groups it carries. The body is principally concerned with making a single porphyrin called protoporphyrin-IX which occurs in hemoglobin. (The red complex of Fe^{2+} ion and protoporphyrin-IX is called *heme,* which is responsible for the red color of hemoglobin.)

The other porphyrins found in the body are largely either precursors of protoporphyrin-IX, or breakdown products of these precursors. Some of these precursors are unstable and tend to constantly break down into side products to a small extent rather than undergo complete conversion to protoporphyrin-IX. The two main groups of breakdown products are uroporphyrins and coproporphyrins. They can be regarded as dropout molecules from the school of protoporphyrin-IX synthesis. Because changes in the relative and absolute amounts of the uroporphyrins and coproporphyrins occur in certain disease states, their quantitation is important. Presented below is a qualitative test used simply to screen for increased levels of these substances. It takes advantage of the fact that both uroporphyrins and coproporphyrins are fluorescent.

When a quantitative procedure is available, such as the one described on page 231, it is preferable to not use a preliminary screening test, such as the one given below, since the quantitative method provides more reliable and useful results.

PORPHYRINS SCREENING TEST [11]

PROCEDURE

To a 250 ml separatory funnel add 25 ml of urine, 10 ml of glacial acetic acid, and 50 ml of ether.

Stopper and shake vigorously for 3 minutes, opening the stopcock (with funnel inverted) occasionally to release the ether fumes.

Allow the 2 layers to separate, remove the stopper, and drain the urine (bottom layer) into one flask and the ether into another.

Return the urine to the separatory funnel, add another 50 ml of ether, and shake for 3 minutes.

Allow the solutions to separate. Then drain the urine into a flask and save it.

Combine the 2 ether extracts in the funnel, add 10 ml of 5% (v/v) hydrochloric acid, and shake the combination for 3 minutes.

Allow the solutions to separate. Then collect the acid (bottom layer) in a test tube.

Pour the urine into another tube and examine both the urine (after extraction) and the acid extract under an ultraviolet light in a dark room. A pink or red fluorescence in the acid indicates the presence of coproporphyrins, while the same fluorescence in the urine is due to the presence of uroporphyrins. A light pink fluorescence signifies a slight increase, while a bright pink to red means a moderate to large increase in porphyrin excretion.

DISCUSSION

The main purpose of the extraction procedure is to remove porphyrins from the urine where other interfering substances might be present. Since coproporphyrins are ether soluble while uroporphyrins are not, this procedure also serves to separate these two groups of substances from each other. The coproporphyrins are ether soluble after the addition of acid because they possess only four carboxyl (COOH) groups on their side chains. The uroporphyrins stay in the water phase because they have eight of these carboxyl groups.

Although this fluorescence method is a useful screening test, a fluorometer or a spectrophotometer and a modified extraction procedure is required for quantitative results. While the quantitative fluorometric method is approximately 1000 times more sensitive than the absorption spectrophotometric method, greater care must be taken to avoid interferences in the fluorometric method. A high-quality spectrophotometer is required for the absorption method because the absorption bands are narrow and therefore require a narrow, accurate band of light. For this reason and because of the necessity (1) to confirm that the absorption peaks occur at the expected wavelengths (that is, that the absorbing substance indeed is the porphyrin of interest) and (2) to obtain absorbance values at more than one wavelength, it is particularly advantageous to scan the wavelength region of interest using a recording spectrophotometer. Such a method is presented here for both coproporphyrins and uroporphyrins.

PORPHYRINS METHOD

METHOD
Sveinsson et al. [15].

MATERIAL
Urine.

REAGENTS
Glacial acetic acid.
Diethyl ether (ethyl ether).
Hydrochloric acid, 0.1N. Dilute 8.3 ml of concd HCl to 1 liter.
Sodium acetate, saturated. Add sodium acetate to water until no more will dissolve. Crystals of sodium acetate should always be visible.
Dibasic sodium phosphate, 5%. Dissolve 5 g of Na_2HPO_4 in water and dilute to 100 ml.
Calcium chloride, 3%. Dissolve 3 g of $CaCl_2$ in water and dilute to 100 ml.
Sodium hydroxide, 1N. Dissolve 20 g of NaOH in water and dilute to 500 ml.
Hydrochloric acid, 0.5N. Dilute 42 ml of concd HCl to 1 liter.
Sodium hydroxide, 0.1N. Dissolve 2 g of NaOH in water and dilute to 500 ml.

COPROPORPHYRINS METHOD
Note the total volume of the 24-hour urine sample.
To a 120 ml separatory funnel add 15 ml of urine, 1 ml of glacial acetic acid, and 30 ml of ethyl ether.
Shake for 1 minute.
Allow the layers to separate. Then run the aqueous (bottom) layer into a tube and save for uroporphyrin. Wash the ether layer with 10 ml of 5% acetic acid. Discard the aqueous layer.
To the ether layer, add 5 ml of 0.1N HCl.
Shake for 1 minute.
Allow the layers to separate. Then run the aqueous (bottom) layer into a tube and centrifuge for 2 minutes.
Scan from 450 nm to 350 nm in a 10 mm cuvet against 0.1N hydrochloric acid in the reference cuvet using a double-beam recording spectrophotometer.
Coproporphyrins are present in detectable quantities *only* if there is a spectral absorption peak at 401 nm.

CALCULATIONS

Obtain the absorbance values at 401, 380, and 430 nm; that is, obtain the A_{401}, A_{380}, and A_{430} values. Substitute these values in the following equation:

$$241 \times [2 \times A_{401} - (A_{380} + A_{430})] = \text{coproporphyrins } (\mu g/\text{liter})$$

$$\mu g/\text{liter} \times \frac{\text{24-h volume (ml)}}{1000} = \text{coproporphyrins } (\mu g/d)$$

UROPORPHYRINS METHOD

Transfer 5 ml of the ether-extracted urine which was saved from the coproporphyrins procedure to a 16 × 125 mm test marked at the 10 ml level.

Add a saturated solution of sodium acetate dropwise and with swirling until the pH is 5.5. (Monitor the pH with a pH meter.)

Dilute the sample to 10 ml with water and mix.

Transfer 2 ml to a 16 × 125 mm test tube.

Add 2 drops of 5% dibasic sodium phosphate solution and mix.

Add 2 ml of 3% calcium chloride solution and mix.

Add 2 ml of 1N sodium hydroxide and mix.

Centrifuge for 5 minutes.

Decant the supernatant and discard it.

Wash the precipitate with 5 ml of 0.1N sodium hydroxide.

Centrifuge for 5 minutes.

Decant and discard the supernatant.

Dissolve the precipitate in 5 ml of 0.5N hydrochloric acid.

Scan from 450 nm to 350 nm in a 10 mm cuvet against 0.1N hydrochloric acid in the reference cuvet. Uroporphyrins are present in detectable quantities *only* if there is a spectral absorption peak with a maximum absorbance at 405 nm.

CALCULATIONS

$$6025 \times [2 \times A_{405} - (A_{380} + A_{430})] = \text{uroporphyrins } (\mu g/\text{liter})$$

$$\mu g/\text{liter} \times \frac{\text{24-h volume (ml)}}{1000} = \text{uroporphyrins } (\mu g/d)$$

NORMAL VALUES

The normal range in adults for coproporphyrin is approximately 30 to 180 μg per day, and that of uroporphyrin is zero to 35 μg per day.

DISCUSSION

As in the screening test in an earlier section in this chapter the organic extraction procedure serves to isolate the coproporphyrins and uroporphyrins from each other and from interfering substances which might be present. The coproporphyrins can be extracted back out of the ether phase with 0.1N aqueous HCl no doubt because the acid protonates one or more of the pyrrole nitrogens to give the porphyrin molecule a net positive charge and therefore aqueous rather than organic solubility. The uroporphyrin is further separated from other interfering substances by adsorbing it to a calcium phosphate precipitate. An excess of Ca^{2+} is used to avoid any free phosphate because the adsorption of the uropor-phyrin to the precipitate is incomplete unless this precaution is taken. Subsequent addition of acid after a washing step dissolves the precipitate and releases the uroporphyrin into solution for spectrophotometric analy-sis. Using three absorbance values rather than one improves the accuracy by making the calculations less sensitive to background absorbance from any residual interfering materials.

PHYSIOLOGICAL SIGNIFICANCE OF PORPHYRINS

Increased excretion of coproporphyrins occurs with certain liver and hematological disorders and with lead poisoning. Increased excretion of both coproporphyrins and uroporphyrins occurs with a disease known as porphyria. Acute porphyria is usually accompanied by an increased excretion of porphobilinogen as well as porphyrins. Porphobilinogen is one of the precursor molecules of protoporphyrin-IX which were men-tioned earlier in this chapter. A simple qualitative test for porphobilino-gen is presented below:

Porphobilinogen Method

METHOD
Watson and Schwartz [18].

REAGENTS
Ehrlich's reagent modified. Dissolve 0.70 g of p-dimethylamino-benzalde-hyde, $(CH_3)_2NC_6H_4CHO$, in 150 ml of concd hydrochloric acid. Add 100 ml of water and mix. Store in a dark bottle.

Saturated sodium acetate (aqueous). Crystals of sodium acetate $(NaC_2H_3O_2 \cdot 3H_2O)$ should always be visible.

Chloroform (CHCl$_3$).

Butanol (C$_4$H$_9$OH).

PROCEDURE
To a large test tube add 5 ml of urine and 5 ml of modified Ehrlich's reagent. Mix and allow to stand for 3 min. Add 10 ml of sodium acetate solution, mix, and allow to stand for 3 minutes again. Add 5 ml of chloroform and shake thoroughly. Porphobilinogen forms a red compound with the aldehyde reagent. The red compound is in-soluble in chloroform. Therefore, a red color in the *aqueous* (top) phase indicates the presence of porphobilinogen. A red color in the

chloroform probably is due to the presence of urobilinogen, and is of no significance in this test. If a large amount of urobilinogen is present, one chloroform extraction may be insufficient. Therefore, the chloroform extraction should be repeated.

If the aqueous phase is pink, it should be shaken with 10 ml of butanol. Since porphobilinogen will not extract into butanol, persistence of the pink color in the aqueous phase confirms the presence of porphobilinogen.

If porphobilinogen is present, the red color should appear after addition of the Ehrlich's reagent. If the color does not appear until after the addition of sodium acetate, the chromogenic compound probably is urobilinogen.

Flocculation Tests

Since the liver is actively engaged in protein metabolism, malfunction of this organ frequently results in qualitative or quantitative changes (or both) in serum globulin fractions. There are several specific liver function tests based on the precipitation or flocculation of proteins under certain abnormal conditions. Two of these tests are discussed here. Although they are still used, they have largely been replaced by some of the more specific protein tests discussed in Chapter 15.

CEPHALIN-CHOLESTEROL FLOCCULATION TEST

This test is commonly referred to as the *ceph-floc test* and is performed as follows:

METHOD
Hanger [2].

MATERIAL
Serum.

REAGENTS
Ceph-floc antigen. The ceph-floc antigen is prepared from sheep brain and may be purchased commercially in a dry form.

Stock antigen. Treat the desiccated material in the following manner: Add 5 ml of ethyl ether per unit to effect solution of the contents. If turbidity persists, add one drop of water to obtain a clear solution. This is the stock ether antigen and is stable for months if kept tightly stoppered at room temperature.

Working antigen. Add, slowly and with stirring, 1 ml of stock ether antigen solution to 35 ml of water previously warmed to 65 C to 70 C on a hot plate. Heat slowly to boiling. Then simmer until the volume is reduced to 30 ml. Cool to room temperature and store in refrigerator. A 50 ml Erlenmeyer flask marked at 30 ml is a suitable vessel.

Note: During the heating, all coarse granular clumps should be dispersed. The result should be a stable, milky, translucent emulsion free of all traces of ether.

PROCEDURE
Add 0.2 ml of serum to 1.0 ml of working antigen and 4 ml of 0.85% (w/v) NaCl in a 15 ml centrifuge tube.
Mix by rotating the tube rapidly between the hands.
Let stand in the dark at room temperature for 48 hours.
Note the flocculation at 24 and 48 hours, and record as 0 to 4 plus, according to the following descriptions:

0   equals turbidity unchanged.
1+ equals $\frac{1}{4}$ of supernatant cleared.
2+ equals $\frac{1}{2}$ of supernatant cleared.
3+ equals $\frac{3}{4}$ of supernatant cleared.
4+ equals complete flocculation: supernatant entirely cleared.

DISCUSSION AND PHYSIOLOGICAL SIGNIFICANCE OF CEPH-FLOC TEST
Serum from healthy adults may produce up to 2+ flocculation in 48 hours. The technic itself is simple and requires no special explanations other than those offered in the above description. The problem of judging the degree of flocculation is best solved with supervised experience.

The physical state of the emulsion is not disturbed in the presence of serum containing normal concentrations of albumin and gamma globulin. However, the concentrations of these protein fractions are altered in parenchymal liver disease. The resultant changes cause flocculation of the antigen. It is thought that both a low albumin concentration and an increased globulin concentration are responsible for causing abnormal flocculation.

The most important application of this test is in the differentiation of parenchymal liver disease from early obstructive or hemolytic jaundice, because the ceph-floc value is elevated in liver disease, and usually unaffected in the two latter disorders.

Thymol Turbidity Test

METHOD
Maclagan [6], modified.

MATERIAL
Serum.

REAGENTS
Thymol buffer (pH 7.7). Weigh out:
    1.03 g of sodium barbital, $(C_2H_5)_2C(CONNaCONH)CO$
    1.38 g of barbital, $(C_2H_5)_2C(CONH)_2CO$
    3 g of crystalline thymol, $CH(CH_3)_2C_6H_3OHCH_3$

Add these to 500 ml of boiling water in an Erlenmeyer flask. Remove
the flask from the heat, mix well, and allow to cool to room temperature. Add a speck of powdered thymol to the solution and allow it to
stand overnight; thymol crystals will settle out around the seedlings.
Shake and filter. The pH, measured at 25 C, should be 7.70 ± 0.05.
If the solution is stored in a refrigerator, it will remain clear, but more
thymol will settle out and thereby reduce the sensitivity of the reagent.
It is therefore recommended that the reagent be kept at room temperature. Although it becomes opaque, this does not appear to affect its
sensitivity.

PROCEDURE

Into a 19 mm cuvet measure 6 ml of thymol buffer and 0.1 ml of serum.
Mix, allow to stand for 30 minutes, mix again immediately, and read
the absorbance at 650 nm in a spectrophotometer, setting the zero
with a cuvet containing 6 ml of thymol buffer. The Test should be
mixed again *just before* reading.

CALCULATIONS

$$\text{Absorbance of Test} \times \frac{\text{concentration of Standard}}{\text{absorbance of Standard}} = \text{Maclagan units}$$

If a turbidity greater than 15 units is encountered, add another 6 ml of
buffer to the Test, mix, read, and multiply the result by 2.

STANDARDIZATION

In Maclagan's original paper [6], the thymol turbidity unit is arbitrarily
defined as a test which, when performed as described above, produces a
turbidity comparable to that of a 10 mg per dl Kingsbury-Clark [4]
urine protein standard. Therefore:

$$\frac{\text{Kingsbury-Clark values (mg/dl)}}{10} = \text{Maclagan units}$$

Several methods have been suggested for standardizing the thymol
turbidity method. One method employs barium sulfate suspensions for
standardization. In that paper [13], the concentration of the barium
solution was given in terms of normality where molarity was intended.
Therefore, when standardization is performed using this procedure, the
units (Shank-Hoagland) are ½ the value of the Maclagan units. However, even when this correction is made, the barium standards do not
agree exactly with serum standards. The procedure recommended here is
simple and follows Maclagan's original definition closely.

Select a sample of human serum with normal total protein, albumin,
and globulin concentrations, and determine the total protein concentra-

tion.[5] Dilute this serum 1:50 with 0.85% (w/v) saline solution, mix, and prepare cuvets as follows:

ml of Dilute Serum	ml of 3% Sulfosalicylic Acid	Thymol Value in Maclagan Units
2.5	7.5	$P^a \times 2$
2.0	8.0	$P \times 1.6$
1.5	8.5	$P \times 1.2$
1.0	9.0	$P \times 0.8$
0.5	9.5	$P \times 0.4$

[a] P = total protein concentration (grams per 100 ml) of the serum being used.

Mix all tubes and read the absorbances at 650 nm against a Blank of 3% (w/v) sulfosalicylic acid.

If a protein concentration of 10 mg/dl equals 1 Maclagan unit, the total protein (TP) value of the serum being used for standardization, in *milligrams per 100 ml,* when divided by 10 equals Maclagan units:

$$\frac{TP\ (g/dl) \times 1000}{10} = \text{Maclagan units}$$

or:

$$TP\ (g/dl) \times 100 = \text{Maclagan units}$$

Since the standardization directions call for a 1:50 dilution of the serum, the formula becomes:

$$TP\ (g/dl) \times \frac{100}{50}\ [\text{or } 2] = \text{Maclagan units}$$

The Kingsbury-Clark standards are based on the use of 2.5 ml of the protein solution plus 7.5 ml of 3% sulfosalicylic acid. Since the first standard in the above tabulation is prepared in this manner, the calculated factor of 2 applies. The other factors are proportional to the amounts of dilute serum used.

DISCUSSION AND PHYSIOLOGICAL SIGNIFICANCE OF THYMOL TURBIDITY TEST

Serum from healthy adults may produce up to 2 units of turbidity. The formation of a precipitate in the thymol turbidity test is dependent upon

---

[5] The serum used for standardizing the protein methods may be used for this method as well.

the nature of the serum β-globulin fraction, which includes a portion of the serum lipoproteins. Certain qualitative changes occur in this protein fraction during liver disease that render the β-globulins capable of precipitating thymol in a buffered solution. The large dilution of the serum (1:60) and the use of the 650 nm wavelength minimize interference due to hemoglobin or bilirubin in the serum.

The pH of the thymol-barbital buffer was originally stated to be 7.8. However, when measured at 25 C, the pH of this solution is closer to 7.7. It has been suggested that the thymol solution be buffered to a pH of 7.55 and that *tris*-hydroxymethyl-aminomethane be used rather than barbital. Although the sensitivity of the method is greater at this pH, the specificity may be less. Furthermore, use of the more sensitive reagent requires reassessment of the normal range. It would seem as though either solution may be used as long as the aforementioned points are considered.

The physiological significance of the thymol turbidity test is essentially the same as that of the ceph-floc test given in the preceding section.

References

1. Ducci, H., and Watson, C. J.   The quantitative determination of the serum bilirubin with special reference to the prompt-reacting and the chloroform-soluble types. *J. Lab. Clin. Med.* 30:293, 1945.
2. Hanger, F. M.   The flocculation of cephalin-cholesterol emulsions by pathological sera. *Trans. Assoc. Am. Physicians* 53:148, 1938.
3. Heirwegh, K. P. M., Meuwissen, J. A. T. P., and Fevery, J.   Critique of the Assay and Significance of Bilirubin Conjugation. In Sobotka, H., and Stewart, C. P. (Eds.), *Advances in Clinical Chemistry.* New York: Academic, 1973. Vol. 16.
4. Kingsbury, F. B., Clark, C. P., Williams, G., and Post, A. L.   The rapid determination of albumin in urine. *J. Lab. Clin. Med.* 11:981, 1926.
5. Kingsley, G. R., Getchell, G., and Schaffert, R. R.   Bilirubin. In Reiner, M. (Ed.), *Standard Methods of Clinical Chemistry.* New York: Academic, 1953. Vol. I.
6. Maclagan, N. F.   Thymol turbidity test: A new indicator of liver dysfunction. *Br. J. Exp. Pathol.* 25:234, 1944.
7. Malloy, H. T., and Evelyn, K. A.   The determination of bilirubin with the photoelectric colorimeter. *J. Biol. Chem.* 119:481, 1937.
8. McGann, C. J., and Carter, R. E.   The effect of hemolysis on the van den Bergh reaction for serum bilirubin. *J. Pediatr.* 57:199, 1960.
9. Nosslin, B.   The direct diazo reaction of bile pigments in serum. *Scand. J. Clin. Lab. Invest.* 12:1, 1960.
10. Schwartz, S., Sborov, V., and Watson, C. J.   Studies of urobilinogen: IV. The quantitative determination of urobilinogen by means of the Evelyn photoelectric colorimeter. *Am. J. Clin. Pathol.* 14:598, 1944.
11. Schwartz, S., Hawkinson, V., Cohen, S., and Watson, C. J.   A micromethod for the quantitative determination of urinary coproporphyrin isomers (I and II). *J. Biol. Chem.* 168:133, 1947.
12. Seligson, D., Marino, J., and Dodson, E.   Determination of sulfobromophthalein in serum. *Clin. Chem.* 3:638, 1957.

13. Shank, R. E., and Hoagland, C. L.   A modified method for the quantitative determination of the thymol turbidity reaction of serum. *J. Biol. Chem.* 162:133, 1946.
14. Smith, C. H.   The magic numbers—"20 milligrams of bilirubin." *J. Pediatr.* 56:712, 1960.
15. Sveinsson, S. L., Rimington, C., and Barnes, H. D.   Complete porphyrin analysis of pathological urines. *Scand. J. Clin. Invest.* 1:2, 1949.
16. Van den Bergh, A. A. H., and Muller, P.   Ueber eine direkte und eine indirekte Diazoreaction auf Bilirubin. *Biochem. Z.* 77:90, 1916.
17. Wallace, G. B., and Diamond, J. S.   The significance of urobilinogen in the urine as a test for liver function with a description of a simple quantitative method for its estimation. *Arch. Intern. Med.* 35:698, 1925.
18. Watson, C. J., and Schwartz, S.   A simple test for urinary prophobilinogen. *Proc. Soc. Exp. Biol. Med.* 47:393, 1941.
19. Watson, C. J., Schwartz, S., Sborov, V., and Bertie, E.   Studies of urobilinogen: V. A simple method for the quantitative recording of the Ehrlich reaction as carried out with urine and feces. *Am. J. Clin. Pathol.* 14:605, 1944.
20. White, D., Haider, G. A., and Reinhold, J. G.   Spectrophotometric measurement of bilirubin concentrations in the serum of the newborn by the use of a microcapillary method. *Clin. Chem.* 4:21, 1958.
21. Winkel, P., Ramsoe, K., Lyngbye, J., and Tygstrup, N.   Diagnostic value of routine liver tests. *Clin. Chem.* 21:71, 1975.

20. Enzymes

Enzymes are proteins which catalyze essential biochemical reactions. This means that they increase the rates of these reactions without being used up. Their activity usually is quite specific.

Enzymes are measured in terms of *activity* rather than by units of weight. Since their activity varies with temperature, pH, substrate concentration, product concentration, the presence of activators or inhibitors, and other factors, the methods of measurement demand careful control of all influencing conditions. A discussion of some of these conditions is included in this chapter. More comprehensive presentations may be found in appropriate sources [3, 5, 7, 10, 17, 20].

The particular substances on which enzymes act are called *substrates*. Each enzyme catalyzes a special reaction for only a certain type of substrate and generally is named after the usual substrate with the suffix *-ase* added. Organic phosphates are substrates of enzymes called phosphatases, for instance, and the enzyme which catalyzes the *transfer* of an *amino* group from *glutamic* acid to *oxalacetic* acid is called *glutamic oxalacetic transaminase*. Some of the enzymes isolated before this system of naming was adopted end in the suffix *-in*; for example, pepsin, trypsin, and rennin. In these cases, the name is unrelated to the type of substrate of the enzyme.

General Considerations

COFACTORS AND ACTIVATORS

Some enzymes require special auxiliary substances, called *cofactors,* for their activity. Cofactors are either metal ions or organic molecules called *coenzymes.* Zinc and magnesium ions are two examples of metal cofactors while many of the vitamins function as coenzymes (or coenzyme precursors). The coenzyme in most common use in clinical laboratories is nicotinamide adenine dinucleotide (NAD), which is derived from the vitamin niacin. The older name for this coenzyme is diphosphopyridine nucleotide (DPN).

ACTIVITY

Enzymes are measured by their activity rather than their concentration not only because activity measurements are convenient but because enzymes exist in very small amounts in biological fluids and also are difficult to isolate from other reactants. The general reaction may be represented as follows when a coenzyme is used:

$$\text{substrate} + \text{coenzyme} \xrightarrow{\text{enzyme}} \text{product} + \text{changed coenzyme}$$

The rate (speed) of the reaction may be determined by measuring any one of the following: (1) decrease in concentration of substrate, (2) decrease in concentration of coenzyme, (3) increase in concentration of product, or (4) increase in concentration of changed coenzyme. It is easiest to relate the rate of the reaction to the enzyme concentration if the enzyme is kept as busy as possible catalyzing the conversion of substrate into product. This requires the presence of a large excess of substrate and coenzyme. As long as this is the case, the reaction is said to be following *zero-order kinetics*. This is because none (zero) of the reactants are undergoing a significant change in concentration. As busy as the enzyme is, it may be able to catalyze only the reaction of a few percent of substrate and coenzyme during the first few minutes. Therefore the concentrations of these two substances are essentially constant (only a few percent decrease) during this period. [Under these conditions, it is preferable to determine the rate of reaction by measuring (3) or (4) rather than (1) or (2).] The concentration of the enzyme is absolutely constant, of course, because it is not being used up. Whenever possible, as it generally is, the enzyme assays in the clinical chemistry laboratory are carried out under zero-order conditions because this makes the enzymatic activity (rate of reaction) directly proportional to the concentration of the enzyme. Over a long period of time, the concentrations of substrate and coenzyme will decrease to the point where the reaction slows down because the enzyme is no longer converting substrate into product at a maximal rate. Therefore, it is important to measure the *initial* rate of a zero-order enzyme reaction in order to allow the activity to serve directly as a measure of the enzyme concentration and not of the substrate concentration too. In some cases, however, the initial time used for timing the reaction must be delayed because it takes a few moments for the reaction to reach full speed. Such reactions are said to exhibit a *lag phase*. The problem typically is that some impurities are being used up, the temperature has not stabilized, or the assay involves a sequence, for example, of two enzymes in which the product of the first enzyme is the substrate of the second enzyme. Although only one of the enzymes is being quantitated, the presence of a second enzyme allows the reaction of the enzyme of interest to be measured more conveniently or reliably. Such a coupled-enzyme assay takes time to reach full activity because the intermediate product (from the first enzyme) is not present to serve as substrate for the second enzyme until the activity of the first enzyme produces this product. Additional time therefore is required for the second enzyme (and therefore the overall two-enzyme process) to attain full activity. A three-enzyme coupled assay for creatine phosphokinase is outlined in this chapter.

Since enzymes catalyze both the forward and reverse directions of their reactions, it is important that the concentration of the product at the start of the reaction be zero or negligible so that the enzyme does not spend part of its time catalyzing the formation of substrate from product (reverse reaction). When this is the case, the enzyme is not being kept

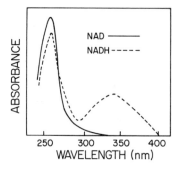

Figure 20-1
Spectral Absorption Curves of NAD and NADH

as busy as possible with the forward reaction. This is another reason why the initial rate should be measured. (The reaction slows down because product has accumulated and started to compete with substrate for the enzyme.) This problem is avoided, of course, when a means is provided for the removal, usually by some further reaction, of product as it is formed.

The rate of an enzymatic reaction may be determined in one of three ways, all of which should be confined to the zero-order phase of the reaction and each of which can involve measurement of the change in concentration of any of the components (substrate, coenzyme, product, or changed coenzyme): (1) single-point assay—stop the reaction, usually by inactivating the enzyme, after a specified period of time and measure one of the above components; (2) multiple-point assay—remove increments of the reaction at specified times, stop the reaction in each increment immediately after it is taken, and measure one of the components; or (3) continuous assay—constantly monitor the change in concentration of one of the components, usually in a recording spectrophotometer.

Enzyme reactions involving the coenzyme NAD (nicotinamide adenine dinucleotide) often are measured by a continuous assay in an ultraviolet recording spectrophotometer because NADH, the changed coenzyme of NAD, absorbs light strongly at 340 nm, while NAD does not, as shown in Figure 20-1. Therefore, the transition from one form to the other, during an enzyme reaction, may be followed by measuring the increase or decrease in absorbance as a function of time.

UNITS OF ACTIVITY

Enzymes are not measured in weight per unit volume, as are most other chemical constituents of the blood, but rather their *activities* are measured. Therefore, the results of enzymatic reactions are reported in various types of arbitrary units. Some units are based on a value assigned to a given quantity of end product formed under certain stated conditions, and the unit frequently carries the name of the author(s) of the method employed. For example, the unit described for one alkaline phosphatase method is equal to "the amount of enzyme activity, per 100 ml of sample, required to produce 1 mg of inorganic phosphorus from β-glycerophosphate in 1 hour at 37 C, pH 9.1"; this is the *Bodansky unit*.

With methods employing the NAD-NADH reaction, the unit is defined as absorbance change per unit of time. One transaminase unit is "the amount of enzyme activity which produces an absorbance change of 0.001 per minute per milliliter of serum at 32 C, pH 7.4"; this is the *Karmen unit*.

Several methods exist for the measurement of each enzyme. The units are not always identical; in fact, sometimes it is not even possible to convert one type of unit to another. This problem is a source of confusion to physicians and hence is of concern to laboratory personnel. In recent years an attempt has been made to establish a universally acceptable method for calculating and expressing enzyme units. The resulting recommendation is the *international unit* [12] (abbreviated U or IU) which is defined as the amount of enzyme catalyzing the transformation of 1 micromole of substrate per minute. The recommended reaction conditions are: temperature 30 C (originally, the temperature was taken at 25 C but was redefined at 30 C), pH and substrate concentration optimal, the initial reaction should be used, and the reaction should follow zero-order kinetics. Actual rates are reported in international units (U) or milli international units (mU) per milliliter or liter of sample. A sample conversion from conventional to international units is performed below.

If 1 Bodansky unit (BU) = 1 mg of phosphorus per hour per 100 ml of sample, then

$$1 \text{ BU} = \frac{1000}{31} \times \frac{1}{60} \times 10 = 5.35 \text{ U/liter}$$

where $\frac{1000}{31}$ converts milligrams to micromoles, $\frac{1}{60}$ converts hours to minutes, and the factor of 10 makes the expression per liter rather than per 100 ml. This is the conversion factor which is used to convert BU to U even though the temperatures for BU and U are different.

The several methods in use for each enzyme do not always result in comparable reaction rates. Therefore, if a sample is analyzed by two different methods and the results are calculated in international units, a different number of units may result. If they are both called by the same name, it may be confusing to the physician.

CONDITIONS AND INHIBITORS

The conditions of temperature and pH, and sometimes ionic strength, are very important in enzyme assays, affecting both the stability and the activity of the enzymes. Enzymes generally are more stable at lower temperatures. Therefore samples which are not analyzed immediately, should be stored in the cold. The stability of some clinically important serum enzymes is given in Table 20-1.

Assays usually are carried out at some specific temperature in the range 25 C to 37 C because of the convenience of performing assays at a temperature close to room temperature and because most enzymes show maximal activity and reasonable stability in this temperature range. Temperatures in the 25–30 C range are difficult to control because, at any given time, room temperature may be either above or below the selected temperature, thereby requiring either heating or cooling. Temperatures in the 32–38 C range require only a heater. Also enzyme activity is greater in this range. Most enzymes, like most proteins, are destroyed by heat, especially above 65 C.

Because the activity of enzymes is sensitive to temperature, precise control of temperature throughout the reaction period is essential to the accuracy of enzyme methods. Such control is difficult to achieve in the cuvet well of a spectrophotometer. Although the temperature of the cuvet compartment may be controlled, the temperature in the cuvet well may vary significantly with the introduction of each sample.

A typical relationship between pH and extent of enzyme activity is shown in Figure 20-2. The pH for maximum activity varies from one enzyme to another, as does the shape of the plateau of the pH curve in this range. Extreme pH values destroy most enzymes just as heat does. A sharp plateau requires fine control of pH while a broad one allows more variation of pH without influencing the rate of reaction. Appropriate buffer solutions are employed in enzyme methods to achieve the proper pH. Ideally, a buffer should bring the ingredients of the reaction

Table 20-1. Stability of Some Serum Enzymes

Enzyme	Stability at 25 C	Stability at 4 C
Amylase	1 week	several months
Lipase	1 week	several weeks
Phosphatase, alkaline	unstable	unstable[a]
Phosphatase, acid	unstable	several days[b]
Transaminase (GO or GP)	several days	several weeks
Lactate dehydrogenase	1 week	1 week
Creatine phosphokinase	unstable	unstable[a]

[a] Store frozen.
[b] If acidified (10 μl of 20% acetic acid per 1 ml of serum).

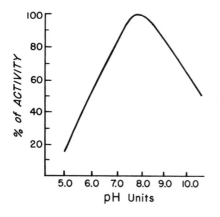

Figure 20-2
Effect of pH on Enzyme Activity

mixture (including sample) to the pH of the buffer solution. However, many buffers in common use are designed to bring the final pH to a value somewhere between that of the buffer and that of the reaction mixture. Some buffers are not strong enough (concentrated enough) always to ensure bringing the solution to the same pH. This situation introduces an uncontrolled variable resulting in spurious results.

Many substances, including metal ions, drugs, detergents, and proteins, can inhibit the activities of enzymes. For this reason, the glassware used for enzyme assays must be meticulously clean. The problem of inhibitors is one of the shortcomings of measuring enzymes based on their activity; the enzyme may be present, but it may be underestimated because its activity is reduced by an inhibitor in the sample.

ISOENZYMES

Some enzymes exist in a number of forms which are molecularly different but act upon the same substrate. These different forms are called *iso-enzymes* or (less preferred) *isozymes*. The total activity observed for such an enzyme, therefore, is contributed by a group of isoenzymes. Usually they can be separated from each other by a technic like electrophoresis or ion-exchange chromatography. Often the different isoenzymes will show different maximal rates for the substrate which they share in common, different susceptibility to extremes of temperature or pH, and be inactivated to different extents by certain inhibitors. Fractionation of lactic dehydrogenase (five isoenzymes) and creatine phosphokinase (three isoenzymes) are commonly performed in clinical laboratories.

A given serum enzyme may derive from several sources in the body. The primary source of an increased amount of enzyme in the blood sometimes may be determined from the isoenzyme composition.

STANDARDIZATION

Standardization technics for enzyme methods are generally unsatisfactory. In some enzyme procedures only the method for measuring the end product is standardized, for example, glucose, phosphate, dyes. In methods where units are expressed as absorbance change, the type and quality of instrument employed is critical. Although reference or control samples (lyophilized reference materials, pooled frozen serum) which have been assayed are quite useful as secondary standards, enzymatic methods cannot be properly standardized until primary standards are available.

Amylase

Methods for the determination of amylase depend upon the ability of this enzyme to catalyze the hydrolysis of starch to simple sugars. A simple amylase method consists in serial dilution of the serum, addition of starch to each tube, incubation, and detection of the disappearance (hydrolysis) of the starch using iodine as an indicator. This (amyloclastic) method is somewhat crude, but may be used as a screening technic.

There are many technics that involve the use of starch solutions and measurement of liberated reducing sugars by the usual glucose methods [15]. With these (saccharogenic) methods, the most serious problem is the instability of starch solutions, which are readily contaminated by molds. Success with these methods therefore depends, to a great extent, upon the control of these solutions. In one method the problem of unstable starch preparations is overcome through the use of β-amylose starch fraction dried into cotton squares [1].

Commercial tablets composed of an insolubilized preparation of starch to which blue dye molecules are attached are available for amylase analysis [19]. The blue tablet disperses (but does not dissolve) when added to an aqueous buffer. When serum is added to this suspension, the amylase hydrolyzes (cleaves) some of the linkages in the blue-starch substrate, liberating small, soluble, blue fragments. After a period of time, the reaction is stopped by the addition of strong acid and centrifuged or filtered. The amount of blue color present in the supernatant (or filtrate) then is a measure of the amylase concentration in the added serum.

The method presented here is a simple, rapid amyloclastic technic that yields satisfactory results.

AMYLASE METHOD

METHOD
Caraway [9].

MATERIAL
Serum.

REAGENTS

Substrate. Dissolve 13.3 g of anhydrous disodium phosphate (Na_2HPO_4) and 4.3 g of benzoic acid (C_6H_5COOH) in approximately 250 ml of water in a 600 ml beaker. Bring the solution to a boil and add 0.200 g of soluble starch (according to Lintner) which has been suspended in 5 ml of cold water. Rinse the container with additional water so as to transfer all of the starch. Continue boiling the solution for 1 minute after addition of starch. After the solution cools to room temperature, transfer it quantitatively to a 500 ml volumetric flask and dilute to 500 ml with water. The pH should be 7.0 ± 0.1. This solution is stable for at least 2 months at room temperature. Sterilization of the reagent bottle to be used for storing the substrate helps prevent growth of molds in the solution.

Stock iodine solution 0.1N. Dissolve 3.567 g of potassium iodate (KIO_3) and 45 g of potassium iodide (KI) in approximately 800 ml of water in a 1 liter volumetric flask. Add slowly and with mixing 9 ml of concd hydrochloric acid and dilute to 1 liter with water. Store this solution in the refrigerator.

Working iodine solution 0.01N. Dilute 50 ml of the stock iodine solution to 500 ml with water. This solution should be renewed each time a new substrate is made.

PROCEDURE

Pipet 5 ml of substrate into each of two 50 ml volumetric flasks.

Put one flask (Test) into a water bath at 37 C for 5 minutes. The other flask (Control) need not be incubated.

Measure exactly 0.1 ml of serum into the bottom of the Test flask, mix by swirling, and return it to the water bath for 7.5 minutes. No serum is added to the Control tube.

After incubation, remove the Test flask from the bath and *immediately* add 5 ml of working iodine solution, with mixing. Add 5 ml of the iodine solution to the Control flask and dilute to 50 ml with water the contents of both flasks. Mix well.

Pour at least 5.5 ml of solution from each flask into 19 mm cuvets and in a spectrophotometer measure against a water Blank the absorbances of both solutions at 660 nm.

CALCULATIONS

$$\frac{\text{absorbance of Control} - \text{absorbance of Test}}{\text{absorbance of Control}} \times 800$$

$$= \text{amylase units per 100 ml}$$

If the amylase activity exceeds 400 units, repeat the test using 0.02 ml of serum and multiply the result by 5.

STANDARDIZATION

One amylase unit for this method is "the amount of enzyme that catalyzes the hydrolysis of 10 mg of starch in 30 minutes to a stage at which
no color is given by iodine." Since the substrate contains 0.4 mg of starch
per milliliter, the 5 ml amount used provides 2 mg of starch. By the
above definition, 2 mg of starch hydrolyzed in 7.5 minutes is equivalent to

$$\frac{2}{10} \times \frac{30}{7.5} \times 1 \text{ (unit)} = 0.8 \text{ unit}$$

Therefore, under the conditions of this procedure, complete hydrolysis
of the substrate represents 0.8 amylase unit per 0.1 ml of sample or 800
units per 100 ml.

The number of absorbance units between zero and the control reading
(A_1) represents 800 amylase units. The number of absorbance units
between the control reading and the reading of the Test (A_2) represents
the amount of substrate hydrolyzed by the sample. Therefore,

$$\frac{A_2}{A_1} \times 800 = \text{amylase units per 100 ml of sample}$$

If the solutions are prepared carefully and the reading of the Control
does not change appreciably, no further standardization of the procedure
is necessary.

NORMAL VALUES

The range of amylase activity in the serum of healthy adults is essentially
the same with this method as with the saccharogenic methods, that is,
up to 160 units per 100 ml.

DISCUSSION

If the analysis cannot be carried out within a short time, the sample
should be stored in a freezer to preserve the amylase activity.

It is well known that, when iodine is added to a solution containing
starch, a blue color results. With the above procedure, the concentration
of the iodine solution is held constant, so that any change in color (or
absorbance) signifies a change in concentration of the starch solution.
It is assumed that this change is proportional to the amylase activity in
the sample. A change in concentration in either the iodine or the starch
solution, even though it might be due to contamination or deterioration,
results in a change in the absorbance of the control. Therefore, although
the readings of the control may vary somewhat from time to time, a large
change in absorbance (for example, over 0.050) or a continuous change
in one direction indicates the need for renewal of one or both solutions.
With the method described above, the control usually reads about 0.700.

Since the control represents the greatest amount of starch to be mixed with the iodine under the conditions of this method, it must necessarily have a higher reading than a test, in which some of the starch will be hydrolyzed. The quantitative aspects of the reaction are considered under Standardization.

The greatest source of error with this method is from contamination with saliva. *Under no circumstances should any pipet be blown out,* since even the slightest spray of saliva will introduce serious contamination with subsequently high values. For measurement of the sample, a 0.1 ml pipet (such as the Mohr type) calibrated *to deliver* should be used, *not a rinse-out (to-contain)* type. If other analyses are to be done with the serum before the amylase test, some of the sample should be *poured* into a separate tube and saved for the amylase determination in order to prevent contamination when excess serum, pipetted for other tests, is blown back into the tube.

While the water is being added to the volumetric flask to make the volume 50 ml, the flask should be swirled to avoid formation of a precipitate. Once the procedure is started, it should be carried through to completion without delay.

AMYLASE IN OTHER BODY FLUIDS

The method described above may be used to determine the amylase activity in other body fluids. Duodenal drainage and pancreatic fluid usually have extremely high amylase activity (approximately 100 times the values in serum). Therefore, such fluids should be diluted 1:100 with water before analysis.

There are conflicting views regarding the usefulness of the urinary amylase (or diastase) determination as a diagnostic aid. However, it has been shown that, if such determinations are to be of value at all, they must be made within a 24-hour urine sample [22]. Normally, about 2000 units of amylase are excreted per 24 hours in the urine. Therefore the concentration should be approximately the same as in serum and no predilution is necessary.

Saliva has a high amylase activity. Since this fluid is readily available, it can be used to check the substrate in any amylase method. If a small drop of saliva is mixed with a few milliliters of serum, a very high amylase activity should result. This sample may then be used to test the capacity of the substrate and the ability of the method to detect elevated values. If the prepared sample yields too high an activity, it may be diluted into a suitable range with 0.85% (w/v) NaCl solution.

PHYSIOLOGICAL SIGNIFICANCE OF AMYLASE ACTIVITIES

Amylase is found chiefly in the saliva and in pancreatic tissue. Normally, small amounts of amylase are present in the blood, but with various forms of pancreatic disturbance (for instance, pancreatitis) large amounts of amylase are secreted into the blood by the pancreas. The onset of pancreatitis may be swift and severe. Since the amylase activity often is a diagnostic criterion, rapid and accurate performance of the test is an important laboratory service.

The blood amylase activity may fluctuate rapidly and dramatically, rising acutely during an attack and subsiding to normal levels shortly afterward. For this reason, a physician may obtain blood from a patient during an attack of acute abdominal pain and request an emergency amylase level with that sample. In these cases the amylase report often is an essential factor in determining the treatment to be used.

The blood amylase activity may be elevated in cases of mumps also, because of secondary pancreatic involvement.

Lipase

Methods for the determination of lipase utilize the ability of this enzyme to liberate fatty acids from true fats. The method presented here is widely used.

LIPASE METHOD

METHOD
Crandall and Cherry [11].

MATERIAL
Serum.

REAGENTS

Olive oil substrate. Dissolve 5 g of acacia and 0.4 g of sodium benzoate (C_6H_5COONa) in 100 ml of water. Pour this solution into a Waring blender and add 100 ml of pure olive oil. Mix for 10 minutes to form a homogeneous emulsion. Store in a refrigerator. Shake vigorously before using.

Phosphate buffer, pH 7.0. Weigh 5.785 g of dibasic sodium phosphate (Na_2HPO_4) and 3.532 g of monobasic potassium phosphate (KH_2PO_4). Dissolve, transfer to a volumetric flask, and dilute with water to 1 liter.

Ethyl alcohol, 95% (v/v).

Sodium hydroxide, 0.05N. Dilute 50 ml of 0.1N sodium hydroxide to 100 ml. The solution should be standardized with acid and adjusted if necessary.

Thymolphthalein indicator. Dissolve 1 g of thymolphthalein in 100 ml of 95% ethyl alcohol. Filter if necessary.

APPARATUS
Buret, 5 ml capacity, calibrated in increments of 0.1 ml.

PROCEDURE

Into each of two 25 ml Erlenmeyer flasks measure 1 ml of serum, 2 ml
of substrate, 3 ml of water, and 0.5 ml of phosphate buffer.

To one of the flasks add 3 ml of 95% alcohol (Blank). Incubate both
flasks at 37 C for 6 hours.

Remove the flasks from the incubator, add 3 ml of 95% alcohol to the
Test flask, and add 5 drops of indicator to both flasks.

Titrate both solutions to a definite blue endpoint with 0.05N sodium
hydroxide.

CALCULATIONS

The lipase units of activity are equal to the number of milliliters of
0.05N sodium hydroxide used for the neutralization of the formed acids.

titration of Test $-$ titration of Blank
$$= \text{lipase units (ml of } \frac{N}{20} \text{ [0.05}N\text{] NaOH)}$$

NORMAL VALUES

With this method, an activity of less than 1.5 units usually is considered
normal, although the range between 1.0 and 1.5 units sometimes is considered equivocal.

DISCUSSION

Lipase is a relatively stable enzyme and will maintain its activity in
serum for several weeks if the sample is stored in a refrigerator. Since
hemoglobin may inhibit lipase activity, grossly hemolyzed serum should
not be used.

Initially it was recommended that the olive oil be treated with a base
before using it in the preparation of substrate in order to neutralize any
acids that might be present in the oil. While this technic results in lower
blank titrations, the final results appear to be comparable to those obtained with substrate prepared from pure but untreated olive oil. Since
all serum blanks have virtually the same value, a test may be performed
without a blank if the quantity of serum is limited. In this case, the
average value of other serum blanks may be subtracted from the test.
The purpose of the acacia is to keep the olive oil emulsified. The sodium
benzoate acts as a preservative.

During the incubation period, lipase catalyzes the hydrolysis of triolein
(olive oil) with subsequent formation of oleic acids. These acids are then
quantitated by titration with a standard base. The alcohol added to the
blank flask inhibits lipase activity during the incubation period. The
blank titration represents the acid initially present in the substrate plus
any acid formed spontaneously during incubation.

The rate of enzyme activity is not linear throughout the incubation period. It has been determined that the reaction is virtually complete after 6 hours. However, if more convenient, the incubation period may be extended to 18 to 20 hours without hampering the clinical usefulness of the test.

PHYSIOLOGICAL SIGNIFICANCE OF LIPASE ACTIVITIES

It is generally agreed that lipase values have approximately the same significance as amylase values in the diagnosis of acute pancreatitis. During pancreatic disturbances, the serum lipase activity may rise more slowly than the serum amylase, but it may remain elevated for a longer period of time and therefore may be more useful in the diagnosis and follow-up of acute pancreatitis some time after clinical symptoms subside. The blood-lipase activity may also be increased with chronic disease of the pancreas.

Phosphatases

The group of enzymes in the blood which split (hydrolyze) phosphate esters are collectively referred to as *phosphatases*. The clinically important phosphatases are divided into two groups depending on the pH at which they are most active. The group of phosphatases most active at approximately pH 10 is called *alkaline phosphatase* while the group most active at approximately pH 5 is called *acid phosphatase*.

There are a number of methods published for the analysis of either or both of the phosphatases. In these methods a common substrate may be used but the pH of the buffer determines which phosphatase will be measured. A few of the substrates recommended and the product measured are shown in Table 20-2.

Methods employing *p*-nitrophenyl phosphate [4] are the ones most commonly used in clinical chemistry laboratories. Both manual and automated versions of these methods are available.

ALKALINE PHOSPHATASE

METHOD
Bessey, Lowry, and Brock [4].

MATERIAL
Serum.

Table 20-2. Some Phosphate Substrates and Products

Substrate	Product (Besides Phosphate)
Glycerophosphate [6]	Glycerol
Phenolphthalein phosphate [16]	Phenolphthalein
β-Naphthyl phosphate [24]	*β*-Naphthol
Phenylphosphate [2, 23]	Phenol
p-Nitrophenyl phosphate [4]	*p*-Nitrophenol

REAGENTS

Sodium hydroxide, approximately 1N. Dissolve 40 g of NaOH pellets in water and dilute to 1 liter. Store in a plastic bottle.

Sodium hydroxide, approximately 0.02N. Dilute 20 ml of 1N NaOH to 1 liter with water. Store in a plastic bottle.

Glycine buffer, 0.1M, pH 10.6. Dissolve 0.2 g of magnesium chloride ($MgCl_2 \cdot 6H_2O$) and 7.5 g of glycine (aminoacetic acid) in approximately 700 ml of water. Add 85 ml of 1N NaOH. Then adjust the pH to 10.6 ± 0.1 using 1N NaOH or 1N HCl as required. Dilute to 1 liter with water. Add a few drops of chloroform as a preservative. Store refrigerated in a plastic bottle.

Buffered substrate. Dissolve 100 mg of disodium *p*-nitrophenyl phosphate in 25 ml of water. Add 25 ml of glycine buffer. Pipet into 16 × 125 mm tubes in 1 ml amounts. Cap and store in a freezer.

Hydrochloric acid, concd.

PROCEDURE

Warm two 16 × 125 mm tubes, each containing 1 ml of buffered substrate, to 37 C.

To one tube add 0.1 ml of water (Blank) and to the other add 0.1 ml of serum (Test).

Mix, and incubate at 37 C for exactly 30 minutes.

Remove the tubes from the incubator and add 10 ml of 0.02N NaOH. Mix well.

Transfer to 19 mm cuvets and read the absorbance of the Test (A_1) in a spectrophotometer at 410 nm, setting the Blank at zero.

Add 0.1 ml of concd HCl to each cuvet, mix, and read the absorbance again (A_2).

CALCULATIONS

Determine the alkaline phosphatase value for each of the absorbance readings from a calibration curve (see Standardization). Then subtract A_2 from A_1 to get the final value in Bessey, Lowry, Brock units.

STANDARDIZATION

Stock standard, 10 mmol/liter. Dissolve 139 mg of *p*-nitrophenol in 100 ml of water. Store frozen.

Working standard, 50 μmol/liter. Dilute 0.5 ml of stock standard to 100 ml with 0.02N NaOH. Prepare fresh on the day of use. Prepare a standard curve in 19 mm cuvets as shown on page 254. Read the absorbances of all cuvets at 410 nm against 0.02N NaOH set at zero. Plot the absorbances against their respective unit values on linear graph paper. This curve should be checked frequently but it need not be performed each time.

Working Standard (ml)	0.02N NaOH (ml)	Bessey, Lowry, Brock Units
1	10	1
2	9	2
4	7	4
6	5	6
8	3	8

One Bessey, Lowry, Brock unit is defined as the amount of enzyme activity which will liberate 1 μmol of p-nitrophenol per hour under the conditions of the method. In the procedure described, 1 ml of working standard contains

$$\frac{50}{1000} = 0.05 \ \mu\text{mol of } p\text{-nitrophenol}$$

Since only 0.5 hour of incubation time is used instead of 1 hour, this value is multiplied by 2 to give 0.1 μmol. This value is then multiplied by 10 to express the result as μmol per ml because only 0.1 ml of serum is used. Hence, the 1 ml of working standard is equivalent to 1 μmol per ml or 1 mmol/liter or 1 Bessey, Lowry, Brock unit of alkaline phosphatase activity.

NORMAL VALUES
Alkaline phosphatase activity in the serum of healthy adults usually is less than 2.5 units. In children values up to 6.5 units may be found.

DISCUSSION
If the analysis is not to be carried out the same day, the serum should be stored in a freezer. However, even when serum is frozen, alkaline phosphatase activity may change over a period of time.

During incubation with alkaline phosphatase, the almost colorless substrate, p-nitrophenyl phosphate, is hydrolyzed to produce p-nitrophenol, which is yellow in an alkaline solution. Addition of the sodium hydroxide stops the enzyme reaction and brings out the color of the p-nitrophenol. After the initial absorbance reading, the test is acidified to decolorize it. Hence the second absorbance reading serves as a blank because it represents any color or tubidity not due to p-nitrophenol.

It is good policy to read the absorbance of the reagent blank each time against 0.02N NaOH. If this reading increases, it may indicate deterioration of the substrate with subsequent formation of p-nitrophenol.

PHYSIOLOGICAL SIGNIFICANCE OF ALKALINE
PHOSPHATASE ACTIVITIES
Alkaline phosphatase is located chiefly in bone and in the liver. Since this enzyme plays an important role in bone formation, the serum alkaline phosphatase activity is greater in growing children and is increased

in any disease characterized by increased bone formation or attempts at bone formation, for example, Paget's disease, hyperparathyroidism, rickets, and metastatic cancer. Alkaline phosphatase is also excreted in the bile. Most types of liver disease, particularly obstructive jaundice, are accompanied by an increase in alkaline phosphatase activity in the blood. In general, the activity is greater in obstructive jaundice than in primary disease of the liver. The persistence of very low levels of activity of serum alkaline phosphatase is found in a hereditary bone disease known as hypophosphatasia.

ACID PHOSPHATASE (TOTAL AND PROSTATIC)

METHOD
Bessey, Lowry, Brock [4]; Fishman et al. [13].

MATERIAL
Serum.

REAGENTS
Citrate buffer. Dissolve 18.9 g of citric acid ($C_6H_8O_7 \cdot H_2O$) in approximately 500 ml of water. Add 180 ml of 1N NaOH and 100 ml of 0.1N HCl. Mix, and dilute to 1 liter with water. If necessary, adjust the pH to 4.90 ± 0.05 using concd HCl or NaOH as required.

Buffered substrate (total acid phosphatase). Dissolve 100 mg of disodium *p*-nitrophenyl phosphate in 25 ml of water. Add 25 ml of citrate buffer and mix. Pipet into 16 × 125 mm tubes in 1 ml amounts. Cap, and store in a freezer.

Tatrate-citrate buffer. Dissolve 1.5 g of tartaric acid ($C_4H_6O_6$) in 250 ml of citrate buffer.
Adjust the pH to 4.90 ± 0.05 using concentrated (50%) NaOH (dropwise). Store in a refrigerator.

Buffered substrate (prostatic acid phosphatase). Dissolve 100 mg of *p*-nitrophenyl phosphate in 25 ml of water. Add 25 ml of tartrate-citrate buffer and mix. Pipet into 16 × 125 mm tubes in 1 ml amounts. Cap, and store in a freezer.

Sodium hydroxide, 0.1N. Dissolve 4.0 g of NaOH in water and dilute to 1 liter.

PROCEDURE
For *total* acid phosphatases, warm two 16 × 125 mm tubes, each containing 1 ml of *total* buffered substrate, to 37 C.
For *prostatic* acid phosphatase, warm an additional tube containing *prostatic* buffered substrate.

To one of the *total* tubes add 0.2 ml of water (Blank).

Add 0.2 ml of serum to the other tube (and to the *prostatic* tube if applicable).

Mix and incubate at 37 C for exactly 30 minutes.

Remove the tubes from the incubator and add 5 ml of 0.1*N* NaOH. Mix well.

Transfer to 19 mm cuvets. Read the absorbances of the Tests in a spectrophotometer at 410 nm, setting the Blank at zero.

If a sample is icteric or turbid, measure another 0.2 ml of the serum into a cuvet. Add 6 ml of 0.1*N* NaOH. Mix and read the absorbance at 410 nm, setting zero with 0.1*N* NaOH.

CALCULATIONS

If a Blank for icterus or turbidity was performed, subtract its absorbance reading from the reading of the *total* tube (and of the *prostatic* tube if applicable).

For *total* acid phosphatase determine the value (units) from a calibration curve (see Standardization).

For *prostatic* acid phosphatase determine the value of the *prostatic* tube reading from the same calibration curve. Then subtract this from the value of the *total* tube to get units of prostatic acid phosphatase.

STANDARDIZATION

The standards and standard curve are the same as given on page 254 for alkaline phosphatase except for the values of the units, which are as follows:

Working Standard (ml)	0.02N NaOH (ml)	Bessey, Lowry, Brock Units
1	10	0.28
2	9	0.56
4	7	1.12
6	5	1.67
8	3	2.23

In this case the 1 ml standard which is 1 unit for alkaline phosphatase must be multiplied by 0.1/0.2 because 0.2 ml of serum is used and by 6.2/11 because the final volume for the test is 6.2 ml whereas the volume in the Standard tube is 11 ml. Hence the value of the first Standard for acid phosphatase is:

$$1 \times \frac{0.1}{0.2} \times \frac{6.2}{11} = 0.28 \text{ units}$$

NORMAL VALUES

Total acid phosphatase activity in the serum of healthy adults usually is less than 0.65 units. Values for prostatic acid phosphatase usually are less than 0.15 units.

DISCUSSION

Acid phosphatase is quite labile. Therefore, if the analysis cannot be started soon after the sample is obtained, the serum should be frozen to preserve the enzyme activity. The activity may also be preserved by adding a few crystals of disodium citrate to the serum. Hemolyzed serum should not be used because the acid phosphatase activity in the red cells is many times higher than in the serum.

The Discussion under Alkaline Phosphatase applies here as well. A stronger sodium hydroxide solution is used here because the buffer in this case is acid and, since the reaction proceeds at an acid pH, the acid blank reading is not made.

Tartaric acid inhibits the activity of acid phosphatase of prostatic origin but has no effect on acid phosphatase from other sources [13]. Therefore prostatic acid phosphatase is determined indirectly by the difference between the *total* and the *tartrate-inhibited* assays.

PHYSIOLOGICAL SIGNIFICANCE OF ACID PHOSPHATASE ACTIVITIES

Acid phosphatase is found chiefly in the erythrocytes and in the prostate gland. The serum acid phosphatase activity is used almost exclusively for diagnosis of cancer of the prostate. Although acid phosphatase is present in the blood of both men and women, the test is generally performed with serum from male patients. In these cases, an increased total acid phosphatase activity is indicative of prostate malignancy. However, the prostatic acid phosphatase method is thought to be more sensitive and specific for diagnosis of prostatic problems. Increased acid phosphatase values will be found in the serum of men who were subjected to digital prostatic examination prior to the collection of a blood sample.

Glutamic Oxalacetic Transaminase (GOT)

Glutamic oxalacetic transaminase is a tissue enzyme that catalyzes the transfer of amino and keto groups between alpha-amino acids and alpha-keto acids; hence, it is a *transferase*. Methods for determination of this enzyme employ aspartic acid and alpha-ketoglutaric acid as a substrate. The reaction, catalyzed by transaminase, results in the formation of glutamic acid and oxalacetic acid. In the classic method [18], the oxalacetic acid that is formed oxidizes the reduced form of the coenzyme nicotinamide adenine dinucleotide (NADH) in the presence of malic dehydrogenase. The NADH, which absorbs strongly at 340 nm, is converted to NAD, which does not absorb at this wavelength (see Figure 20-1). Thus the conversion results in reduced absorbance at 340 nm.

This procedure is tedious and time consuming, employs unstable reagents, and requires the use of an ultraviolet spectrophotometer.

In a somewhat simpler method [25], the oxalacetic acid is decarboxylated with aniline citrate to form pyruvic acid. The pyruvic acid then combines with dinitrophenylhydrazine to form a pyruvate-dinitrophylhydrazone, which is then extracted with toluene and measured colorimetrically in an alkaline medium.

The simplest version of the method involves the direct combination of oxalacetic acid with dinitrophenylhydrazine and measurement of the color in an alkaline solution [21]. Although the ultraviolet procedure is the reference method, the colorimetric method eliminates the need for an ultraviolet spectrophotometer and lends itself more readily to multiple analysis, while giving results which compare favorably with the ultraviolet technic.

GLUTAMIC OXALACETIC TRANSAMINASE (GOT) METHOD

METHOD
Reitman and Frankel [21].

MATERIAL
Serum or cerebrospinal fluid.

REAGENTS
Phosphate buffer, pH 7.4. Dissolve 11.92 g of anhydrous dibasic sodium phosphate (Na_2HPO_4) and 2.18 g of anhydrous monobasic potassium phosphate (KH_2PO_4) in water and dilute to 1 liter. Store in a refrigerator.

GOT substrate. Weigh 0.0584 g of alpha-ketoglutaric acid and 5.32 g of DL-aspartic acid, and place in a 250 ml beaker. Add 40 ml of 1N sodium hydroxide and stir until solution is complete. Using a pH meter, adjust the pH of the solution to 7.4 ± 0.1 by adding 1N sodium hydroxide dropwise, with stirring.

Transfer the solution quantitatively to a 200 ml volumetric flask with phosphate buffer solution (pH 7.4). Dilute to 200 ml with the buffer, add 2 ml of chloroform, and store in the refrigerator.

Color reagent. Dissolve in 200 ml of 1N hydrochloric acid 0.0396 g of dinitrophenylhydrazine, $(NO_2)_2C_6H_3NHNH_2$; use a magnetic stirrer. Store in the refrigerator.

Sodium hydroxide, 0.4N (approximately). Dissolve 16 g of sodium hydroxide (NaOH) in 1 liter of water.

PROCEDURE
Into each of two 19 mm cuvets, measure 1 ml of substrate.
Warm to 37 C by immersing the tubes in a water bath for a few minutes.

Measure 0.2 ml of sample into one cuvet (Test) and 0.2 ml of water into the other (Blank).

Mix by swirling, and incubate at 37 C for 1 hour.

Remove the tubes from the incubator and add 1 ml of color reagent. Swirl to mix, and allow to stand for 20 minutes.

Add 10 ml of 0.4N sodium hydroxide, stopper, and invert to mix. Allow to stand for at least 5 minutes.

Set the Blank at 0.25 absorbance in a spectrophotometer at 505 nm.

Read the absorbance of the Test. If more than one test is to be read, reset the Blank to 0.25 absorbance between readings.

CALCULATIONS

The results are calculated using a calibration curve (see Standardization). If a value over 200 units is obtained, the test should be repeated using 0.2 ml of a 1:5 dilution of the serum.[1] The result then must be multiplied by 5.

STANDARDIZATION

The transaminase unit is defined, in terms of the ultraviolet procedure, as *the amount of enzyme activity in 1 ml of serum that will lower the absorbance by 0.001 in 1 minute under the conditions of that method.* Most other methods, including the one described here, have been designed so that the units are comparable to the ones in the original ultraviolet method. By definition, then, the proper way to calibrate an instrument for this procedure is to analyze a number of samples of various activities by the colorimetric method *and* by the reference ultraviolet procedure. Then, using linear graph paper, plot the *units of activity* as determined by the reference method against their respective *absorbances* obtained with the colorimetric procedure. This curve then may be used for calculating the tests. An alternative procedure has been suggested whereby an intermediate standard (sodium pyruvate) is evaluated in terms of the reference method. This procedure results in assigning transaminase unit values to several concentrations of pyruvate. The pyruvate solutions then may be used as standards for the colorimetric method. Pure sodium pyruvate or prepared transaminase standard are commercially available.[2]

Pyruvate Standard.[3] Dissolve 20.0 mg of pure sodium pyruvate in 100 ml of phosphate buffer (pH 7.4).

Set up a standard curve in 19 mm cuvets as shown on page 260.

---

[1] Dilutions of serum do not always give values in exact proportion to the magnitude of the dilution. This has also been noted with the spectrophotometric method.
[2] Sigma Chemical Company, St. Louis, Missouri.
[3] Experimentally it has been found that this standard, when used as described, gives a curve which agrees more closely with the reference method than does the 22 mg per 100 ml standard recommended by Reitman and Frankel [21].

Standard (ml)	GOT Substrate (ml)	H₂O (ml)	GOT Units	GPT Units
0	1.0	0.2	Blank	Blank
0.1	0.9	0.2	24	28
0.2	0.8	0.2	61	57
0.3	0.7	0.2	114	97
0.4	0.6	0.2	190	—

No incubation is necessary. Add 1 ml of color reagent, mix, and proceed exactly as for the Tests. Plot the curves on linear graph paper using the units given above and their respective readings. The Blank is plotted at zero concentration and 0.25 absorbance. The curve is quite stable but should be checked periodically.

NORMAL VALUES
Values up to 40 units of GOT activity in serum are considered normal, but the range 40 to 50 units is considered equivocal.

DISCUSSION
As has been mentioned, most methods for the determination of trans-aminase activity employ a substrate containing alpha-ketoglutaric acid and DL-aspartic acid. Transaminase catalyzes the *transfer* of the *amino* group as follows:

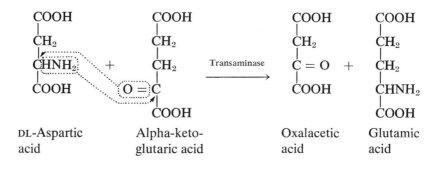

| DL-Aspartic acid | Alpha-keto-glutaric acid | | Oxalacetic acid | Glutamic acid |

The amount of oxalacetic acid formed in this reaction is proportional to the transaminase activity. The various methods differ from each other primarily in the manner in which the oxalacetic acid is measured. In the colorimetric method described above, the oxalacetic acid combines with dinitrophenylhydrazine to form an oxalacetic-dinitrophenylhydrazone. When sodium hydroxide is added, a brown color results, the intensity of which is proportional to the concentration of the hydrazone.

As with any enzyme method, the details concerning time, tempera-ture, and pH should be carefully observed. After addition of sodium hydroxide, the color stabilizes in 5 minutes and remains constant for at least 30 minutes. If desired, a serum blank may be set up, following the same procedure but omitting the 1 hour incubation step and adding the color reagent immediately after adding the serum. The authors of the method found this to be unnecessary. However, the serum blank should be set up if the serum has a very high bilirubin concentration, if it is turbid or hemolyzed, or if it contains a dye sensitive to the alkaline pH (for example, BSP, PSP).

If kept refrigerated, the substrate is stable for at least 1 month. It is good policy to run a control sample each day, since deterioration of the substrate may be recognized by a continuously rising value for the control. If many analyses are performed each day, it may be advantageous to measure 1 ml amounts of substrate into a large number of 16 × 125 mm tubes. These tubes may be stoppered and stored in a freezer for long periods of time. Each day the required number of tubes may be removed, the analyses run in them, and the contents transferred to cuvets for reading.

The tests may be read against water set at zero absorbance if desired. However, the technic of running a reagent blank each time and setting this at 0.25 absorbance helps to compensate for small changes that might occur in the reagents or instrument.

The precision of the reference transaminase method is about ±5 units in the normal or moderately elevated range. If properly standardized, the performance of the colorimetric method is comparable to that of the ultraviolet technic. Although these methods are not very precise as judged by the usual analytical standards, they are adequate to serve the purpose for which they are designed.

The transaminases are relatively stable enzymes and will maintain their activity for several weeks if the serum is stored in a refrigerator.

Glutamic Pyruvic Transaminase (GPT)

Glutamic pyruvic transaminase is similar to GOT and is determined by the same methods except that the substrate contains DL-alanine instead of aspartic acid and the product measured is pyruvic acid rather than oxalacetic acid.

GLUTAMIC PYRUVIC TRANSAMINASE (GPT) METHOD

METHOD
Reitman and Frankel [21].

MATERIAL
Serum.

REAGENTS
All reagents are the same as for GOT except the substrate.

GPT substrate. Weigh 0.0292 g of alpha-ketoglutaric acid and 1.78 g of
DL-alanine, and place in a 100 ml beaker. Add 20 ml of water and

stir until solution is complete. Using a pH meter, adjust the pH of the solution to 7.4 ± 0.1 by adding 1N sodium hydroxide dropwise, with stirring. The adjustment requires approximately 10 drops.

Transfer the solution quantitatively to a 100 ml volumetric flask with phosphate buffer solution (pH 7.4). Dilute to 100 ml with the buffer, add 2 ml of chloroform, and store in the refrigerator.

PROCEDURE

The procedure is the same as for GOT (page 258) except that the incubation time is *30 minutes* rather than 1 hour.

CALCULATIONS

The results are calculated using a calibration curve (see Standardization).

STANDARDIZATION

The standardization procedure is the same as for GOT (page 259) except that, if serum is used for standardization, the GPT values should be determined with the ultraviolet method and, if the pyruvate standard[4] is used, the values assigned for GPT should be used in plotting the calibration curve (see page 260).

NORMAL VALUES

In the absence of disease, GPT values in serum usually are less than 35 units.

DISCUSSION

The discussion presented for GOT (page 260) also applies to GPT. However, in the reaction shown on page 260, the amino group is transferred from DL-alanine (CH_3CHNH_2COOH), and the product is pyruvic acid ($CH_3CHCOCOOH$), rather than oxalacetic acid. The colored compound measured is pyruvate-dinitrophenylhydrazone.

PHYSIOLOGICAL SIGNIFICANCE OF TRANSAMINASE ACTIVITIES

The group of enzymes called transaminase exist in the tissues of many organs. Necrotic activity in these organs causes a release of abnormal quantities of enzyme into the blood. Since heart tissue is rich in glutamic oxalacetic transaminase, myocardial infarction results in high GOT activities in the serum. GOT activity is also increased with various types of liver disease and sometimes with renal disease.

The liver is especially rich in GPT. This enzyme measurement is used primarily as a test for hepatitis. With infectious hepatitis the GPT activity in serum is greater than that of GOT, but both activities usually are increased. Thus, generally speaking, the GOT value is used for diagnosis of myocardial infarction, while the GPT value is useful in diagnosing infectious hepatitis. Neither test is specific.

---

[4] The GPT curve may be determined using the GOT substrate (see page 258).

Lactate Dehydrogenase (LDH)

Lactate dehydrogenase is a tissue enzyme that catalyzes the oxidation of lactic acid to pyruvic acid in the presence of nicotinamide adenine dinucleotide (NAD) or the reduction of pyruvic acid to lactic acid with the reduced form (NADH) of this coenzyme. One method [27] employs the pyruvate-to-lactate reaction, and measures the decrease in ultraviolet absorption as NADH is converted to NAD (see page 242). This method requires preliminary reduction of endogenous pyruvate.

Colorimetric methods are available which measure the formation of a pyruvic-dinitrophenylhydrazone [8], but these procedures require reagents that are difficult to standardize.

The method presented here involves the lactate-to-pyruvate reaction and measures the increase in ultraviolet absorption as NAD is converted to NADH.

LDH Method

METHOD
Wacker et al. [26].

MATERIAL
Serum.

REAGENTS

LDH substrate. Into a 600 ml beaker measure 1.25 ml of L-lactic acid ($CH_3CHOHCOOH$) and approximately 200 ml of water. Dissolve 5.97 g of sodium pyrophosphate ($Na_4P_2O_7 \cdot 10H_2O$) in this solution. Add 9.5 ml of $1N$ sodium hydroxide and mix. Add 0.89 g of nicotinamide adenine dinucleotide (NAD) and stir until dissolved. Using a pH meter, adjust the pH to 8.8 by adding $1N$ sodium hydroxide or hydrochloric acid as necessary.

Measure 3 ml amounts of this substrate into 16×100 mm test tubes, stopper tightly, and store in a freezer.

PROCEDURE

Set the spectrophotometer[5] at 340 nm. Set zero absorbance with water in a 10 mm cuvet.

Bring the contents of a substrate tube to 32 C.

Add 0.2 ml of serum, mix well, and pour the mixture into a 10 mm cuvet.

Read the absorbance of the test (A_1).

Immediately start a timer that has been set for 5 minutes.

Recheck the zero with the Blank and after exactly 5 minutes read the test again (A_2).

---

[5] Use of a recording spectrophotometer is recommended.

CALCULATIONS

With this method 1 unit of activity results in an increase in absorbance of 0.001 per minute per milliliter of serum.

$A_2 - A_1$ = the increase in absorbance in 5 minutes

$\dfrac{A_2 - A_1}{5}$ = increase per minute

Since 0.2 ml of serum was used:

$\dfrac{A_2 - A_1}{5} \times \dfrac{1}{0.2}$ = increase per minute per milliliter

This quantity times 1000 gives Wacker units. The formula thus becomes:

$A_2 - A_1 \times 1000$ = Wacker units

If a recording spectrophotometer is used the absorbance change between any two time points may be used to calculate the increase in absorbance per minute.

NORMAL VALUES

In the absence of disease, LDH values in serum usually are less than 170 Wacker units.

DISCUSSION

Lactate dehydrogenase catalyzes the reaction

$$
\begin{array}{ccc}
\mathrm{CH_3} & & \mathrm{CH_3} \\
| & & | \\
\mathrm{HCOH} & + \quad \mathrm{NAD} \underset{\longleftarrow}{\overset{\mathrm{LDH}}{\rightleftharpoons}} & \mathrm{C{=}O} \quad + \quad \mathrm{NADH} \\
| & & | \\
\mathrm{COOH} & & \mathrm{COOH}
\end{array}
$$

Lactic acid   Pyruvic acid

Since NADH absorbs ultraviolet light at 340 nm while NAD does not (see Figure 20-1), the increase in absorbance of ultraviolet light is a measure of the production of NADH which, under the conditions described, is directly related to LDH activity.

Strict adherence to details concerning time and temperature are essential if reproducible results are to be obtained. Since the reaction rate is linear for at least 5 minutes, the technic involving only 2 readings may be used for the sake of convenience. However, a continuous monitoring of the reaction with a recording spectrophotometer should be made wherever possible.

If a serum is very turbid so as to give a high initial reading, it may be determined against a blank consisting of 0.2 ml of the serum in 3 ml of water. Since hemolyzed serum may yield spuriously high results, it should not be used for this test. Samples with very high LDH activities should be repeated using 0.1 ml of serum. The result must then be multiplied by 2.

Lactate dehydrogenase is a relatively stable enzyme and will maintain its activity for at least 1 week if the serum is stored in a refrigerator.

PHYSIOLOGICAL SIGNIFICANCE OF LDH ACTIVITIES

Lactate dehydrogenase is found chiefly in the liver, heart, skeletal muscle, and erythrocytes. Its activity is used primarily for diagnosis of myocardial infarction, although increased activities also may be found with liver disease and various malignant diseases. Usually this test is used in conjunction with the transaminases and creatine phosphokinase (CPK).

Creatine Phosphokinase (CPK)

The enzyme creatine phosphokinase (CPK) is measured in some clinical chemistry laboratories. The method most commonly used [14] is based on the following sequence of three coupled reactions, the first of which is catalyzed by CPK.

$$ADP + phosphocreatine \xrightarrow{\text{CPK}} ATP + creatine$$

$$ATP + glucose \xrightarrow{\text{hexokinase}} ADP + G\text{-}6\text{-}P$$

$$G\text{-}6\text{-}P + NADP \xrightarrow{\text{G-6-PD}} 6\text{-phosphogluconate} + NADPH$$

where: ADP = adenosine diphosphate
          ATP = adenosine triphosphate
          G-6-P = glucose 6-phosphate
          G-6-PD = glucose 6-phosphate dehydrogenase
          NADP = nicotinamide adenine dinucleotide phosphate
          NADPH = reduced nicotinamide adenine dinucleotide phosphate

As described earlier in this chapter for NAD and NADH (page 242), the reduction of NADP to NADPH results in an increase of absorbance at 340 nm. The complexity and instability of the reagents makes this enzyme more difficult to measure reliably than some of the others.

Since CPK is unstable in serum, samples should be frozen until analyzed.

CPK is one of the enzyme tests used to diagnose and follow the course of myocardial infarction. However, like the transaminases and LDH, it is not clinically specific for the disorder and may be elevated in other conditions such as muscular dystrophy, chronic alcoholism, cellulitis, or pneumonia.

CPK Isoenzymes

Greater specificity and sensitivity for diagnosing myocardial infarction can be achieved by analyzing for the isoenzymes of CPK. Electrophoresis or an ion-exchange technic allows separation of three CPK isoenzymes. Subsequent exposure of the bands or fractions to a suitable substrate solution allows the individual isoenzymes to be quantitated, or at least visualized. The three bands are called CPK-1 (or BB), CPK-2 (or MB), and CPK-3 (or MM). Normally only CPK-3, which remains near the point of application, is detectable in the serum. Soon after the onset of a myocardial infarction CPK-2, which migrates with alpha-2 globulin, becomes visible.

References

1. Andersch, M. A.   The determination of serum amylase, with particular reference to the use of beta-amylose as the substrate. *J. Biol. Chem.* 166:705, 1946.
2. Benotti, J., Rosenberg, L., and Dewey, B.   Modification of the Gutman and Gutman method of estimating "acid" phosphatase activity. *J. Lab. Clin. Med.* 31:357, 1946.
3. Bergmeyer, H. V.   *Methods of Enzymatic Analysis.* New York: Academic, 1974. Vol. 1–4.
4. Bessey, O. A., Lowry, O. H., and Brock, M. J.   A method for the rapid determination of alkaline phosphatase with five cubic millimeters of serum. *J. Biol. Chem.* 164:321, 1946.
5. Blume, P., and Freier, F. (Eds.).   *Enzymology in the Practice of Laboratory Medicine.* New York: Academic, 1974.
6. Bodansky, A.   Phosphatase studies; determination of serum phosphatase. Factors influencing the accuracy of the determination. *J. Biol. Chem.* 101:93, 1933.
7. Boyer, P. D. (Ed.).   *The Enzymes* (3rd ed.). New York: Academic, 1970–1973. Vol. 1–9.
8. Cabaud, P. G., and Wroblewski, F.   Colorimetric measurement of lactic dehydrogenase activity in body fluids. *Am. J. Clin. Pathol.* 30:234, 1958.
9. Caraway, W. T.   A stable starch substrate for the determination of amylase in serum and other body fluids. *Am. J. Clin. Pathol.* 32:97, 1959.
10. Coodley, E. L. (Ed.).   *Diagnostic Enzymology.* Philadelphia: Lea & Febiger, 1970.
11. Crandall, L. A., and Cherry, I. S.   Studies on the specificity and behavior of blood and tissue lipases. *Proc. Soc. Exp. Biol. Med.* 28:570, 1930–31.
12. *Enzyme Nomenclature—Recommendations (1972) of the International Union of Pure and Applied Chemistry and the International Union of Biochemistry.* Amsterdam: Elsevier, 1973.
13. Fishman, W. H., and Lerner, F.   A method for estimating serum acid phosphatase of prostatic origin. *J. Biol. Chem.* 200:89, 1953.
14. Henry, R. J., Cannon, D. C., and Winkelman, J. W.   *Clinical Chemistry Principles and Technics* (2nd ed.). New York: Harper & Row, 1974. P. 899.
15. Henry, R. J., and Chiamori, N.   Study of the saccharogenic method for the determination of serum and urine amylase. *Clin. Chem.* 6:434, 1960.
16. Huggins, C., and Talalay, P.   Sodium phenolphthalein phosphate as a substrate for phosphatase tests. *J. Biol. Chem.* 159:399, 1945.

17. Jencks, W. P. *Catalysis in Chemistry and Enzymology.* New York: McGraw-Hill, 1969.
18. Karmen, A. A note on the spectrophotometric assay of glutamic oxalacetic transaminase in human blood serum. *J. Clin. Invest.* 34:131, 1955.
19. Klein, B., Foreman, J. A., and Searcy, R. L. New chromogenic substrate for determination of serum amylase activity. *Clin. Chem.* 16:32, 1970.
20. Mahler, H. R., and Cordes, E. H. *Biological Chemistry* (2nd ed.). New York: Harper & Row, 1971.
21. Reitman, S., and Frankel, S. A colorimetric method for the determination of serum glutamic oxalacetic and glutamic pyruvic transaminases. *Am. J. Clin. Pathol.* 28:56, 1957.
22. Saxon, E. I., Hinkley, W. C., Vogel, W. C., and Zieve, L. Comparative value of serum and urinary amylase in the diagnosis of acute pancreatitis. *A.M.A. Arch. Intern. Med.* 99:607, 1957.
23. Schwartz, M. K., Kessler, G., and Bodansky, O. Comparison of serum alkaline phosphatase activities determined with sodium β-glycerophosphate and sodium phenylphosphate as substrates. *Am. J. Clin. Pathol.* 33:275, 1960.
24. Seligman, A. M., Chauncy, H. H., Nachlas, M. M., Hanheimer, L. H., and Ravin, H. A. The colorimetric determination of phosphatases in human serum. *J. Biol. Chem.* 190:7, 1951.
25. Tonhazy, N. E., White, N. G., and Umbreit, W. W. A rapid method for the estimation of the glutamic-aspartic transaminase in tissues and its application to radiation sickness. *Arch. Biochem.* 28:36, 1950.
26. Wacker, W. E. C., Ulmer, D. D., and Vallee, B. L. Metalloenzymes and myocardial infarction. *N. Engl. J. Med.* 255:449, 1956.
27. Wroblewski, F., and LaDue, J. S. Lactic dehydrogenase activity in blood. *Proc. Soc. Exp. Biol. Med.* 90:210, 1955.

21. Lipids

The term *lipids* applies to all fatty compounds present in the blood. These substances, in their free form, are virtually insoluble in water, but are freely soluble in solvents such as alcohol, ethyl ether, petroleum ether, and chloroform. They can be classified [4] as shown in Figure 21-1. In serum the lipids are soluble due to their combination with protein. These combinations are called *lipoproteins*.

The lipid compounds most often quantitated in clinical chemistry are total lipids, phospholipids, total fatty acids, cholesterol (free and esterified), and triglycerides. Methods for the determination of each of these compounds are presented here; notes on the clinical significance of lipid concentrations are at the end of the chapter.

Total Lipids

The total lipids include all of the materials shown in Figure 21-1. The classic procedure for their quantitation has been the gravimetric method [19]. In this method, a serum sample is combined with a solvent mixture such as ethanol-ether (Bloor's reagent [3]) or methanol-chloroform (Folch's reagent [10]). This extraction step disrupts the lipoprotein complexes, precipitates the proteins, and dissolves the lipids in the organic phase. If the alcohol component is not present, the extraction of lipid is much less complete because the lipoprotein complexes are not fully disrupted. After washing or reextraction to remove nonlipid components, the organic phase is evaporated, and the total lipid residue is quantitated by weighing it. In the gravimetric method, 7 to 12 mg of lipid is obtained from 2 ml of serum.

Total serum lipids also may be measured by turbidimetric methods [7, 11] in which lipids are precipitated with sulfuric acid or phenol plus salt. However, generally the results do not correlate well with the gravimetric method. Therefore turbidimetric methods are not widely used.

Total lipids can be estimated by determining each of the lipid components by separate methods and then summing the result. However, because there is some overlap of lipid components in the analyses, the sum may not correlate exactly with the determined total lipid value.

The phosphovanillin colorimetric method for total lipids gives results which correlate well with the gravimetric method, and it is simple and rapid. It is the method presented here.

Total Lipids Method

METHOD
Frings et al. [12].

MATERIAL
Serum.

REAGENTS
Concentrated sulfuric acid.

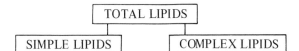

SIMPLE LIPIDS	COMPLEX LIPIDS
Simple lipids contain no ester or amide linkages and therefore are resistant to cleavage by acidic or basic hydrolysis conditions.	Complex lipids contain one or more ester or amide linkages and therefore are cleaved by acidic or basic hydrolysis conditions.

1. Fatty acids

2. Steroids
   a. Cholesterol
   b. Steroid hormones
      and vitamin D

3. Terpenes (includes carotene
   and vitamins A, E, and K)

4. Prostaglandins

1. Acylglycerols (includes triglycerides,
   i.e., triacylglycerols)

2. Acylsterols (includes cholesterol
   esters)

3. Phosphoglycerides (phosphoglycerides +
   phosphorous-containing sphingolipids =
   phospholipids)

4. Sphingolipids

5. Waxes (some waxes are acylsterols)

Figure 21-1
Classification of Lipids

Phosphovanillin reagent. Dissolve 1.2 g of vanillin in 200 ml of water.
  With constant stirring, add 800 ml of concentrated phosphoric acid.
  Store in a dark bottle.

PROCEDURE

Add 0.1 ml of serum and 2 ml of concentrated sulfuric acid to a 16 ×
  100 mm tube (Test).
Add 0.1 ml of the standard and 2 ml of concentrated sulfuric acid to
  another tube (Standard).
Mix all tubes well, and heat in a boiling water bath for 10 minutes.
Cool to room temperature. Then transfer 0.1 ml of each mixture to 19
  mm cuvets. Measure 0.1 ml of concentrated sulfuric acid to another
  cuvet (Blank). To all cuvets add 6 ml of phosphovanillin reagent.
Mix all cuvets well and incubate at 37 C for 15 minutes.
Cool to room temperature and measure the absorbances at 540 nm using
  the Blank to set zero absorbance.

CALCULATIONS

$$\frac{\text{absorbance of Test}}{\text{absorbance of Standard}} \times 700 = \text{total lipids (mg/dl)}$$

Note: If a value greater than 1000 mg per dl occurs, repeat the test,
starting not at the beginning but at the step after the boiling-water bath

treatment, using 12 ml of phosphovanillin reagent instead of 6 ml. Multiply the final result by 2.

STANDARDIZATION

Standard, 700 mg/dl. Dissolve 700 mg of olive oil, corn oil, or triolein in 100 ml of *absolute* ethanol and store in a refrigerator.

NORMAL VALUES

Serum from healthy adults usually contains between 400 and 1000 mg per dl of total lipids.

DISCUSSION

The exact chemistry involved in this method is not completely under-stood. Apparently it is the carbon-to-carbon double bonds (C=C) of lipids which react with the reagent to form the chromophore. Heating the sample in the presence of sulfuric acid probably not only forms such double bonds in many of those lipids which do not already possess them (by a dehydration type of reaction), but also reacts with these carbon-to-carbon double bonds to form intermediate compounds which later react to give a color when vanillin and phosphoric acid are added. The method is reasonably specific for lipids because only very small amounts of non-lipid substances occur in serum which possess reactive carbon-to-carbon double bonds or the ability to form them.

Samples with lipid concentrations greater than 1000 mg per dl must be diluted because Beer's law is followed (a plot of absorbance vs. total lipid concentration is linear) only up to 1000 mg per dl at 540 nm, even though the absorbance is only about 0.5 in 19 mm cuvets at this con-centration.

Occasionally, a lipid concentration greater than 2,000 mg per dl is encountered. In this case, the test should be repeated using a smaller sample size or a dilution of the serum.

Triglycerides

As shown below, three fatty acids in ester linkage with glycerol comprise a triglyceride. These components of triglycerides can be released by bases (saponification), acids, or enzymes (lipases).

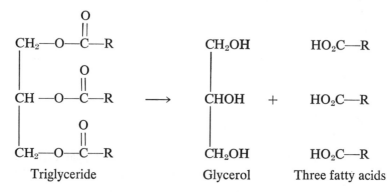

| Triglyceride | Glycerol | Three fatty acids |

The three R groups are long hydrocarbon chains (for example, $C_{16}H_{33}$) and may differ from each other.

Triglycerides sometimes are referred to as *neutral fat*. They represent the storage form of lipid in the body and constitute the major lipid found in both the pre-beta lipoproteins and chylomicrons, two of the four basic types of lipid-protein complexes which occur in the blood.

Triglycerides are measured based either on their fatty acid content or their glycerol content, although measurements of the latter are more widely used. The glycerol is released by a saponification reaction as illustrated above, then quantitated either chemically or enzymatically. Most of the chemical methods start with liberation of formaldehyde ($H_2C{=}O$) from glycerol using periodic acid, then form a colored product from the formaldehyde using diacetylacetone and ammonia [28], chromotropic acid [17, 34] or phenylhydrazine and potassium ferricyanide [13]. The enzymatic method [30] involves the use of three enzymes, glycerol kinase, pyruvate kinase, and lactate dehydrogenase, in the following reactions:

$$\text{glycerol} + \text{ATP} \xrightarrow{\text{glycerol kinase}} \text{glycerol phosphate} + \text{ADP}$$

$$\text{ADP} + \text{phosphoenol pyruvate} \xrightarrow{\text{pyruvate kinase}} \text{ATP} + \text{pyruvate}$$

$$\text{pyruvate} + \text{NADH} \xrightarrow{\text{lactate dehydrogenase}} \text{lactate} + \text{NAD}$$

The last reaction is monitored by the decrease in absorbance at 340 nm accompanying the loss of NADH (see page 242). Since the reaction rate depends on the amount of glycerol present (more glycerol yields more ADP, which yields more pyruvate), this rate serves as a measure of the glycerol and, therefore, the triglyceride concentration.

TRIGLYCERIDE METHOD

METHOD
Gottfried and Rosenberg [14]; Levy [23].

MATERIAL
Serum.

REAGENTS
Heptane, $CH_3(CH_2)_5CH_3$.

Isopropanol ($CH_3CHOHCH_3$).

Sulfuric acid, 0.08N. Dilute 2.2 ml of concentrated sulfuric acid (H_2SO_4) to 1 liter with water.

Potassium hydroxide, 6.25M. Dissolve 17.5 g of potassium hydroxide (KOH) in water and dilute to 50 ml.

Periodate reagent. Add 5 ml of glacial acetic acid to 95 ml of water. Dissolve 0.6 g of sodium metaperiodate in this solution.

Ammonium acetate, 2M. Dissolve 154 g of ammonium acetate (CH_3COONH_4) in water and dilute to 1 liter.

Acetylacetone reagent. Dilute 1.5 ml of acetylacetone (2,4-pentanedione; $CH_3COCH_2COCH_3$) to 200 ml with 2M ammonium acetate. Store refrigerated; stable for at least 2 weeks.

PROCEDURE

To a 16 × 125 mm tube add 0.5 ml of serum (Test). Add 0.5 ml of water to another tube (Blank) and treat the same as the Test.

Add 2 ml of heptane, 3.5 ml of isopropanol, and 1 ml of 0.08N sulfuric acid.

Mix for 20 seconds on a vortex mixer. Then let stand for at least 5 minutes.

Carefully transfer the heptane (upper) layer to a clean tube.

Measure 0.2 ml of this extract into a 13 × 100 mm tube.

Add 2 ml of isopropanol and 1 drop of 6.25M KOH.

Mix well on a vortex mixer (at least 15 seconds).

Heat the tubes for 10 minutes in a heating block at 70 C.

Remove the tubes from the block and add 0.2 ml of periodate reagent and 1 ml of acetylacetone reagent.

Mix well on a vortex mixer.

Heat again for 10 minutes at 70 C.

Remove the tubes from the block and pour the contents into 12 mm cuvets.

Read the absorbance in a spectrophotometer at 425 nm against the Blank set at zero.

CALCULATIONS

Determine the concentration of triglycerides in the Test from a calibration curve (see Standardization).

Triglyceride concentration may be expressed as mg/dl or mmol per liter. If the latter expression is used, a molecular weight of 885 is assumed for triolein. Therefore, 1 mmol = 885 mg; so

$$\frac{mg/dl \times 10}{885} = mmol/liter$$

or

$$mg/dl \times 0.0113 = mmol/liter$$

STANDARDIZATION

Triglyceride Stock Standard, 1000 mg/dl. Weigh 1000 mg of triolein and dilute to 100 ml with isopropanol. Store refrigerated.

Table 21-1. Normal Serum Triglyceride Values

Age (yr)	Upper Limit (mg/dl)
0–29	140
30–39	150
40–49	160
50–59	190

Triglyceride Working Standards. Dilute the following amounts (ml) of stock standard to 50 ml with isopropanol, 2.5, 5.0, 10.0, 15.0, 17.5. The concentrations of these standards are 50, 100, 200, 300, and 350 mg per dl respectively. Store refrigerated.

These standards are analyzed exactly like the tests except that 3.0 ml of isopropanol (instead of 3.5 ml) and 0.5 ml of water are used in the extraction step in addition to the heptane and sulfuric acid.

A standard curve should accompany each set of analyses.

NORMAL VALUES

The commonly accepted range of normal values for triglyceride in serum is 40–150 mg per dl. However, even in healthy people triglyceride levels in serum vary with age. The usual upper limits of normal for various age groups are shown in Table 21-1. It is important that a person fast for at least 14 hours before a blood sample is obtained for triglyceride analysis.

DISCUSSION

The first step in this method serves to precipitate the serum proteins and simultaneously extract the triglycerides, leaving behind potential interfering substances. Although centrifugation is not necessary, the solvent phases and precipitate separate more clearly if the tubes are centrifuged. The upper (heptane) layer should be transferred to a clean tube before the 0.2 ml is measured in order to avoid contamination with precipitate or water-alcohol phase.

During heating with KOH, the triglycerides are saponified to free glycerol. Oxidation of this glycerol with periodate releases formaldehyde which subsequently is condensed with acetylacetone and ammonia to give a yellow fluorescent chromophore. Quantitation may be made colorimetrically or fluorometrically.

With some versions of this method, a turbidity may develop as the tubes cool. However, this problem does not occur with the modification presented here.

The color reaction does not follow Beer's law through the range of concentrations covered in the method. Therefore a standard curve should be plotted each time and used for calculations.

Phospholipids

The complex lipids containing phosphorus are called phospholipids. Methods for the determination of phospholipids involve extraction, digestion, and the determination of phosphorus. Extraction is performed using lipid solvents such as alcohol, ether, or chloroform. As an alternative to extraction, phospholipids may be precipitated from the sample with trichloroacetic acid [35]. Although dry-ashing has been suggested, digestion with mineral acids is preferred. The final determination of phosphorus may be performed using any of the several satisfactory methods available.

The method described below involves extraction with Bloor's reagent, digestion with sulfuric acid and hydrogen peroxide, and determination of phosphorus using the method described in Chapter 17.

PHOSPHOLIPID METHOD

METHOD
Bloor [3]; Baumann [2]; Fiske and Subbarow [9].

MATERIAL
Serum.

REAGENTS
Bloor's reagent. Mix 600 ml of ethyl alcohol with 200 ml of ethyl ether. Store in a glass-stoppered bottle.

Sulfuric acid, 5N (approximately). Dilute 14 ml of concd sulfuric acid (H_2SO_4) to 100 ml with water.

Hydrogen peroxide, 30% (w/w).

Urea, 5% (w/v). Dissolve 5 g of urea in 100 ml of water.

Molybdate reagent. Dissolve 25 g of ammonium molybdate, $(NH_4)_6Mo_7O_{24} \cdot 4H_2O$, in water and dilute with water to 1 liter.

Sulfonic acid reagent. See page 208.

PROCEDURE
To a 25 ml volumetric flask add approximately 15 ml of Bloor's reagent. Add 0.5 ml of serum with continuous swirling.

Bring the solution to a boil on a hot plate set at medium heat. Then remove from the heat and cool to room temperature.

Dilute to 25 ml with Bloor's reagent, mix well, and filter through Whatman no. 1 (11 cm) filter paper.

To a 25 × 200 mm digestion tube (Folin-Wu nonprotein-nitrogen tube), add 10 ml of filtrate and 2 glass beads.

Evaporate the contents of the tube to dryness in a 60 C water bath with the aid of a stream of air.

Add 1 ml of 5N sulfuric acid and set the tube at an angle over a micro burner.

Allow the contents to boil until the solution turns black.

Remove the heat and allow the tube to cool for 1 minute.

Add 1 drop of hydrogen peroxide directly to the residue and heat the
solution again. If the residue does not become colorless, add another
drop of peroxide.

After the residue becomes colorless, cool it slightly. Then add 2 drops of
urea solution for each drop of peroxide used.

Boil the solution again for 1 minute. Then cool it to room temperature.

Dilute the solution to 12.5 ml with water and mix.

In another tube dilute 1 ml of 5N sulfuric acid to 12.5 ml with water
(Blank).

To a third tube add 1 ml of working standard and 1 ml of 5N sulfuric
acid. Dilute to 12.5 ml with water (Standard).

To all tubes add 0.5 ml of sulfonic acid reagent, mix, and allow to stand
for 1 minute.

Add 1.5 ml of molybdate reagent, mix, and allow to stand for 10
minutes.

Transfer to 19 mm cuvets and read the absorbances of the Test and
Standard at 660 nm, setting the Blank at zero.

CALCULATIONS

$$\text{absorbance of Test} \times \frac{\text{concentration of Standard (10 mg/dl)}}{\text{absorbance of Standard}} \times 25$$
$$= \text{phospholipid (mg/dl)}$$

STANDARDIZATION

Stock Phosphorus Standard (20 mg/dl). See page 210.

Working Standard. Dilute 5.0 ml of stock standard to 50 ml with water.
Store in a refrigerator.

Only the *inorganic phosphorus* portion of the method is standardized.
The concentration of phosphorus in the working standard is

$$\frac{5.0}{50} \times 20 = 2.0 \text{ mg/dl}$$

Therefore the 1 ml of standard used in the test contains 0.02 mg of
phosphorus. In the method, 0.5 ml of sample is diluted to 25 ml and
10 ml of this dilution are used. This contains

$$0.5 \times \frac{10}{25} = 0.2 \text{ ml of sample}$$

Therefore, the standard is equivalent to 0.02 mg of phosphorus per
0.2 ml (10 mg per 100 ml of serum).

The calculations give *lipid phosphorus;* phospholipids are calculated on
the assumption that 4% (by weight) of the phospholipid molecule is
phosphorus. Therefore lipid phosphorus is multiplied by 25 to get
phospholipids.

NORMAL VALUES

Serum from healthy adults usually contains between 150 mg and 275 mg of phospholipids per 100 ml. This is 6 to 11 mg of lipid phosphorus per 100 ml.

DISCUSSION

Treatment with Bloor's reagent extracts the phospholipids and precipitates proteins. Inorganic phosphorus normally present in the serum is not extracted with this reagent.

During digestion, the phospholipids are decomposed and their phosphoric acid liberated. Hydrogen peroxide facilitates oxidation of the organic materials, and urea is added to decompose the excess peroxide, preventing it from interfering with the final color reaction. A discussion of the color reaction used in the determination of phosphorus may be found on page 210.

The calculated results actually represent phosphorus of phospholipid origin. However, the several types of phospholipids contain very nearly the same proportion of phosphorus (4% by weight). Therefore, no significant error is incurred when a factor of 25 is assumed in calculating phospholipid from lipid phosphorus.

Fatty Acids

The fatty acids are long-chain carboxylic acids present in human plasma primarily as neutral fat, phospholipids, and cholesterol esters. Most of the fatty acids are esterified, with only about 5% of the total present in the free form.

Analytic methods for the determination of *total* fatty acids involve extraction, saponification, separation of the free acids, and titration. Methods for determining free fatty acids also are available [16], but these methods are less reliable because of the very low concentration and relative instability of the free fatty acids.

TOTAL FATTY ACIDS METHOD

METHOD

Stoddard and Drury [33].

MATERIAL

Serum.

REAGENTS

Bloor's reagent. Mix 600 ml of ethyl alcohol with 200 ml of ethyl ether. Store in a glass-stoppered bottle.

Sodium hydroxide, concentrated (50% w/w).

Thymol blue indicator. Dissolve 1 g of thymol blue in 100 ml of ethyl alcohol.

Hydrochloric acid, 4N. Dilute 33 ml of concentrated hydrochloric acid (HCl) to 100 ml with water.

Sodium chloride, 5%. Dissolve 50 g of sodium chloride (NaCl) in water and dilute to 1000 ml.

Phenolphthalein indicator. Dissolve 1 g of phenolphthalein in 100 ml of ethyl alcohol.

Sodium hydroxide, 0.020N. Dilute 20 ml of standard 0.100N sodium hydroxide (NaOH) to 100 ml with water. Verify the normality by titrating with a standard acid.

Purified sea sand.

PROCEDURE
To a 50 ml volumetric flask add approximately 35 ml of Bloor's reagent.
Add 3 ml of serum with continuous swirling (Test).
Bring the solution to a boil on a hot plate set at medium heat. Then remove from the heat and cool to room temperature.
Dilute to 50 ml with Bloor's reagent, mix well, and filter through Whatman no. 1 (11 cm) filter paper into a 50 ml cylinder.
Measure 40 ml of the filtrate into a 100 ml beaker.
Add a few grains of purified sea sand and, with the aid of a stream of air, evaporate to approximately $\frac{1}{2}$ volume on a hot plate set at medium heat.
Add 0.1 ml of concentrated NaOH. Cover with a watch glass, and warm gently for 30 minutes.
Remove the watch glass, increase the heat, and evaporate the contents of the beaker to dryness.
Remove from the heat, add 15 ml of hot purified water, and mix gently.
Cool to room temperature. Then add 3 drops of thymol blue indicator. Add 4N HCl dropwise (with stirring) until the solution turns pink.
Place the beaker in ice or in a refrigerator for at least 30 minutes (or overnight if more convenient).
Filter through Whatman no. 1 (11 cm) filter paper.
Rinse the beaker 3 times with 5 ml volumes of 5% NaCl, pouring the rinsings through the filter paper.
Wash the precipitate on the filter paper at least 10 times with 5% NaCl. The washing should be continued until a 5 ml amount of wash solution containing 1 drop of phenolphthalein indicator turns pink when 1 drop of 0.020N NaOH is added.
Discard the washings and place a 50 ml Erlenmeyer flask under the funnel.
Add five 4 ml amounts of *hot* 95% ethyl alcohol to the precipitate, and collect the filtrate in the Erlenmeyer.
Add 20 ml of ethyl alcohol to another flask to serve as a Blank.

To the Blank and Test add 3 drops of thymol blue and titrate to a blue-
green endpoint with 0.020N NaOH, using a 2 ml buret.
Alternatively, the indicator may be omitted and the Test and Blank
titrated to pH 9.0 using a pH meter.

CALCULATIONS

$$\text{titration of Test} - \text{titration of Blank} \times 0.02 \times \frac{5}{12} \times 1000$$
$$= \text{total fatty acids, mmol (or meq) per liter}$$

$$\frac{\text{mmol (or meq)per liter}}{10} \times 277 = \text{mg/dl}$$

The concentration of fatty acids may be expressed in terms of millimoles (mmol) or milliequivalents (meq) of NaOH used in the titration.
The titration of the Test (minus the Blank correction) times the concentration of NaOH gives millimoles (or milliequivalents) of acid present
in the final solution. Since 40 ml of a 3:50 dilution of sample are used,
this represents

$$\frac{40}{50} \times 3 = \frac{12}{5} \text{ ml}$$

Therefore, the result is multiplied by

$$\frac{5}{12} \times 1000$$

to get millimoles (or milliequivalents) per *liter*.
To convert millimoles to milligrams, the molecular weight of a compound must be known. Since the fatty acids are a group of compounds,
there is no single molecular weight that is applicable. However, an average molecular weight of 277 can be assumed which introduces an error
of not more than 2%. Dividing millimoles per liter by 10 gives millimoles per 100 ml, and this figure, multiplied by the average molecular
weight (or equivalent weight) of 277, gives milligrams per 100 ml.

NORMAL VALUES
Serum from healthy adults usually contains 250–400 mg per dl or 9–16
mmol (or meq) per liter of fatty acids.

DISCUSSION
Bloor's reagent extracts the fatty acids and precipitates proteins. Since
the fatty acids are present largely as esters, they must be saponified; this
is accomplished by heating with sodium hydroxide. During this procedure, the esters are split and the fatty acids converted to sodium salts,
or soaps. These soaps are dissolved in hot water, and, when the solution
is acidified, they become insoluble. Precipitation of the fatty acids is

facilitated by lowering the temperature of the solution. The test may be left overnight at this point if desired.

When the solution is filtered, the precipitated fatty acids are retained on the filter paper. It is very important that the washing be thorough since any mineral acid remaining will be titrated along with the fatty acids, giving a spuriously high result.

The fatty acids are dissolved in hot ethanol and titrated with standard base. The indicator thymol blue (endpoint approximately pH 9) ensures complete titration of the acids.

Cholesterol

Cholesterol is the best known and most widely determined lipid and is an important intermediate in the synthesis of steroid hormones.

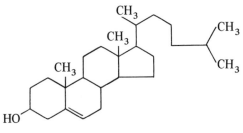

Cholesterol

In the blood, cholesterol occurs both free and esterified. The latter form is a mixture of many cholesterol esters, each of which has a different fatty acid bound in ester linkage to cholesterol at the OH group. Usually, only the *total* cholesterol concentration is of clinical interest, but occasionally separate estimation of the free and esterified forms is of value.

A large number of methods have been devised for cholesterol determination. Both the reference [31] and routine procedures involve the development of a color when cholesterol is treated with a hot acid or a metal plus hot acid. The acids used are sulfuric acid and acetic acid. The metal used most frequently is iron. The color results because cholesterol is dehydrated and oxidized to various polyene derivatives [5]. (A succession of conjugated, carbon-to-carbon double bonds is introduced into the fused rings of cholesterol.) Depending upon the exact reagents and conditions employed, a variety of intermediate and final colors ranging from purple to green to red may be obtained because various mixtures of polyenic products can be generated.

An important feature of any colorimetric cholesterol method is whether any separation steps are involved. This relates not only to the separation of total cholesterol from the serum prior to a colorimetric reaction, but also to the separation of the free and esterified cholesterol components from each other prior to a colorimetric reaction. The usual choices are as follows:

1. No separation step [36]: Total cholesterol can be determined directly in serum using either a mixture of ethyl acetate, sulfuric acid, and ferric perchlorate, or a mixture of ferric acetate, uranium acetate, sulfuric acid, and ferrous sulfate.
2. One separation step [6]: Total cholesterol is extracted from the serum with an organic solvent and a colorimetric reaction is performed.
3. Two separation steps: After an organic extraction, either the free and esterified cholesterol components are separated by digitonin [31] treatment or by a chromatographic technic like column chromatography on silicic acid [8, 15]. (Digitonin is a naturally occurring substance which forms an insoluble complex with free but not esterified cholesterol, thereby providing a means for their separation.)

One reference method for cholesterol is the *method of Abel* [1], which involves the following steps:

1. Heat with KOH plus ethanol to liberate the cholesterol from protein complexes (lipoproteins) and convert the cholesterol esters into free cholesterol.
2. Extract the cholesterol into petroleum ether.
3. Perform the Lieberman-Burchard colorimetric reaction after evaporation of the petroleum ether. This last step involves treating the cholesterol residue with a mixture of sulfuric acid and acetic anhydride to develop a green color.

The method presented here uses a mixture of ferric chloride, acetic acid, and sulfuric acid to develop a red color from an isopropanol extract of cholesterol, both with (free cholesterol) and without (total cholesterol) a digitonin precipitation step.

CHOLESTEROL METHOD
(Total, Free, and Esters)

METHOD
Loeffler and McDougald [25].

MATERIAL
Serum.

REAGENTS
Isopropanol, reagent grade.

Sulfuric acid, concentrated.

Acetone.

Ferric chloride reagent. Dissolve 700 mg of ferric chloride hexahydrate ($FeCl_3 \cdot 6H_2O$) in 1 liter of glacial acetic acid.

Digitonin solution. Dissolve 1 g of digitonin in 57 ml of absolute ethanol. Dilute to 100 ml with water. If some of the digitonin precipitates, redissolve it by warming it to 56 C and mixing.

PROCEDURE

A. *Total Cholesterol*

Add 4.8 ml of isopropanol to each of two 15 ml conical centrifuge tubes.

To one add 0.2 ml of serum (Test) and to the other add 0.2 ml of standard (Standard).

Mix, allow to stand for 5 minutes, then centrifuge for 5 minutes.

Transfer 1 ml of each clear supernatant to appropriately labeled 25 × 100 mm tubes.

Save the remainder of the supernatant for the free and ester method given below.

Add 1 ml of isopropanol to another 25 × 100 mm tube (Blank).

To one tube at a time, add 2 ml of ferric chloride reagent, mix, immediately add 2 ml of concentrated sulfuric acid, and mix well.

Pour into 13 mm cuvets and read the absorbances at 550 nm using the Blank to set zero absorbance.

CALCULATIONS

$$\frac{\text{absorbance of Test}}{\text{absorbance of Standard}} \times 200 = \text{total cholesterol (mg/dl)}$$

B. *Free Cholesterol and Cholesterol Esters*

To a 15 ml conical centrifuge tube add 2 ml of the isopropanol supernatant which was saved from the total cholesterol method given above.

Add 4 ml of acetone and 2 ml of digitonin solution and mix well.

Allow to stand in an ice bath for 30 minutes.

Centrifuge for 10 minutes. Then decant and wipe off the mouth of the tube with a tissue.

Add 5 ml of acetone and mix.

Centrifuge for 10 minutes, decant, and wipe off the mouth of the tube as before.

Add 1 ml of isopropanol and mix well.

Add 1 ml of isopropanol to a 15 ml centrifuge tube (Blank).

To one tube at a time, add 2 ml of ferric chloride reagent, mix, then immediately add 2 ml of concentrated sulfuric acid, and mix well.

Pour into 13 mm cuvets and read the absorbances at 550 nm, using the Blank to set zero absorbance.

CALCULATIONS

$$\frac{\text{absorbance of Test}}{\text{absorbance of Standard}} \times 100^1 = \text{free cholesterol (mg/dl)}$$

$$\text{total} - \text{free} = \text{cholesterol esters (mg/dl)}$$

$$\frac{\text{mg/dl esters}}{\text{mg/dl total}} \times 100 = \text{percent cholesterol esters}$$

STANDARDIZATION

Standard, 200 mg/dl. Dissolve 200 mg of pure cholesterol[2] in isopro-
    panol and dilute to 100 ml with isopropanol. Keep tightly stoppered.
This standard is used as described under Procedure.

NORMAL VALUES

Although the range of concentrations for total serum cholesterol in
healthy adults usually is stated as being approximately 140 to 250 mg
per dl, this range applies only to persons in younger age groups. In older
age groups, serum cholesterol levels are significantly higher [20]. The
approximate upper limits of normal for cholesterol levels in various age
groups are shown in Table 21-2.

The normal range for cholesterol esters is between 70% and 75% of
the total, with free cholesterol making up the remaining 25% to 30%.

DISCUSSION

The isopropanol serves two essential purposes: It liberates the cholesterol from its protein complexes (lipoproteins), and it precipitates these
and other proteins so they do not interfere in the color reaction later in
the procedure. Bilirubin and hemoglobin also are coprecipitated in this
step. The addition of concentrated sulfuric acid to the isopropanol solution produces enough heat to activate the colorimetric reaction. Transfer
of the solution while it is still hot to the cuvet prevents the accumulation
of bubbles. The solutions should be cooled to room temperature before
the absorbances are read, however, and should be protected during this
period from direct exposure to light, which may cause the color to fade.

Digitonin is derived from plants. It is composed of 5 monosaccharides
(4 of which are glucose) and a steroid. Somehow it forms a 1:1, highly
insoluble complex with cholesterol but not with cholesterol esters. Addition of acetone in the above method enhances the precipitation of
cholesterol as the digitonide. After a washing step with acetone, to
further remove cholesterol esters, the cholesterol digitonide precipitate
is dissolved in isopropanol.

---

[1] Since 2 ml of extract are used for the free and 1 ml for the total, the value of the
200 mg per dl standard in this case is 100 mg per dl.
[2] Available as a Standard Reference Material from the U.S. National Bureau of
Standards.

Table 21-2. Normal Serum Cholesterol Values[a]

Age (yr)	Upper Limit (mg/dl)
0–19	230
20–29	240
30–39	270
40–49	310
50–59	330

[a] Some clinicians maintain that, even though serum cholesterol values over 250 mg per dl are found in a large percentage of apparently healthy persons, these levels should not be considered normal (desirable).

Lipoproteins

A number of the proteins in serum are called *lipoproteins* because their function is to transport lipids [26, 29]. Without them, the lipids could not exist in a dissolved state in the blood. The four major classes of lipoproteins are chylomicrons, pre-beta lipoproteins, beta lipoproteins, and alpha lipoproteins. This classification is based on paper electrophoresis, which separates the lipoproteins into these four groups. They can be separated also by ultracentrifugation, and, although basically the same groups are observed, different names are applied. These alternate names as well as the lipid content of the lipoprotein fractions are summarized in Table 21-3.

Lipoproteins not only are named but also are measured in terms of the separation provided by electrophoresis [22] or ultracentrifugation [18]. These measurements are carried out principally in cases of increased concentrations of blood lipids (hyperlipidemia), which, of course, means that the lipoprotein concentrations also must be increased (hyperlipoproteinemia) since they involve these same lipids. Various combinations of lipoproteins can be increased in these disorders. Correlation of hyper-

Table 21-3. Lipoproteins

Categories Based on Paper Electrophoresis	Categories Based on Ultracentrifugation	Major Lipid Component[a]
Chylomicrons	Chylomicrons	Triglycerides
Pre-beta lipoproteins	Very low density lipoproteins (VLDL)	Triglycerides
Beta lipoproteins	Low density lipoproteins (LDL)	Cholesterol
Alpha lipoproteins	High density lipoproteins (HDL)	Phospholipid

[a] Each of the four categories of lipoproteins contains variable quantities of triglycerides, cholesterol, and phospholipids. The other common type of lipid in the blood, fatty acids, is transported mostly by albumin.

lipidemia, hyperlipoproteinemia, and clinical disorders is under intensive study [21, 24], but no universally acceptable explanations are available.

Hyperlipidemia currently is receiving considerable attention because it has been positively correlated with atherosclerotic cardiovascular disease, which has a high mortality and morbidity rate.

Physiological Significance of Lipid Concentrations

The concentration of *total lipids* in serum reflects the presence or absence of hyperlipidemia, but analysis of the various lipid fractions usually is indicated. Changes in *phospholipid* concentrations usually parallel those of cholesterol. *Fatty acid* concentrations have essentially the same significance as do triglycerides. The concentration of *triglycerides* is increased following a meal, maximum concentration usually being reached at approximately 4 hours. This is the lipid fraction that, when present in increased concentration, imparts a lactescence (milky appearance) to the serum. Increased concentrations of the other lipid fractions are not apparent in this way.

The concentrations of total cholesterol, phospholipids, and triglycerides in serum are elevated in nephrosis, diabetes mellitus, and hypothyroidism. However, the increases usually are not clinically diagnostic of these diseases.

The concentrations of all lipid fractions in the serum are increased in essential hyperlipidemia, a disease which may be familial or acquired.

Cholesterol exists in the human blood as a free sterol and also in an esterified or combined form. In some diseases, total cholesterol concentration is affected, while in others only the ester fraction is disturbed. Jaundice of the obstructive type usually is accompanied by an elevated total serum cholesterol with a normal ester fraction. Diabetes, hypothyroidism, and certain types of kidney disease are other disorders that may exhibit the same type of cholesterol disturbance. Low total cholesterol values with normal ester fractions are noted mainly in malnutrition and hyperthyroidism.

Total serum cholesterol concentrations may vary markedly in healthy individuals; changes in concentration of ±50 mg per dl over the course of a few hours have been noted [27].

Since the liver cells esterify cholesterol, infectious hepatitis, cirrhosis, and most types of liver damage are accompanied by a decreased percentage of cholesterol esters.

Only the ester *percentage* is significant, not the absolute ester concentration [32]. In the example on page 285, both patients have the same ester concentration and normal total cholesterol levels. However, when the percentages are calculated, it is clear that patient A has a normal percentage, while patient B probably has liver disease.

In recent years, interest in lipid metabolism has increased sharply, primarily as a result of the possible relationship between lipid metabo-

Patient	Total Cholesterol (mg/dl)	Free Cholesterol (mg/dl)	Esters (mg/dl)	Esters Percentage of Total
A	160	46	114	71
B	228	114	114	50

lism and atherosclerosis. It is probable that more intensive clinical research and improved analytical methods will enhance the importance of lipid measurements in clinical medicine.

References

1. Abell, L. L., Levy, B. B., Brodie, B. B., and Kendall, F. E.   A simplified method for the estimation of total cholesterol in serum and demonstration of its specificity. *J. Biol. Chem.* 195:357, 1952.
2. Baumann, E. J.   On the estimation of organic phosphorus. *J. Biol. Chem.* 59:667, 1924.
3. Bloor, W. R.   A method for the determination of fat in small amounts of blood. *J. Biol. Chem.* 17:377, 1914.
4. Bloor, W. R.   Biochemistry of the fats. *Chem. Rev.* 2:243, 1925.
5. Burke, R. W., Diamondstone, B. I., Velapoldi, R. A., and Menis, O.   Mechanisms of the Liebermann-Burchard and Zak color reactions for cholesterol. *Clin. Chem.* 20:794, 1974.
6. Carr, J. J., and Drekter, I. J.   Simplified rapid technic for the extraction and determination of serum cholesterol without saponification. *Clin. Chem.* 2:353, 1956.
7. Cheek, C. S., and Wease, D. F.   A summation technic for serum total lipids. *Clin. Chem.* 15:102, 1969.
8. Creech, B. G., and Sewell, B. W.   A rapid micro procedure for determination of free and esterified cholesterol in blood serum using silicic acid separation. *Anal. Biochem.* 3:119, 1962.
9. Fiske, C. H., and Subbarow, Y.   The colorimetric determination of phosphorus. *J. Biol. Chem.* 66:375, 1925.
10. Folch, J., Lees, M., and Sloane-Stanley, G. H.   A simple method for the isolation and purification of total lipids from animal tissues. *J. Biol. Chem.* 226:497, 1957.
11. Fosbrooke, A. S., and Rudd, B. T.   The determination of particulate fat in serum and its use in clinical studies. *Clin. Chim. Acta* 13:251, 1966.
12. Frings, C. S., Fendley, T. W., Dunn, R. T., and Queen, C. A.   Improved determination of total serum lipids by the sulfo-phospho-vanillin reaction. *Clin. Chem.* 18:673, 1972.
13. Galletti, F.   An improved colorimetric micromethod for the determination of serum glycerides. *Clin. Chim. Acta* 15:184, 1967.
14. Gottfried, S. P., and Rosenberg, B.   Improved manual spectrophotometric procedure for determination of serum triglycerides. *Clin. Chem.* 19:1077, 1973.
15. Henry, R. J., Cannon, D. C., and Winkelman, J. W.   *Clinical Chemistry Principles and Technics* (2nd ed.). New York: Harper & Row, 1974. P. 1441.

16. Henry, R. J., Cannon, D. C., and Winkelman, J. W.   *Clinical Chemistry Principles and Technics* (2nd ed.). New York: Harper & Row, 1974. P. 1449.
17. Henry, R. J., Cannon, D. C., and Winkelman, J. W.   *Clinical Chemistry Principles and Technics* (2nd ed.). New York: Harper & Row, 1974. P. 1456.
18. Henry, R. J., Cannon, D. C., and Winkelman, J. W.   *Clinical Chemistry Principles and Technics* (2nd ed.). New York: Harper & Row, 1974. P. 1509.
19. Jacobs, S. L., and Henry, R. J.   Studies on the gravimetric determination of serum lipids. *Clin. Chim. Acta* 7:270, 1962.
20. Keys, A., Mickelsen, O., Miller, E.v.O., Hayes, E. R., and Todd, R. L. The concentration of cholesterol in the blood serum of normal man and its relation to age. *J. Clin. Invest.* 29:1347, 1950.
21. Lees, R. S.   A progress report on lipoprotein phenotyping. *J. Lab. Clin. Med.* 82:529, 1973.
22. Lees, R. S., and Hatch, F. T.   Sharper separation of lipoprotein species by paper electrophoresis in albumin-containing buffer. *J. Lab. Clin. Med.* 61:518, 1963.
23. Levy, A. L.   Triglycerides by nonane extraction and colorimetry, manual and automated. *Ann. Clin. Lab. Sci.* 6:474, 1972.
24. Levy, R. I., Morganroth, J., and Rifkind, B. M.   Treatment of hyperlipidemia. *N. Engl. J. Med.* 290:1295, 1974.
25. Loeffler, H. H., and McDougald, C. H.   Estimation of cholesterol in serum by means of improved technics. *Am. J. Clin. Pathol.* 39:311, 1963.
26. Nelson, G. J. (Ed.).   *Blood Lipids and Lipoproteins: Quantitation, Composition and Metabolism.* New York: Wiley, 1972.
27. Peterson, J. E., Wilcox, A. A., Haley, M. I., and Keith, R. A.   Hourly variation in total serum cholesterol. *Circulation* 22:247, 1960.
28. Sardesai, V. M., and Manning, J. A.   The determination of triglycerides in plasma and tissues. *Clin. Chem.* 14:156, 1968.
29. Scanu, A. M., and Ritter, M. C.   The Proteins of Plasma Lipoproteins: Properties and Significance. In Sobotka, H., and Stewart, C. P. (Eds.), *Advances in Clinical Chemistry.* New York: Academic, 1973. Vol. 16.
30. Schmidt, F. H., and von Dahl, K.   Zur methode der enzymatischen neutral fettbestimmung in biologischem material zeitschrift. *Z. Klin. Chem. Klin. Biochem.* 6:156, 1968.
31. Schoenheimer, R., and Sperry, W. M.   A micromethod for the determination of free and combined cholesterol. *J. Biol. Chem.* 106:745, 1934.
32. Sperry, W. M.   The relationship between total and free cholesterol in human blood serum. *J. Biol. Chem.* 114:125, 1936.
33. Stoddard, J. L., and Drury, P. E.   A titration method for blood fat. *J. Biol. Chem.* 84:741, 1929.
34. Van Handel, E., and Zilversmit, D. B.   Micromethod for the direct determination of serum triglycerides. *J. Lab. Clin. Med.* 50:152, 1957.
35. Zilversmit, D. B., and Davis, A. K.   Microdetermination of plasma phospholipids by trichloroacetic acid precipitation. *J. Lab. Clin. Med.* 35:155, 1950.
36. Zlatkis, A., Zak, B., and Boyle, A. J.   A new method for the direct determination of serum cholesterol. *J. Lab. Clin. Med.* 41:486, 1953.

22. Hormones

The endocrine (ductless) glands include the pituitary, thyroid, para-thyroid, adrenals, pancreatic islets of Langerhans, ovaries, testes, and placenta. These glands secrete active chemical substances called *hormones* which are transported by the blood to other tissues where they exert specific metabolic regulatory effects. The hormones of the human body are numerous and their chemistry complex. A small amount of the more pertinent information is presented in this chapter; for more exhaustive accounts of the chemistry and methodology of hormones the reader is referred to appropriate textbooks [2–5, 14, 16, 26].

Structurally the hormones are polypeptides or proteins, aromatic amines, or steroids. The polypeptide or protein hormones are elaborated by the pituitary, parathyroids, and pancreatic islets and are quantitated by bioassay or immunochemical procedures. The hormonal aromatic amines—thyroxine and triiodothyronine—are secreted by the thyroid gland. Methods for determining the concentration of these hormones in the blood are given in Chapter 23.

Steroids

Steroids are elaborated by the ovaries, testes, placenta, and adrenal cortex. Structurally they possess in common a cyclopentanoperhydro-phenanthrene nucleus. This structure consists of 4 rings (labeled A through D) and 17 carbon atoms numbered as shown below:

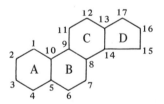

The activity of a steroid is related to the number and location of side chains and oxygen substituents.

Based on the number of carbons, the steroid hormones fall into three categories:

18 carbons: estrogens (female sex hormones)
19 carbons: androgens (male sex hormones)
21 carbons: corticosteroids and progestins

The three most abundant and important *estrogens* are estrone, estra-diol-17-β, and estriol. Chemically, they are characterized by a phenolic A-ring and a hydroxy or keto group at C-17 (carbon number 17). They are responsible for the development and maintenance of the female sex organs and secondary sex characteristics. The ovary in the female is the primary site of estrogen formation except in pregnancy, when the placenta becomes the predominant organ for estrogen production.

The two most important *androgens* are testosterone and dehydro-epiandrosterone. Testosterone is secreted chiefly by the testes of the male and is responsible for the development and maintenance of male secondary sex characteristics. Dehydroepiandrosterone is secreted by the adrenal gland and is thought to be a testosterone precursor. Like the estrogens, the androgens have either a hydroxyl or a keto group at C-17, but they have also an extra CH_3 (methyl) group at C-10, giving them 19 instead of 18 carbon atoms.

Aldosterone, cortisol, and corticosterone are the three most important *corticosteroids*. All three are secreted by the adrenal gland. Aldosterone accounts for at least 95% of all mineralocorticoid activity (regulation of Na and K levels), and cortisol accounts for about 90% of all gluco-corticoid activity (increased blood glucose). Corticosterone provides a small percentage of both activities. Chemically, the corticosteroids are characterized by a $-COCH_2OH$ side chain at C-17, plus an -OH (hydroxyl) group at C-11. The structure of cortisol is shown on page 291.

Progesterone is a female pregnancy hormone. It has a $-COCH_3$ group at C-17 and is one of the precursors in the body of both corticosteroids and testosterone. In nonpregnant women it is secreted mainly by the ovaries while in pregnant women, mainly by the placenta. It prepares the uterus for pregnancy and helps maintain the conditions favorable for continuance of pregnancy.

Most commonly, the steroid hormones are measured in groups to screen for adrenal, pituitary, or sex-gland abnormalities. For instance, the urine tests for 17-hydroxycorticosteroids and 17-ketogenic steroids both measure primarily cortisol and its metabolites, with the latter test measuring a larger portion of these metabolites. Similarly, the urine test for 17-ketosteroids measures the androgens. Although these types of methods remain popular because they allow the relatively simple measurement of the total output of a class of steroid, technics such as gas chromatography (Chapter 7) and competitive binding assays (Chapter 8) are rapidly providing practical methods for measurement of individual steroids.

17-Ketosteroids

Methods for the determination of 17-ketosteroids (17-KS) usually involve hydrolysis, extraction, purification, and a colorimetric reaction [6, 7]. Numerous methods have been developed in an attempt to increase the efficiency and specificity of the determination. The method presented here is one of several satisfactory technics that employ the general principles stated above.

Total 17-Ketosteroid Method

METHOD
Kraushaer et al. [13].

MATERIAL
Urine.

REAGENTS
Sodium hydroxide, 50% (w/w). Obtain commercially or prepare as described on page 11.

Hyamine 1622 solution, 2.5%. Dissolve 25 g of Hyamine 1622 (Rohm and Haas) in water and dilute to 1 liter.

Meta-dinitrobenzene solution. Dissolve 0.18 g of purified *meta*-dinitrobenzene in 100 ml of 2.5% Hyamine 1622 solution. Store refrigerated and renew monthly.

Potassium hydroxide 10N. Dissolve 280 g of KOH in water and dilute to 500 ml. Store in a plastic bottle.

Formalin solution, 10%. Dilute 10 ml of formalin to 100 ml with water.

Dichloromethane (methylene chloride).

Methanol.

Hydrochloric acid (concd).

PROCEDURE
Collect a complete 24-hour sample of urine, storing the container in a refrigerator during the collection period if possible. Measure the total volume.

Measure 5 ml of urine, 0.1 ml of formalin solution, and 0.5 ml of concentrated hydrochloric acid into a 25 × 150 mm tube (Test).

Mix. Then heat in a boiling water bath for 15 minutes.

Cool. Then add 20 ml of dichloromethane and shake for 20 minutes.

Centrifuge for 2 minutes. Then suction off the urine (top layer) and discard.

Add 10 ml of 50% sodium hydroxide and shake for 2 minutes. Centrifuge for 2 minutes.

Pipet 5 ml of the dichloromethane extract into a 16 × 125 mm tube and evaporate to dryness at 60 C.

Add 0.5 ml of methanol.

Add 0.5 ml of methanol to another 16 × 125 mm tube (Blank).

Add 0.5 ml of working standard to another 16 × 125 mm tube (Standard).

Add 0.2 ml of *meta*-dinitrobenzene solution to each tube.

Add 1 ml of 10N potassium hydroxide and mix. Let stand for 10 minutes. Then add 1.5 ml of 2.5% Hyamine 1622 solution and mix.

Read the absorbances of the Test and Standard in 12 mm cuvets in a spectrophotometer at 525 nm against the Blank set at zero.

CALCULATIONS

$$\text{absorbance of Test} \times \frac{\text{concentration of Standard (16 mg/liter)}}{\text{absorbance of Standard}}$$

$$\times \frac{\text{24-h urine volume (ml)}}{1000} = 17\text{-KS (mg/d)}$$

STANDARDIZATION

Stock Standard, 1 mg/ml. Dissolve 100 mg of dehydroepiandrosterone in 100 ml of methanol. Store in a refrigerator.

Working Standard, 0.04 mg/ml. Dilute 1 ml of the stock standard to 25 ml with methanol. Store in a refrigerator.

In this method the hormone is extracted from 5 ml of urine into 20 ml of solvent, but only 5 ml of the extract are used in the subsequent analysis. Therefore, the amount of hormone in the final reaction tube is from

$$\frac{5}{20} \times 5 = 1.25 \text{ ml of urine}$$

The 0.5 ml of working standard used in the method contains 0.02 mg of ketosteroid. Hence the value of this standard relative to the test is

$$\frac{0.02 \text{ mg}}{1.25 \text{ ml}} \quad \text{or} \quad 16 \text{ mg/liter}$$

NORMAL VALUES

Healthy men excrete approximately 10 to 25 mg of 17-ketosteroids per 24 hours, while women excrete 5 to 15 mg per 24 hours.

DISCUSSION

The principal compounds measured in the method are dehydroepiandrosterone, androsterone, and etiocholanolone. The 17-ketosteroids are excreted largely in a conjugated form bound to sulfate or glucuronic acid. Heating with acid splits the conjugate so that the free steroids may be measured. The acid concentration and heating time are important since insufficient hydrolysis results in incomplete splitting of the conjugates while conditions that are too vigorous may produce other chromogens which will interfere with the subsequent determination.

The steroids are extracted into dichloromethane to remove them from other interfering compounds in the urine. However, a small amount of water and other undesirable substances also go into the dichloromethane. Shaking with sodium hydroxide removes the interfering phenolic compounds (estrogens).

When ketosteroids are treated with *meta*-dinitrobenzene and potassium hydroxide, a reddish purple dinitrobenzene complex is formed that has an absorption maximum at 520 nm; this is the Zimmerman reaction [27]. It can be illustrated as follows for the reactive part of 17-ketosteroids:

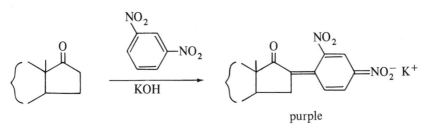

purple

Hyamine 1622 is a detergent which is used to solubilize the *meta*-dinitrobenzene and the colored product of the Zimmerman reaction.

All steps involving dichloromethane should be carried out in a fume hood since this substance can be toxic if inhaled.

PHYSIOLOGICAL SIGNIFICANCE OF 17-KETOSTEROID CONCENTRATIONS
In women virtually all of the 17-ketosteroids come from the adrenal cortex, whereas in men approximately two-thirds come from the adrenal cortex and the remainder from the testes. Decreased excretion of 17-ketosteroids usually accompanies hypopituitarism, hypogonadism, and hypoadrenalism (Addison's disease). With congenital adrenal hyperplasia and virilizing adrenocortical tumors the 17-ketosteroid excretion usually is increased.

17-Hydroxycorticosteroids
The corticosteroids are 21-carbon steroids secreted by the adrenal cortex. In terms of biological activity the three most important corticosteroids are aldosterone, cortisol, and corticosterone. Cortisol has the following structure:

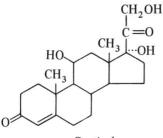

Cortisol

Cortisol and related corticosteroids (and their metabolites) frequently are referred to as 17-hydroxycorticosteroids (17-OHCS). Methods for the measurement of 17-hydroxycorticosteroids involve hydrolysis, extraction, purification, and a colorimetric reaction. Although several methods are available, none is considered specific. Some methods require enzymatic hydrolysis [20] (to remove sulfate or glucuronate groups) while others employ mineral acid [18] for this purpose.

One indirect method [15] recommends the oxidation of 17-OHCS to 17-ketosteroids (17-KS) and subsequent measurement of the 17-KS by the Zimmerman reaction. The preformed 17-KS are also measured. When this measurement is subtracted from the total, a value called the 17-ketogenic steroids (17-KGS) is obtained. The 17-hydroxycorticosteroid method presented here is relatively simple but gives results comparable to those obtained with more complex procedures.

TOTAL 17-HYDROXYCORTICOSTEROID METHOD

METHOD
Kornel [12].

MATERIAL
Urine.

REAGENTS

Sulfuric acid 62% (v/v). To 380 ml of water in a Pyrex flask add 620 ml of concentrated sulfuric acid slowly, with mixing and intermittent cooling. Cool to room temperature. Then dilute to 1 liter with water.

Phenylhydrazine-sulfuric acid reagent. Dissolve 125 mg of purified phenylhydrazine crystals in 200 ml of 62% sulfuric acid. Store in a tightly stoppered bottle in a refrigerator.

Chloroform-butanol solvent. To 1 liter of chloroform add 100 ml of *n*-butanol and mix well.

Sulfuric acid, 5N. Dilute 14 ml of concentrated sulfuric acid to 100 ml with water.

Ammonium sulfate.

PROCEDURE

Collect a 24-hour urine sample, storing the container in a refrigerator during the collection period if possible. Measure the total volume.

Measure 3 ml of urine and 0.1 ml of 5N sulfuric acid into a 25 × 150 mm screw-capped tube. (If the correct amount of 5N sulfuric acid is added, the resultant pH will be 1. Check this for each new preparation of 5N sulfuric acid.)

Add a scoop (3 g) of ammonium sulfate and mix for 1 minute. Some crystals should remain undissolved. If not, add more ammonium sulfate and mix again.

Add 30 ml of chloroform-butanol solvent and shake for 15 seconds.

Centrifuge for 5 minutes. Then discard the top (aqueous) layer by suctioning it off with a blunt needle or glass tube connected to an aspirator trap.

Measure two 10-ml portions of the remaining solvent layer into two 16 × 150 mm tubes (Test Blank and Test).

Measure two 10-ml portions of the chloroform-butanol solvent into two 16 × 150 mm tubes (Reagent Blank and Reagent).

Measure two 9.9-ml portions of chloroform-butanol solvent and two 0.1-ml portions of the working standard into two 16 × 150 mm tubes (Standard Blank and Standard).

To each Blank (Test Blank, Reagent Blank, and Standard Blank) add 2 ml of 62% sulfuric acid and mix.

To the Test, Reagent, and Standard add 2 ml of phenylhydrazine-sulfuric acid reagent and mix.

Centrifuge all tubes for 5 minutes. Then carefully pipet the top layers into 10 mm cuvets.

Heat the cuvets at 55–60 C for 45 minutes.

Read the absorbances of the Reagent, Standard, and Test in a spectrophotometer at 410 nm against their respective Blank tubes set at zero.

CALCULATIONS

There should be absorbance readings for the Reagent, Standard, and Test. Subtract the Reagent reading from the Test and subtract the Reagent reading from the Standard to obtain corrected Test and Standard readings. Then, letting A_x = corrected Test absorbance and A_s = corrected Standard absorbance, substitute these values in the following formula:

$$A_x \times \frac{16.6}{A_s} \times \frac{\text{24-h urine volume (ml)}}{1000} = \text{17-OHCS (mg/d)}$$

STANDARDIZATION

Stock Standard. Dissolve 100 mg of cortisone acetate in *n*-butanol and dilute to 100 ml with the same solvent. Store in a refrigerator. The resultant concentration of cortisone is 83.3 mg per dl.

Working Standard. Dilute 2 ml of the stock standard to 10 ml with *n*-butanol and store in a refrigerator. The concentration of cortisone is 16.6 mg per dl.

In this method the hormones are extracted from 3 ml of urine into 30 ml of solvent, but only 10 ml of the extract is used in the subsequent analysis. Therefore, the reaction tube contains

$$\frac{10}{30} \times 3 = 1.0 \text{ ml of urine}$$

The 0.1 ml of working standard used in the method contains 0.0166 mg of 17-OHCS. Hence the value of this standard relative to the test is

$$\frac{0.0166 \text{ mg}}{1.0 \text{ ml}} \quad \text{or} \quad 16.6 \text{ mg/liter}$$

NORMAL VALUES

Healthy adults usually excrete between 1 and 10 mg of 17-hydroxy-corticosteroids per 24 hours in the urine.

DISCUSSION

The principal compounds measured with this method are cortisol and its metabolites. However, only about 50% of the urinary metabolites of cortisol retain the dihydroxyacetone side chain necessary for measurement with the phenylhydrazine reaction.

Acidification of the sample and addition of ammonium sulfate facilitate the extraction of total (free and conjugated) 17-OHCS into the chloroform-butanol solvent. The extraction is incomplete unless the urine is acidified to pH 1 and unless ammonium sulfate is used. The low pH enhances the organic solubility of the conjugated 17-OHCS by protonating and thereby quenching all or some of the charge of the conjugated groups (glucuronate or sulfate). The ammonium sulfate dissolves only in the water, making this phase much more ionic (full of ions). This reduces the solubility of the 17-OHCS, which are organic substances, so that they partition into the organic phase.

Some versions of this method use n-butanol instead of chloroform-butanol as the extracting solvent. This more polar extracting solvent is a source of greater interference problems from polar substances like ascorbic acid and sugars, which partly extract into the butanol along with the 17-OHCS, then later react with phenylhydrazine to give a yellow color. The presence of a small amount of chloroform in the butanol provides a solvent which is sufficiently nonpolar to reject these polar interfering substances, but not so nonpolar as to reject the conjugated 17-OHCS. Thus, the chloroform-butanol solvent constitutes a significant improvement of the method.

The reaction of phenylhydrazine with the dihydroxyacetone side chain of the 17-OHCS in sulfuric acid-alcohol solution to form a yellow color (absorption maximum at 410 nm) is known as the Porter-Silber reaction [17]. The naturally occurring, active corticosteroids possessing this side chain are cortisol and compound S (11-deoxycortisol). Other biologically active corticosteroids lack the hydroxy group at the 17 position and do not undergo the Porter-Silber reaction.

During the heating process (in acid) the conjugates are split (glucuronate and sulfate groups are released). A color reaction takes place between free 17-OHCS and phenylhydrazine, as illustrated below (for just the reactive part of a 17-OHCS): The color is quite stable and does not have to be measured immediately.

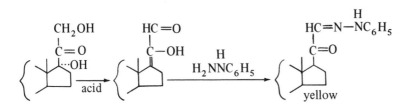

PHYSIOLOGICAL SIGNIFICANCE OF 17-HYDROXYCORTICOSTEROID CONCENTRATIONS

The concentration of corticosteroids in urine reflects the activity of the adrenal glands. With hyperadrenalism (Cushing's syndrome), abnormally large amounts of 17-OHCS appear in the urine. However, with hypoadrenalism (Addison's disease) diminished excretion of 17-OHCS may *not* be apparent. To test for primary hypoadrenalism 24-hour urine samples should be collected before and after administration of adrenocorticotropic hormone (ACTH). Normally, ACTH stimulates the adrenal glands, resulting in a large increase in excretion of 17-OHCS. However, with primary hypoadrenalism there is little or no change in 17-OHCS excretion following administration of ACTH.

Epinephrine and Norepinephrine (Catecholamines)

The hormones epinephrine (adrenaline) and norepinephrine (noradrenaline, arterenol) are secreted by the adrenal medulla. Since these compounds are amines and also have hydroxyl groups in ortho positions on the benzene ring (catechol), they are called *catecholamines*. The structures of these two compounds are as follows:

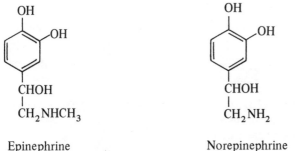

Epinephrine   Norepinephrine

Chemical methods for quantitative determination of catecholamines in urine are relatively crude but, since only large increases in excretion are of clinical significance, several of these methods are considered adequate. Most procedures involve adsorption of the catecholamines with aluminum oxide, washing, elution, and conversion to a fluorescent form [25]. One simple screening test [10] requires that the sample, so treated, be examined visually under ultraviolet light. More refined methods involve photofluorometric determination, and some require column chromatography. Potassium ferricyanide [21] and manganese dioxide [24] have been proposed as oxidizing agents.

The catecholamines are excreted both in a free form and in conjugation with glucuronic acid or sulfate. Usually the concentration of the conjugate in urine is approximately twice that of the free form. To determine both forms (total catecholamines), the sample first must be heated with acid to hydrolyze the conjugate. However, during this hydrolysis other chromogens usually are formed which interfere with the subsequent determination. Furthermore, when catecholamines are excreted in abnormally large amounts, they occur mostly in the free form. For these reasons, the hydrolysis step frequently is omitted.

A comprehensive review of the methodology and clinical significance of catecholamines has been published [9, 22]. The method described here employs column chromatography, oxidation with ferricyanide, formation of lutin, and fluorometric measurement.

CATECHOLAMINES METHOD

METHOD
Jacobs et al. [11].

MATERIAL
Urine.

REAGENTS

Sodium hydroxide, 5N. Dissolve 200 g of sodium hydroxide in water and dilute to 1 liter. Store in a plastic bottle.

EDTA reagent, 1% (w/v). Dissolve 10 g of disodium ethylenediamine tetraacetate in water and dilute to 1 liter.

Boric acid, 0.7M. Dissolve 43.3 g of boric acid (H_3BO_3) in water and dilute to 1 liter.

Zinc sulfate, 0.25% (w/v). Dissolve 0.25 g of zinc sulfate heptahydrate ($ZnSO_4 \cdot 7H_2O$) in water and dilute to 100 ml.

Potassium ferricyanide, 0.25% (w/v). Dissolve 0.25 g of potassium ferricyanide, $K_3Fe(CN)_6$, in water and dilute to 100 ml. Store in a refrigerator and renew each week.

Phosphate buffer, 0.2M, pH 6.5. Dissolve 9.03 g of sodium phosphate, dibasic (Na_2HPO_4) and 18.55 g of potassium phosphate, monobasic (KH_2PO_4) in water and dilute to 1 liter. Adjust the pH to 6.5 $\pm$ 0.1 with HCl or NaOH if necessary.

Phosphate buffer, pH 7.0. Dissolve 71.0 g of dibasic sodium phosphate (Na_2HPO_4) and 43.3 g of monobasic potassium phosphate (KH_2PO_4) in water and dilute to 1 liter. Adjust the pH to 7.0 $\pm$ 0.1 with HCl or NaOH if necessary.

Resin Preparation. Suspend 1 lb of Bio-Rex 70[1] (50–100 mesh, sodium form) ion-exchange resin in water, allow it to settle briefly and suction off the supernatant cloudy water containing the fines. Repeat this procedure until the supernatant is clear. Decant the water and add a volume of phosphate buffer, pH 6.5, equal to the volume of the settled resin. Mix, decant the buffer, then repeat this washing procedure. Add buffer a third time and allow to stand overnight. Mix. Then check to see that the pH is 6.5 ± 0.1. Store the resin in buffer.

Ascorbic acid.

Chromatography columns. Plastic chromatography columns, 10 mm I.D. (no. 003),[2] with funnels (no. 00E3).[2]

PROCEDURE

To a 50 ml beaker add 5 ml of urine (Test).

To another beaker add 5 ml of working standard (Standard).

Add 14 ml of EDTA to each beaker and adjust the pH to 6.5 ± 0.1 using 0.5N NaOH.

Add resin to chromatography columns to a height of 35 mm.

Pour the Test and Standard through the columns.

After all the solutions have run through, fill the reservoirs of the columns with water and allow it to run through.

Place each column in a 16 × 150 mm tube, add 7.5 ml of boric acid to the columns, and collect the eluates.

Mix each Test eluate. Then pipet duplicate 3.5 ml amounts into 16 × 150 mm tubes (Test Blank and Test).

Mix each Standard eluate. Then pipet duplicate 3.5 ml amounts into 16 × 150 mm tubes (Standard Blank and Standard).

Prepare alkaline ascorbate reagent just before use as follows:

1. Dissolve 0.1 g of ascorbic acid in 5 ml of water.
2. Mix 2 ml of the ascorbic acid solution with 9 ml of 5N NaOH.

To all the eluates add 1 ml of phosphate buffer, pH 7.0, and mix.

To the Test Blank and Standard Blank add 1 ml of alkaline ascorbate reagent. Mix and allow to stand for 2 minutes.

To *all* tubes add 0.2 ml of zinc sulfate and mix.

Add 0.2 ml of ferricyanide, mix, and allow to stand for 2 minutes.

To the Test and Standard, add 1 ml of alkaline ascorbate reagent and mix.

Transfer to 10 mm fluorometry cuvets.

Using an excitation (primary) wavelength of 400 nm and an emission (secondary) wavelength of 500 nm, set the reading of the Standard at 100 (full scale). Read the fluorescence of the Standard Blank, the Test, and the Test Blank.

[1] Bio-Rad Laboratories, Richmond, California.
[2] Whale Scientific, Denver, Colorado.

If a Test reads off scale, set its reading at 100. Then reread the Test Blank, Standard Blank, and Standard and use these readings to calculate this particular test.

CALCULATIONS

$$\frac{200}{S - SB} \times (T - TB) = \text{catecholamines } (\mu\text{g/liter})$$

$$\frac{\mu\text{g}}{\text{liter}} \times \frac{\text{24-h volume (ml)}}{1000} = \text{mg/d}$$

where: S = fluorescence of the Standard
SB = fluorescence of the Standard Blank
T = fluorescence of the Test
TB = fluorescence of the Test Blank

NORMAL VALUES
Healthy individuals usually excrete less than 150 mg of catecholamines in the urine per 24 hours.

STANDARDIZATION
Stock Standard Norepinephrine, 100 mg/liter. Dissolve 19.9 mg of levarterenol bitartrate (monohydrate) in 100 ml of 0.1N hydrochloric acid and store in a refrigerator.

Working Standard, 200 μg/liter. Dilute 0.1 ml of the stock standard to 50 ml with water. This solution is stable for approximately one week in a refrigerator.

The molecular weight of levarterenol (norepinephrine) bitartrate monohydrate is 337 while the weight of norepinephrine is 169. Therefore, 10 mg of this hormone are contained in

$$\frac{337}{169} \times 10 \quad \text{or} \quad 19.9 \text{ mg of the bitartrate}$$

The stock standard is diluted 1:500. So the concentration of the working standard is

$$\frac{1}{500} \times 100 = 0.2 \text{ mg/liter} \quad \text{or} \quad 200 \ \mu\text{g/liter}$$

DISCUSSION
As presented here, this method determines only *free* catecholamines. Total catecholamines may be determined if the sample is acidified to pH 2 with hydrochloric acid and heated in a boiling-water bath for 20 minutes before alkalinizing it. However, for reasons presented on page 296, this procedure is not recommended.

The resin column should be packed just before use. Once the procedure is begun, the solutions should be added in rapid succession, and the column should not be allowed to run dry.

The pH of the urine is adjusted to 6.5 because at this pH catecholamines are adsorbed by resin. The amino groups of the catecholamines are protonated at this pH. This renders these groups positively charged ($-NH_3^+$ or $-NH_2^+CH_3$). Binding of the catecholamines to the resin occurs because this resin contains negatively charged, carboxylate groups ($-CO_2^-$), that is, because of ionic binding of the positively charged catecholamines to the negatively charged resin. The catecholamines are somewhat labile at pH 6.5; so the procedure should be carried out as rapidly as possible.

When the column is washed with water, the excess buffer and some of the undesirable urinary constituents are removed. Boric acid elutes the catecholamines from the resin.

For fluorometric analysis, 2 tubes are set up and the eluate in each is buffered to pH 7.0. When the alkaline solution is added to the blank, the catecholamines are rapidly destroyed. The zinc sulfate and ferricyanide solutions in the test oxidize the catecholamines to adrenochrome. Alkalinization converts the adrenochrome to a fluorescent form called *adrenolutin*. Inclusion of ascorbic acid in the alkaline reagent retards oxidation long enough to make possible fluorometric readings.

For this method, the maximum excitation (primary) wavelength is approximately 400 nm, while the maximum emission (secondary) wavelength is approximately 500 nm. The fluorometric method is quite sensitive, but not specific; fluorescent drugs such as quinine and quinidine may give falsely high values [19].

PHYSIOLOGICAL SIGNIFICANCE OF
CATECHOLAMINE CONCENTRATIONS

Measurement of urinary catecholamines is used almost exclusively as an aid in the diagnosis of pheochromocytoma, which is a tumor of the chromaffin cells of the adrenal medulla characterized by greatly increased excretion of catecholamines. The importance of this test lies in the fact that it is a reliable guide to the diagnosis of one of the few curable forms of hypertension. Although slightly increased catecholamine excretion may occur with psychiatric disorders, pheochromocytoma usually results in excretions of 500 μg or more per 24 hours. The method described here is free of interference from the usual drugs used in the treatment of hypertension. However, the drug alpha-methyl-3,4-dihydroxyphenylalanine (alpha-methyl-dopa or Aldomet) is a catecholamine and therefore is measured with this method. The usual dosages of this drug result in very high urinary catecholamine values [8], but the drug is not measured with VMA methods, a discussion of which follows.

Vanilmandelic Acid (VMA)

Vanilmandelic acid (3-methoxy-4-hydroxymandelic acid; vanillylman-delic acid; VMA) is the major urinary metabolite of the catecholamines epinephrine and norepinephrine. Its structure is as follows:

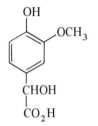

Vanilmandelic Acid

A variety of methods have been used to measure VMA. None of them is simple because various interfering substances in urine must be elimi-nated. These interfering substances may be eliminated by extraction or chromatographic steps prior to a colorimetric reaction for VMA. For instance, a common method [23] is to extract acidified urine with an organic solvent, wash this VMA-containing extract with aqueous base, oxidize the VMA to vanillin with periodate, and then measure the vanillin spectrophotometrically after extracting it into another organic solvent.

The method given here uses a combination of organic extraction and thin layer chromatography to isolate VMA prior to a colorimetric re-action with a diazonium dye.

VANILMANDELIC ACID (VMA) METHOD

METHOD
Annino et al. [1].

MATERIAL
A portion of a 24-hour urine specimen which has been collected and combined with 20 ml of 6N [50% (v/v)] HCl.

REAGENTS
Developing solvent. Mix 90 ml of butanol, 20 ml of glacial acetic acid, and 40 ml of water. Renew every 2 weeks or sooner if used daily.

Borate buffer, 0.1M. Dissolve 38.1 g of sodium borate ($Na_2B_4O_7 \cdot 10H_2O$) in 950 ml of water. Adjust the pH to 10.2 ± 0.1 using 50% NaOH (requires approximately 8 ml). Dilute to 1 liter with water.

Eluting solution. Mix 15 ml of borate buffer with 10 ml of methanol. Prepare just before use.

Saturated potassium carbonate. Add an excess of potassium carbonate to water.

Concentrated hydrochloric acid.

Florisil.

Absolute ethanol.

Silica gel G.

Stock dye solution. Dissolve 140 mg of *p*-nitroaniline in 0.5*N* hydrochloric acid and dilute to 100 ml with the acid. Store in a refrigerator.

Sodium nitrite 1M. Dissolve 6.9 g of sodium nitrite ($NaNO_2$) in water and dilute to 100 ml. Store in a refrigerator.

Working dye solution. To 2.5 ml of stock dye solution add 1*M* sodium nitrite until the solution is colorless (2–3 drops).
Bring the dye solution to pH 2.5 ± 0.1 with 1*N* NaOH. Then dilute to 50 ml with water. This solution should be prepared fresh before each use.

APPARATUS
Glass plates (200 × 200 mm) and applicator system for coating them with silica gel.
Tank for developing these plates.
Flat-edged spatula.
Air blower (warm air).

PROCEDURE
Preparation of plates.[3] To a 250 ml Erlenmeyer flask add 40 g of silica gel G and 80 ml of water. Stopper *immediately* and shake vigorously for one minute. Pour into an applicator and apply the slurry onto a series of 200 × 200 mm glass plates mounted on an applicator board. Use an even motion. The applicator setting should be 0.35 mm to give a layer of silica gel of this thickness. Allow the plates to air dry overnight. Then store them in a cabinet.

Preparation of sample.
Into a 25 × 150 mm tube, measure 25 ml of urine and 2.5 ml of concd HCl. Swirl to mix.
Place in a boiling water bath for 10 minutes.
Cool to room temperature, add 2 g (4.5 ml) of Florisil, stopper, and shake for 10 minutes.
Centrifuge for 5 minutes.
To another 25 × 150 mm tube add 4 ml of the treated urine and 10 ml of ethyl acetate.
Stopper. Shake for 2 minutes. Then centrifuge for 2 minutes.
Transfer the upper layer to a 50 ml wide-mouthed centrifuge tube and

---

[3] Commercially prepared plates may be used (Q4 Silica gel plates no. 2019, Quantum Industries, Fairfield, N.J.). However, separations are not as good as for the plates described here and development time is slower.

place the tube in a 60 C water bath under a stream of air to facilitate
evaporation.

To the urine, add another 10 ml of ethyl acetate and extract as before.
Add this extract to the first one.

When the extracts have evaporated to dryness, remove the tubes from
the bath and add 0.4 ml of *absolute* ethanol down the sides of the
tube.

Mix well to redissolve the VMA. Then transfer the solution to a 10 mm
tube, leaving behind any residue. Cap the tube and store in a refrigerator until chromatography is to be performed.

Chromatography.

Line the chromatography tank with filter paper. Then pour in the developing solvent, wetting the paper. Keep covered at all times.

With a pencil, score a line into the gel on a plate 11.5 cm from one edge.

Using a syringe pipet, apply 50 μl of sample 1.5 cm from the same edge.
The sample should be applied as a streak 2 cm long; use a template as
a guide. Each sample requires several applications. Use a heat gun to
dry between each application.

Apply 50 μl of the working standard in the same fashion. At least one
standard should be placed on each plate.

Place the plate in the tank, cover, and allow to develop until the solvent
reaches the line on the plate (about $1\frac{1}{4}$ hours).

Remove the plate from the tank and dry with a heat gun (in a hood) for
15 minutes. Do not place the heat gun closer than about 18 inches
from the plate.

Quantitation.

Spray the plate with the working dye solution. Dry with the air blower.

Spray with saturated potassium carbonate solution and dry again.

Outline the VMA (purple) spots (Test and Standard) and 1 extra spot
below the Standard to serve as a Blank. Scrape the excess gel from
around the spots. Then scrape the spots into 16 × 100 mm tubes.

To each tube add 2.5 ml of freshly prepared eluting solution.

Mix well with a vortex mixer to break up the gel.

Add 0.5 ml of dye solution and mix again.

Centrifuge for 5 minutes. Then decant into 10 mm cuvets. Within 20
minutes after the dye has been added, read the absorbances of the
Standard and Test in a spectrophotometer at 510 nm against the
Blank.

CALCULATIONS

$$\text{absorbance of Test} \times \frac{\text{concentration of Standard (11 mg/liter)}}{\text{absorbance of Standard}}$$

$$\times \frac{\text{24-h urine volume (ml)}}{1000} = \text{VMA (mg/d)}$$

STANDARDIZATION

Stock standard, 250 mg/liter. Dissolve 25 mg of 3-methoxy-4-hydroxy-mandelic acid (VMA)[4] in 100 ml of absolute ethanol. Store in a refrigerator.

Working standard. Dilute 4 ml of the stock standard to 10 ml with absolute ethanol. Store in a refrigerator and prepare fresh once a month.

The actual concentration of this standard is 100 mg per liter. However, as used in this method it is compared with a urine sample which has been concentrated 10 times (4 ml to a final volume of 0.4 ml). In addition, the urine was diluted slightly with HCl (25 ml + 2.5 ml) for hydrolysis. Therefore, the value of the standard in the method is

$$100 \times \frac{0.4}{4} \times \frac{27.5}{25} = 11 \text{ mg/liter}$$

NORMAL VALUES

In the absence of disease, normal adults excrete up to 10 mg of VMA per 24 hours in the urine.

DISCUSSION

It is important to collect the urine specimen in acid because VMA is labile in neutral or alkaline solutions. In untreated urine, VMA exists in both an ionized and an unionized form, that is, with its carboxyl group in a charged (-COO$^-$) and uncharged (-COOH) condition. The addition of acid to the sample converts all of the VMA into its unionized form, facilitating its extraction into the organic solvent, ethyl acetate. The acid serves also to keep in the aqueous phase certain interfering substances, as does the Florisil (magnesium silicate), which adsorbs some of these substances. Additional interfering substances are eliminated in the step in which the ethyl acetate extract is evaporated to dryness, followed by redissolving the VMA in absolute ethanol. The final purification is made with thin layer chromatography during which the VMA is isolated. Good sensitivity is achieved because the VMA is dissolved in only a small volume of ethanol and because a broad heavy band of this solution is applied to the thin layer plate.

The plate is sprayed with dye to locate the VMA spot. Its position should be verified with reference to a standard on the same plate. This spot should be pink to purple, depending on its intensity. Also, it should be outlined and the excess gel removed from the plate before the gel containing the VMA spot is quantitatively scraped into a tube.

When the gel is agitated with eluting solution, the VMA is redissolved. Addition of the dye solution produces full color development. Care must

[4] Available as a Standard Reference Material from the U.S. National Bureau of Standards.

be taken not to resuspend any gel when transferring the solution to a cuvet because the slight turbidity will cause spurious instrument readings.

PHYSIOLOGICAL SIGNIFICANCE OF VMA CONCENTRATIONS

The physiological significance of VMA is largely the same as that of epinephrine and norepinephrine. Its excretion is increased in pheochromocytoma. (See Physiological Significance of Catecholamine Concentrations in the preceding section on Epinephrine and Norepinephrine in this chapter.) Since it is the principle urinary metabolite of these catecholamines, it serves as a sensitive measure of their levels in the body.

The analysis for VMA therefore often is used instead of, or in conjunction with, urinary catecholamines. The major advantage of measuring VMA is that its concentration in urine is significantly higher than that of total catecholamines. False increased results may be obtained after ingestion of coffee, tea, chocolate, vanilla, or bananas.

References

1. Annino, J. S., Lipson, M., and Williams, L. A.   Determination of 3-methoxy-4-hydroxymandelic acid (VMA) in urine by thin-layer chromatography. *Clin. Chem.* 11:905, 1965.
2. Berson, S. A.   *Methods in Investigative and Diagnostic Endocrinology.* New York: American Elsevier, 1972, 1973. Vols. 1, 2.
3. Butt, W. R.   *Hormone Chemistry.* London: Van Nostrand, 1967.
4. Dale, S. L.   *Principles of Steroid Analysis.* Philadelphia: Lea & Febiger, 1967.
5. Dorfman, I.   *Methods in Hormone Research.* New York: Academic, 1962. 2 vols.
6. Drekter, I. J., Heisler, A., Scism, G. R., Stern, S., Pearson, S., and McGavack, T. H.   The determination of urinary steroids: I. The preparation of pigment-free extracts and a simplified procedure for the estimation of total 17-ketosteroids. *J. Clin. Endocrinol.* 12:55, 1952.
7. Drekter, I. J., Pearson, S., Bartczak, E., and McGavack, T. H.   A rapid method for the determination of total urinary 17-ketosteroids. *J. Clin. Endocrinol.* 7:795, 1947.
8. Gifford, R. W., Jr., and Tweed, D. C.   Spurious elevation of urinary catecholamines during therapy with alpha-methyl-dopa. *J.A.M.A.* 182:493, 1962.
9. Hermann, H., and Mornex, R.   *Human Tumors Secreting Catecholamines: Clinical and Physiological Study of Pheochromocytomas.* New York: Pergamon, 1964.
10. Hingerty, D.   Thirty-minute screening test for pheochromocytoma. *Lancet* 1:766, 1957.
11. Jacobs, S. L., Sobel, C., and Henry, R. J.   Specificity of the trihydroxyindole method for determination of urinary catecholamines. *J. Clin. Endocrinol. Metab.* 21:305, 1961.
12. Kormel, L.   Corticosteroids in human blood: I. Determination of free and conjugated 17-hydroxycorticosteroids in plasma. *J. Clin. Endocrinol. Metab.* 22:1079, 1962.
13. Kraushaer, L. A., Epstein, E., and Zak, B.   Characteristics of a 17-ketosteroid reaction. *Clin. Chem.* 12:282, 1966.

14. Loraine, J. A. *Hormone Assays and Their Clinical Application* (3rd ed.). Baltimore: Williams & Wilkins, 1971.
15. Norymberski, J. K., Stubbs, R. D., and West, H. F. Assessment of adrenocortical activity by assay of 17-ketogenic steroids in urine. *Lancet* 1:1276, 1953.
16. O'Malley, B. W., and Hardman, J. G. (Eds.). Hormone Action. In *Methods in Enzymology.* New York: Academic, 1974. Vols. 36–40.
17. Porter, C. C., and Silber, R. H. A quantitative color reaction for cortisone and related 17,21-dihydroxy-20-ketosteroids. *J. Biol. Chem.* 185: 201, 1950.
18. Reddy, W. J. Modification of the Reddy-Jenkins-Thorn method for the estimation of 17-hydroxycorticoids in urine. *Metabolism* 3:489, 1954.
19. Sax, S. M., Waxman, H. E., Aarons, J. H., and Lynch, H. J. Effect of orally administered quinine and quinidine on apparent values for urinary catecholamines. *Clin. Chem.* 6:168, 1960.
20. Silber, R. H., and Porter, C. C. The determination of 17,21-dihydroxy-20-ketosteroids in urine and plasma. *J. Biol. Chem.* 210:923, 1954.
21. Sobel, C., and Henry, R. J. Determination of catecholamines (adrenalin and noradrenaline) in urine and tissue. *Am. J. Clin. Pathol.* 27:240, 1957.
22. Straus, R., and Wurm, M. Catecholamines and the diagnosis of pheochromocytoma. *Am. J. Clin. Pathol.* 34:403, 1960.
23. Sunderman, F. W., Jr., Cleveland, P. D., Law, N. C., and Sunderman, F. W. A method for the determination of 3-methoxy-4-hydroxymandelic acid (vanillylmandelic acid) for the diagnosis of pheochromocytoma. *Am. J. Clin. Pathol.* 34:293, 1960.
24. Thomas, L. E. The chemical determination of catecholamines in the diagnosis of pheochromocytoma. *Am. J. Clin. Pathol.* 28:605, 1957.
25. Von Euler, U. S., and Floding, I. Diagnosis of pheochromocytoma by fluorimetric estimation of adrenalin and noradrenaline in urine. *Scand. J. Clin. Lab. Invest.* 8:288, 1956.
26. Williams, R. H. (Ed.). *Textbook of Endocrinology* (4th ed.). Philadelphia: Saunders, 1968.
27. Zimmerman, W. Eine Farbreaktion der Sexualhormone und ihre Anwendung zur quantitativen colorimetrischen Bestimmung. *Z. Physiol. Chem.* 233:257, 1935.

23. Thyroid Function Tests

The principal hormones secreted by the thyroid gland are thyroxine (tetraiodothyroxine or T4) and liothyronine (triiodothyronine or T3), both of which are amino acids. The terms T4 and T3 derive from the number of iodine atoms in each compound. These two hormones act similarly and are necessary for normal growth and maturation, although T3 is somewhat more active than T4.

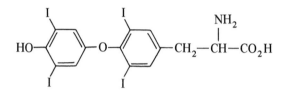

Tetraiodothyronine (Thyroxine)
(T4)

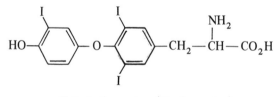

Triiodothyronine (Liothyronine)
(T3)

After release by the thyroid gland into the blood, these hormones are mostly (more than 99%) reversibly bound by plasma proteins, especially by the protein thyroxine-binding globulin (TBG). This protects them from rapid excretion by the kidneys. In each case, however, the active hormone is the small fraction which is free, that is, the fraction which circulates in the blood not bound to TBG or other proteins. Both T3 and T4 compete for the same binding sites on TBG.

Abnormal levels of T3 and T4 in plasma occur in numerous disease processes, most of which directly involve the thyroid gland. Measurements of T4 are more common, largely because it is present in the plasma in about 50 times higher concentration than T3 is. Therefore it is easier to measure accurately. Both the free and the total levels of these substances are of clinical interest. The levels of each depend on both the status of the thyroid gland and on the amount of TBG (thyroxine-binding globulin) in the blood.

306

In reviewing thyroid function methodology, it helps to recognize that testing for T3 and T4 is somewhat unique because of the iodine atoms on these hormones. These atoms facilitate measurement of T3 and T4 not only because they can be removed chemically as iodide ions (I^-), which are easily quantitated, but also because they can be displaced by radioactive iodine atoms to give radioactive T3 and T4, allowing measurement of the unlabeled hormones by isotopic technics. At the same time, the tests which measure T3 and T4 based on their iodine content suffer from a large number of interferences from iodine-containing materials (for example, iodine antiseptics, cosmetics, cough syrups, multivitamin preparations, certain drugs, diagnostic scanning agents, and radiographic contrast media). Most of these materials remain in the blood in interfering concentrations for at least a week. Some remain for as long as 30 years.

Another important consideration in some of the T3 and T4 tests has been the use of two readily available binding agents for these hormones. One is TBG, the natural plasma-binding protein for T3 and T4. The other is any of the strongly basic anion-exchange resins (for example, one of the Dowex-1 resins), which bind T3 and T4 quite strongly. (Basic groups such as $-N^+(CH_3)_3$ attached to these resins bind the $-CO_2^-$ group of T4 and T3.) TBG is used in assays of the competitive-binding type (Chapter 8), while the resins are used both in this manner and for separation purposes. In the latter case, T3 and T4 temporarily are bound by the resin while interfering substances are washed away.

Protein-Bound Iodine

The measurement of protein-bound iodine (PBI) was the principle diagnostic test of thyroid function for many years. This test measures protein-bound T3 and T4, which accounts for essentially all (more than 99%) of each of these hormones in the blood. Three basic steps are involved:

1. Serum proteins are precipitated with zinc hydroxide, trichloroacetic acid, or chloric acid. The precipitate is washed to remove the iodide ions (I^-) that normally are present in blood, as well as other iodine-containing materials (except the T3 and T4, of course, which are bound to TBG in the protein precipitates).

2. The precipitated proteins are digested by dry-ashing in a muffle furnace [2] or by wet digestion with a corrosive agent like chloric acid [1, 17]. The digestion liberates the iodine atoms of T3 and T4 as iodide ions (I^-).

3. Colorimetric measurement is made of the amount of iodide ions based on their ability to quantitatively catalyze the reduction of yellow ceric ions to colorless cerous ions by arsenite ions. This reaction occurs in two steps. Note that I^- is regenerated just like a true catalyst.

$$2I^- + 2Ce^{4+} \longrightarrow 2Ce^{3+} + I_2$$
$$I_2 + As^{3+} \longrightarrow As^{5+} + 2I^-$$

More complete details of this reaction may be found in the literature [7]. Although this test is analytically sound, it requires special working conditions and equipment, is susceptible to interferences from a wide variety of iodinated materials and mercurial compounds, and may be invalidated by increases or decreases in the level of TBG or in the relative amounts of T4 and T3 in the sample.

Sometimes a total iodine analysis is carried out along with the PBI. Normally there is a small amount of free iodine in the serum, so the total iodine value will be slightly higher (0–2 μg per dl) than the PBI. A difference greater than 2 μg per dl suggests the presence of inorganic iodine contamination and the validity of the PBI is questionable.

Butanol-Extractable Iodine and T4 by Column
To circumvent the interferences from the nonhormonal iodine-containing materials occasionally found in the blood, tests were developed for butanol-extractable iodine (BEI) and for separating T4 by column chromatography. Unfortunately, these methods did not solve all of the interference problems and generally have been less precise than the PBI determination. Both tests actually measure total T3 plus T4. The result approximates the T4 value because the T3 concentration is only a small fraction of the T4 concentration.

In the BEI method [9], the T3 and T4 are extracted from the serum into butanol. Then some of the interfering iodine-containing materials are reextracted from the butanol with aqueous alkali. The butanol is evaporated, and the subsequent analysis is similar to that for PBI. The butanol extraction eliminates interferences from inorganic iodides and iodoproteins but not from organic iodine compounds and mercurial diuretics.

In the method for T4 by column [11], the total T3 and T4 are extracted from the serum with aqueous alkali. They are then bound to an anion-exchange resin. After some of the iodine-containing contaminants are rinsed from the column, the T4 and T3 are eluted with acid. After digestion, the iodide is quantitated by the ceric-arsenite reaction.

T3 Uptake
The resin T3-uptake test [14] is used to determine the unused binding capacity of naturally occurring TBG. In this test, radioactive T3 (^{125}I-T3) is added to serum. Some of it binds to sites on TBG not already occupied by T3 and T4. After equilibration, the excess labeled T3 which is not bound is adsorbed by an ion-exchange resin fabricated as a sponge. The radioactivity in the sponge then is an inverse measure

of labeled T3 bound to TBG. Results of this test are reported in terms of the percentage of added T3 adsorbed by the resin, not that bound by TBG.

The T3-uptake value is an inverse measure of unused TBG and therefore is an inverse, indirect measure of TBG. This is because unused TBG tends to increase when TBG is elevated, which results in a decreased T3-uptake value. For instance, the TBG elevations caused by pregnancy or the pill, even though accompanied by a rise in T4, cause an increased unused TBG level and therefore a decreased T3-uptake value. A decreased T3-uptake value also tends to occur whenever TBG is constant but T4 decreases because this causes the unused binding capacity of TBG to be greater.

It is important to realize that this test does not measure T3 either directly or indirectly. It is called the T3-uptake test because labeled T3 is used as one of the key reagents.

One shortcoming of the method is that it is valid only between certain limits so that very high or low values are recorded only as being at the limits of the method.

The basic purpose of the test is to permit correction of T4 values for the effect of changes in TBG levels. One such application requires the use of the T3-uptake value to calculate an index, as discussed below.

Free Thyroxine Index

Not infrequently, an elevated T4 value is seen in a patient with normal thyroid function. This is especially true for women who are pregnant or taking birth-control pills. These and certain other conditions cause a primary elevation of TBG, the blood protein which binds T4. To compensate for the extra TBG, the thyroid secretes more T4 into the blood. The result is a normal level of free (unbound) T4, but the amount of TBG-bound T4, and therefore of total T4, is elevated. It is important to distinguish this type of total-T4 elevation from one involving a primary thyroid problem. Three ways of accomplishing this are: (1) determine free T4, which usually is normal as long as thyroid function is normal, or (2) determine a normalized T4 (discussed below), or (3) calculate the free thyroxine index [4] (also called *T4-T3 uptake index* or *T-7*).

Free thyroxine index = T3 uptake (percent) $\times$ T4 (μg/dl)

The free thyroxine index shows a reasonable correlation with free-T4 values and therefore also is helpful in determining whether the cause of an elevated T4 is a TBG elevation (TBG is not produced by the thyroid) or a primary thyroid problem. A primary TBG elevation causes a decreased T3-uptake value. However, T4 tends to increase in this case in order to keep free T4 constant. Since the free thyroxine index multiplies the increased T4 value by the decreased T3-uptake value, this index also will tend to stay constant. Although the free thyroxine index is clinically useful, it does require two separate analyses to calculate it.

Normalized T4 [12]

The normalized T4 (NT4), also called the "effective thyroxine ratio" (ETR) or "compensated T4" is another way, besides the free thyroxine index, to compensate for a primary elevation of TBG. In this technic, T4 is extracted from the serum with ethanol. This extracted T4 and an untreated portion of the same serum are added to a solution of radioactively labeled T4 bound to exogenous TBG (*T4 · TBG).[1] The extracted T4 competes with free *T4 for binding sites on the exogenous TBG. This tends to increase the amount of free *T4 in the mixture. Endogenous TBG (from the serum) tends to re-form *T4·TBG, thereby lowering the amount of free *T4 in the mixture. After a suitable incubation period free *T4 is separated from *T4·TBG and either the free or bound material is counted. Since both the T4 and the TBG from the serum react, the test is sensitive to concentrations of either or both. If the T4 alone is elevated it will occupy more binding sites on the exogenous TBG, leaving a greater amount of free *T4 in the mixture. If the endogenous TBG alone is elevated, it will bind more *T4, leaving less free *T4 in the mixture. If both T4 and TBG are increased, as happens in pregnancy, a normal amount of free *T4 will remain in the mixture.

Total T3

The only practical method for measuring circulating T3 is the radioimmunoassay technic [3], which involves the principles discussed in Chapter 8. It is difficult to achieve the same degree of accuracy with this test as that achieved with T4 because the concentration of circulating T3 is so much lower than that of T4. When thyroid function tests are requested, total T3 should not be confused with T3 uptake, which is discussed in a preceding section.

Free T4

Since the free form of T4 (not bound by proteins) is responsible for the physiological effects of that hormone, there is interest in the measurement of the small fraction (0.02–0.04% of the total) of T4 which is free in plasma. The equilibrium dialysis method [6] involves addition of radioactive T4 (*T4) and dialysis of the sample to separate the free T4 plus the free *T4 fraction (dialyzable fraction) from the protein-bound or nondialyzable fraction. The ratio of the dialyzable radioactive hormone to the total radioactive hormone reflects the percentage of free T4.

[1] The asterisk designates radioisotopically labeled material (see Chapter 8).

Multiplication of this value by the total T4 concentration (determined separately) gives the free T4 level.

Another method [8] involves addition of *T4 to the sample, adsorption of some of the free T4 and *T4 on Sephadex resin, washing the resin, then counting the *T4 adsorbed. When compared with appropriate standards, the percentage of free T4 may be calculated. Again, a total-T4 analysis is also required.

Because of the low levels of free T4 that are being measured and the double methodology involved, these methods usually are less accurate and more demanding than the total-T4 analysis.

Free T4 concentration is independent of rising levels of TBG and, therefore, is not influenced by oral contraceptives or estrogens. However, free T4 levels can increase in serious illness where there is a fall in TBG concentration. Also free T4 levels can be increased by the administration of heparin.

Thyroxine-Binding Globulin

If thyroxine-binding globulin (TBG) is measured in the clinical laboratory, usually an indirect method [16] is used. The sample is equilibrated with radioactive T4 so as to saturate the TBG. After separation of the free and bound forms, determination of the amount of labeled T4 that is bound gives an indirect measure of TBG concentration. Under normal circumstances, approximately 75% of the T4 in circulating plasma is bound to TBG while the rest, except for the small amount of free T4, is bound to albumin and prealbumin.

Thyroid-Stimulating Hormone

The thyroid-stimulating hormone (TSH) is a glycoprotein hormone, secreted by the anterior pituitary which stimulates the synthesis and secretion of T4 and T3 by the thyroid. Secretion of TSH is regulated by a negative feedback mechanism. Hence, if the serum-T4 level falls, the TSH level increases to restore the T4 level to normal. Conversely, increased serum-T4 levels result in suppression of TSH levels, sometimes to undetectable amounts. This reciprocal relationship between TSH and thyroid hormone renders the TSH a valuable test in the diagnosis of primary hypothyroidism. This is because the increased TSH levels may be measured more accurately than decreased levels of T4 or T3.

Usually TSH is measured by a radioimmunoassay technic [5, 15], the principles of which are discussed in Chapter 8.

Thyroxine (T4)

Thyroxine-binding globulin (TBG) can be used as the binding protein in a competitive binding assay for thyroxine. This technic, which was developed by Murphy and Pattee, is widely used and is essentially free of interferences. Other names for this test besides *T4* are *Total-T4, T4-by-displacement, T4 by competitive-protein binding, Murphy-Pattee thyroxine,* and *Thyroxine.* After the T4 is extracted, with ethanol, from the serum, it is allowed to compete with radioactive T4 (*T4) for binding to TBG. Free and bound *T4 then are separated by addition of an anion-exchange resin, which binds the free but not the bound T4 and *T4. Subsequently, the radioactivity in either the free or bound fraction (or both) may be counted. The method is presented below.

T4 (THYROXINE)

METHOD
Murphy [10].

MATERIAL
Serum.

REAGENTS

Barbital buffer, pH 8.6. Dissolve 14.7 g of barbital and 82.2 g of sodium barbital in water and dilute to 4 liters.

TBG-*T4 solution. To a 500 ml volumetric flask add 4.0 ml of propylene glycol, 100 μl of ^{125}I-T4, and 15 ml of normal pooled serum. Dilute to 500 ml with barbital buffer, pH 8.6. Store refrigerated.

Resin treatment. Suspend 2 lb of rexyn (202, mixture of Cl and SO_4 forms)[2] in 3 liters of barbital buffer in a 4-liter beaker. Stir for 30 minutes with a magnetic stirrer. Let settle and decant. Repeat the washing process two additional times. Spread the resin in aluminum pans and dry overnight at 105 C.

Phosphate buffer, 0.1M, pH 7.6. Dissolve 13.92 g of dibasic potassium phosphate (K_2HPO_4); 2.76 g of monobasic sodium phosphate hydrate ($NaH_2PO_4 \cdot H_2O$); and 8.76 g of NaCl in approximately 900 ml of water. Adjust the pH to 7.6 $\pm$ 0.05 with 1% (v/v) phosphoric acid or 0.1N NaOH. Dilute to 1 liter with water and mix well. Check the pH and adjust if necessary. Store at 4 C.

PROCEDURE

Measure 1 ml of 95% ethanol into each of a series of 13 $\times$ 100 mm glass tubes.

While mixing each tube on a vortex mixer, slowly add 0.5 ml of serum or standard to each tube (Tests and Standards).

Centrifuge for 10 minutes at 2500 rpm.

Measure 300 μl of each clear supernatant into 16 $\times$ 150 mm counting tubes (in duplicate).

[2] Fisher Scientific Co., Pittsburgh, Pa.

Evaporate the contents of all tubes to dryness in a 45 C bath under a stream of air.

Label 2 additional tubes for total counts (TC).

To all tubes, including TC, add 1 ml of the TBG-*T4 solution and mix on a shaker for 1 minute.

Incubate at 45 C for 10 minutes, mixing by hand 3 or 4 times during this period.

Let stand at room temperature for 10 minutes.

Place in an ice bath for 30 minutes.

As rapidly as possible, add 0.4 ml of dry resin to all tubes except TC.

Place the tubes with resin on a shaker for 1 minute.

Return to the ice bath for 5 minutes.

Add 10 ml of *cold* water (forcefully) to all tubes except TC.

Allow the resin to settle. Then aspirate the supernatant to approximately 10 mm above the resin, taking care *not* to remove any resin.

Repeat the washing and aspiration steps twice more.

Count the radioactivity in all tubes in a gamma counter for 60.0 seconds.

CALCULATIONS

The concentrations of the Standards are 0, 2.5, 5.0, 7.5, 10.0, 12.0, and 15.0 μg per dl. Plot these against their respective counts on linear graph paper. Then determine the values of the Tests from the standard curve. A typical curve is shown in Figure 23-1.

If the T4-iodine value is desired, multiply the T4 value by 0.65.

STANDARDIZATION

To make the treatment and recovery of the thyroxine standards similar to the serum samples, the thyroxine standards are dissolved in an albumin-phosphate-saline solution and are processed through the entire procedure.

Standard diluent. Dilute 100 ml of phosphate buffer (see p. 312) with 800 ml of water. Add 7.9 g of NaCl and 10 g of crystalline bovine albumin (fraction V). Adjust the pH to 7.4 $\pm$ 0.05 with 0.3% phosphoric acid or 0.1N NaOH. Dilute to 1 liter with water. Recheck the pH and adjust if necessary. Store at 4 C.

Thyroxine stock standard, 1 mg/ml. Add 28.5 mg of L-thyroxine (sodium salt, pentahydrate) and 2.5 ml of propylene glycol to a 25 ml volumetric flask. Dissolve by adding 0.1M NaOH in 2 ml amounts, with mixing, until the solution is clear. Dilute to 25 ml with water and mix well. Transfer 2.5 ml amounts to polystyrene tubes, cap, and store frozen.

Thyroxine intermediate standard, 1 mg/dl. Dilute 1.0 ml of thyroxine stock standard to 100 ml with standard diluent. Measure 7 ml portions into polystyrene tubes, cap, and store frozen.

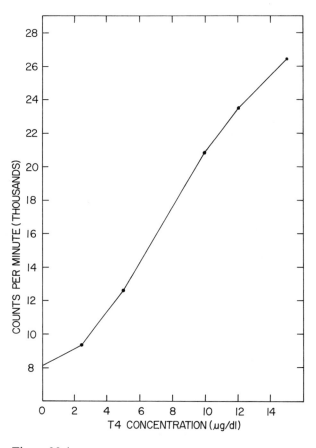

Figure 23-1
Typical Standard T4 Curve

Thyroxine working standards. Dilute 0.25, 0.50, 0.75, 1.00, 1.20, and
1.50 ml of thyroxine intermediate standard to 100 ml with standard
diluent. These standards are 2.5, 5, 7.5, 10, 12, and 15 µg per dl
respectively. The standard diluent itself serves as a zero standard.
Store at 4 C.

An entire set of working standards, including a zero standard, should
accompany each set of analyses. A calibration curve is constructed by
plotting the concentration of each standard against its respective cpm
on linear graph paper. A typical standard curve is shown in Figure 23-1.
The purpose of the propylene glycol is to increase the solubility and
stability of the T4. All standards should be kept dark and cold as much
as possible. They should be allowed to warm to room temperature before

use. Shaking should be avoided since it increases adsorption of T4 to glass.

NORMAL VALUES

Normal serum thyroxine levels fall into the approximate range 4.0 to 11.0 μg per dl. The corresponding range, expressed as T4-iodine, is 2.6 to 7.1 μg per dl. A precision of approximately ± 0.5 μg per dl may be expected with this method.

DISCUSSION

The basic principles of this type of assay are presented in Chapter 8. In this particular case, since T4 is largely bound by some of the proteins (TBG, prealbumin, and albumin) in the untreated serum sample, it is necessary to liberate the T4 and separate it from these proteins prior to addition of known amounts of a T4-binding agent and radioactive T4 for the competitive-assay reaction. This is accomplished by mixing the serum sample with ethanol, which extracts the T4 and simultaneously precipitates the proteins. It is advisable to process the standards through this ethanol precipitation step because the recovery of T4 is only about 80%. However, the standards do not compensate for all of the variation in T4 recovery because some of this variation derives from the individual nature of each serum sample.

The ethanol is removed by evaporation. The dried thyroxine residue then is dissolved in a solution of TBG from normal pooled serum to which radioactive T4 has been added. Although the normal pooled serum as a source of the TBG also contains the T4-binding substances prealbumin and albumin, the strongest binding occurs with TBG. Binding with prealbumin and albumin are minimized through dilution and the use of the barbital buffer.

The T4 from the sample and the added radioactive T4 (*T4) compete for binding sites on the added TBG. The more T4 that is present, the more it, rather than *T4, binds to the TBG. A determination of the amount of *T4 bound by TBG (or a reciprocal determination of the amount of unbound or free *T4) therefore provides a measurement of the amount of T4 present in the sample, as discussed in general in Chapter 8.

Separation of the free and the TBG-bound *T4 is accomplished by the addition of an anion-exchange resin. This insoluble agent pulls the free *T4 (and the unlabeled free T4) out of solution, leaving the TBG-bound *T4 (and unlabeled, TBG-bound T4) in solution. In this method, the solution containing the TBG-bound material is removed by aspiration and washing. The labeled, resin-bound *T4 is counted after the resin is washed. This provides a measure of the amount of T4 in the sample.

The exact duration of the shaking step with the resin is important. Excessive shaking will allow some of the TBG-bound T4 to be taken up by the resin, while insufficient shaking will not allow complete uptake of free *T4 by the resin.

The TC tube provides a measure of the amount of radioactivity added to all tubes. This value should be reasonably constant from day to day (except for the normal decay of *T4). A large change indicates a measurement, technical, or reagent problem.

If the counts for the zero standard are divided by the total counts (TC tube) and this fraction is multiplied by 100, the resultant value is the *percent binding*. This is a measure of the amount of binding which takes place between the TBG and the *T4 when no T4-containing sample is present. The percent binding should be approximately 80%. If the value is significantly different, the integrity of the TBG is suspect.

The method is unaffected by interferences from inorganic and organic iodine compounds as well as mercurial diuretics. Severely hemolyzed samples may give falsely high values. Diphenylhydantoin (Dilantin) and salicylates can cause low values.

Physiological Significance of Thyroid Function Tests

The thyroid gland is not essential for life because its hormones can be taken orally, but, with reduced activity (low T3 and T4 values), called *hypothyroidism,* there is susceptibility to cold, mental and physical slowing, and, in children, the mental and physical slowing of cretinism. Conversely, excess thyroid secretion (high T3 and T4 levels) is called *hyperthyroidism* and results in body wasting, nervousness, tachycardia, tremor, exophthalmos (protruding eyeballs), and excess heat production. The state produced by excessive levels of endogenous (originating within) or exogenous (originating without, that is, ingested) thyroid hormones is called thyrotoxicosis (for example, T3 toxicosis, T4 toxicosis). Normal secretion of thyroid hormones is referred to as euthyroidism.

Goiter is a thyroid disease characterized by an enlarged thyroid gland. The usual cause is an iodine deficiency from an inadequate dietary intake of iodine. The ingestion of substances which inhibit the synthesis of thyroid hormones also causes goiter. These substances may be drugs or naturally occurring inhibitors (goitrogens) as found in cabbage and cauliflower. Goiter also may be associated with hyperthyroidism, for instance, Grave's disease.

Serum thyroxine levels are raised in women who receive oral contraceptives or other estrogen preparations because these substances increase thyroxine-binding globulin (TBG). Conversely, serum thyroxine levels are reduced in acute severe prostrating (exhausting) illnesses because of low TBG levels.

Table 23-1 summarizes normal values, interferences, and the clinical significance of results of some of the thyroid function tests discussed in this chapter. Further information about the use of thyroid function tests is available [13].

Table 23-1. Information on Thyroid Function Tests

Test	Normals	Hypothyroidism			Hyperthyroidism		Interferences			
		Primary	Genetic	Secondary	T4	T3	Organic I	Inorganic I	Mercury	Estrogens
T4	4.0–11.0 µg/dl	−[a]	−	−	+[a]	0[a]	0	0	0	+
Free T4	1.0–2.2 ng/dl	−	−	−	+	0	0	0	0	0
Total T3	60–200 ng/dl	−	−	−	+/0	+	0	0	0	+
TBG	15–25 µg/dl (T4)	0	0	0	0	0	0	0	0	+
T3 Uptake	25–35%	−	−	−	+	+	0	0	0	−
TSH	Up to 10 µU/ml	+	+	−	−	−	0	0	0	0
PBI	3.5–8.0 µg/dl	−	−	−	+	0	+	+	−	+
BEI	3.2–6.4 µg/dl	−	−	−	+	0	+	0	−	+

[a] Key: 0 = no change; − = decrease; + = increase.

References

1. Bodansky, O., Buena, R. S., and Pennacchia, G.   A rapid procedure for determination of total and protein-bound iodine. *Am. J. Clin. Pathol.* 30:375, 1958.
2. Brown, H., Reingold, A. M., and Samson, M.   The determination of protein-bound iodine by dry ashing. *J. Clin. Endocrinol.* 13:444, 1953.
3. Burday, S. Z., and Blum, M.   Clinical and laboratory observations in cases of triiodothyronine toxicosis confirmed by radioimmunoassay. *Lancet* 1:609, 1972.
4. Clark, F., and Hron, D. B.   Assessment of thyroid function by the combined use of the serum protein-bound iodine and resin uptake of 131-triiodothyronine. *J. Clin. Endocrinol.* 25:39, 1965.
5. Hall, R.   The immunoassay of thyroid-stimulating hormone and its clinical applicaiton. *Clin. Endocrinol.* 1:115, 1972.
6. Ingbar, S. H., Braverman, L. E., Dawber, N. A., and Lee, G. Y.   A new method for measuring the free thyroid hormone in human serum and an analysis of the factors that influence its concentration. *J. Clin. Invest.* 44:1679, 1965.
7. Lein, A., and Schwartz, N.   Ceric sulfate-arsenious acid reaction in microdetermination of iodine. *Anal. Chem.* 23:1507, 1951.
8. Levinson, S. S., and Rieder, S. V.   Parameters affecting a rapid method in which Sephadex is used to determine the percentage of free thyroxine in serum. *Clin. Chem.* 20:1568, 1974.
9. Man, E. B., Kydd, D. M., and Peters, J. P.   Butanol-extractable iodine of serum. *J. Clin. Invest.* 30:531, 1951.
10. Murphy, B. E. P., and Pattee, C. J.   Determination of thyroxine utilizing the property of protein-binding. *J. Clin. Endocrinol.* 24:187, 1964.
11. Passen, S., and von Saleski, R.   A semiautomated method which does not require digestion for the determination of serum thyroxine iodine. *Am. J. Clin. Pathol.* 51:166, 1969.
12. Plaut, D.   Timely topics in clinical chemistry: In vitro thyroid testing. *Am. J. Med. Technol.* 40:434, 1974.
13. Schneider, P. B.   Laboratory examination of the thyroid: I. *In vitro* tests. *Postgrad. Med.* 56:91, 1974.
14. Sisson, J. C.   Principles of, and pitfalls in, thyroid function tests. *J. Nucl. Med.* 6:853, 1965.
15. Spierto, F. W., and Sman, B. T_3 Solid phase radioimmunoassay. *Clin. Chem.* 20:631, 1974.
16. Wahner, H. W., and Walsen, A. H.   Measurements of thyroxine-plasma protein interactions. *Med. Clin. North Am.* 56:849, 1972.
17. Zak, B., Willard, H. H., Myers, G. B., and Boyle, A. J.   Chloric acid method for determination of protein-bound iodine. *Anal. Chem.* 24:1345, 1952.

24. Vitamins

Animals require not only carbohydrate, fat, protein, inorganic salts, and water to live. They need vitamins also. These substances originally were thought to be amines, that is, *vital amines,* thus the term *vitamins.* Although many of them are not amines, they all are vital constituents of the diet and organic in nature. Only small amounts of vitamins are required. The vitamins usually are divided into two major groups, the fat-soluble vitamins (A, D, E, and K) and the water-soluble vitamins (B series and C). The water-soluble ones function primarily as coenzymes (enzyme helpers) or coenzyme precursors in the body. A few vitamins of common interest in clinical chemistry are discussed in this chapter. More comprehensive accounts of this subject are available [2, 8].

Vitamin A and Carotene

Neither vitamin A nor carotene is a single substance. Vitamin A and β-carotene, shown below, are typical members of the vitamin A and carotene groups of compounds, respectively.

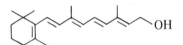

Vitamin A (Retinol; Vitamin A Alcohol)

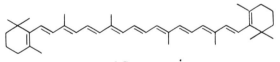

β-Carotene

Carotene is orange colored and occurs naturally in plants along with related compounds, the xanthophylls. Together, these two groups of substances comprise the yellow carotenoids, which give carrots and yellow vegetables their color. When these fat-soluble substances are eaten by animals, they either are stored as such (animal fat always is more or less yellow because of the carotenoids which are present) or converted to vitamin A. For instance, symmetrical, hydrolytic cleavage of β-carotene at the middle double bond affords two molecules of vitamin A. The two effective sources of vitamin A for man are plant foods, which provide carotene, and animal foods, which provide carotene along with preformed vitamin A.

Vitamin A and carotene usually are measured together because of their similar chemical and biological properties, although carotene itself functions only as a vitamin precursor and not as a vitamin. Most analytic methods take advantage of the fat solubility of vitamin A and carotene

by extracting them from serum with petroleum ether. The orange carotene usually is measured directly at this point using a spectrophotometer
[3]. Actually what is measured and reported as carotene is total carotenoids. The nearly colorless vitamin A generally is determined next by
formation of a blue or violet color with antimony trichloride (Carr-
Price reaction) [4], glycerol dichlorohydrin (1,3-dichloro-2-propanol)
(Sobel and Snow reaction) or by the loss in absorbance at 326 nm after
irradiation with long wavelength ultraviolet light, which destroys vitamin
A (Bessey reaction) [3]. A disadvantage of the Carr-Price and Sobel
and Snow reactions is that the blue or violet color has limited stability
and is sensitive to traces of moisture.

VITAMIN A AND CAROTENE METHODS

METHOD
Bessey, et al. [3].

MATERIAL
Serum.

REAGENTS
Absolute ethanol.
Heptane.

PROCEDURE FOR CAROTENE
To a 16 × 125 mm screw-cap tube add 2 ml of serum, 2 ml of absolute
    ethanol, and 5 ml of heptane.
Shake for 5 minutes on a laboratory shaker (fast speed). Then centri-
    fuge for 10 minutes at 2000 rpm.
Transfer approximately 3 ml of the heptane layer to a 10 mm cuvet
    and read the absorbance in a spectrophotometer against heptane at
    450 nm.

CALCULATIONS FOR CAROTENE

$$\text{absorbance of Test} \times \frac{\text{concentration of Standard } (100 \ \mu g/dl)}{\text{absorbance of Standard}}$$
$$= \text{carotene } (\mu g/dl)$$

STANDARDIZATION FOR CAROTENE
Carotene stock standard, 8 μg/ml. Dissolve 40 mg of β-carotene in 500
    ml of petroleum ether. Dilute 5 ml of this solution to 50 ml with
    heptane. Store in a freezer.

Carotene working standard, 40 μg/dl. Dilute 1 ml of stock standard to
    20 ml with heptane. Store in a refrigerator and renew each week.

The absorbance of this standard is read directly at 450 nm against
    heptane. The absolute concentration of the standard is 40 μg/dl.

However, since 2 ml of serum is extracted with 5 ml of heptane, the value of the standard as used in the method is

$$40 \times \frac{5}{2} = 100 \ \mu g/dl$$

This figure is used in the calculation formula.

PROCEDURE FOR VITAMIN A

Read the absorbance (A_1) of the heptane layer from the carotene method at 326 nm against heptane in a 10 mm *silica* cuvet.

Stopper the cuvet and expose it to long-wavelength ultraviolet light (for example, 6 inches away from a Blak-Ray long-wavelength ultraviolet lamp) for one hour.

After one hour, read the absorbance again at 326 nm (A_2).

CALCULATIONS FOR VITAMIN A

$$(A_1 - A_2) \times \frac{\text{concentration of Standard } (125 \ \mu g/dl)}{\text{absorbance of Standard}}$$
$$= \text{vitamin A } (\mu g/dl)$$

STANDARDIZATION FOR VITAMIN A

Vitamin A stock standard, 10 $\mu g/ml$. Dissolve 50 mg of crystalline vitamin A in absolute ethanol and dilute to 100 ml with the ethanol. Dilute 1 ml of this solution to 50 ml with heptane. Store in a freezer.

Vitamin A working standard, 50 $\mu g/dl$. Dilute 1 ml of stock standard to 20 ml with heptane. Store in a refrigerator and renew each week.

The absorbance of this standard is read at 326 nm before and 1 hour after exposure to ultraviolet (UV) irradiation without prior treatment.

The absolute concentration of the standard is 50 μg per dl. However, since 2 ml of serum is extracted with 5 ml of heptane, the value of the standard in the method is

$$50 \times \frac{5}{2} = 125 \ \mu g/dl$$

NORMAL VALUES (VITAMIN A AND CAROTENE)

The normal carotene level (actually total carotenoids) in serum is 25 to 250 μg per dl, while the normal vitamin A level is 15 to 60 μg per dl. Lower values occur in infants.

DISCUSSION (VITAMIN A AND CAROTENE)

Since the standards are not processed through the precipitation and extraction it is advisable to employ a serum control for carotene and vitamin A as a check on these steps.

Vitamin A and carotene are extracted into heptane to remove them from other interfering substances in the serum. Because of their water insolubility, these substances normally bind to the serum proteins. Ethanol is added to break up these complexes and permit vitamin A and carotene to partition into the heptane. Because vitamin A is nearly colorless, while carotene is orange, the absorbance reading at 450 nm measures only the carotene.

Vitamin A is destroyed by ultraviolet light. Therefore the change in absorbance at 326 nm, after UV irradiation, is a measure of the loss of vitamin A.

Samples and standards of both substances should be protected from light. This may be accomplished by keeping them in a refrigerator or cabinet until analyzed. Serum samples may be kept 4 days at room temperature or 2 weeks in a refrigerator with no change in the vitamin A and carotene values, as long as the samples are protected from light.

The necessary irradiation time can be determined with a given lamp by irradiating samples and standards of vitamin A to constant absorbances at 326 nm. It is recommended that the irradiation time be 6 to 8 times the period necessary to destroy 50% of the vitamin A in a standard solution.

PHYSIOLOGICAL SIGNIFICANCE OF
VITAMIN A AND CAROTENE CONCENTRATIONS

Vitamin A and carotene occur in the diet dissolved in other lipids. After they are released from these lipids by the action of lipase (a lipid-hydrolyzing enzyme from the pancreas) and bile (a detergent from the liver) in the small intestine, they are absorbed through the intestinal walls into the blood. These water-insoluble substances then bind to serum proteins, which transport them to the liver. Carotene is converted to vitamin A in the liver and released in small amounts into the blood (as a lipoprotein complex once again) for use by the rest of the body.

Vitamin A deficiencies occur either because of a deficient diet or because of some interference with the intestinal absorption, blood transport, or liver storage of the vitamin.

Generally carotene and vitamin A have the same clinical significance, but sometimes the levels do not correlate. For instance, a dietary excess of only one of them will tend to elevate that particular substance. Also a failure to convert carotene to vitamin A will result in an elevated carotene but lowered vitamin A level.

Vitamin A also may be requested in conjunction with a vitamin A tolerance test. In this procedure, the patient is given a dose of vitamin A in oil by mouth, and the subsequent blood level of the vitamin is monitored. Low levels indicate that the patient's ability to digest and absorb lipids is impaired at some point. This is a typical finding in cystic fibrosis.

Vitamin C (Ascorbic Acid)
Vitamin C, or ascorbic acid, is a carbohydrate-like water-soluble vitamin.

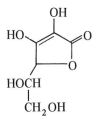

Some animals make vitamin C *in vivo* (for example, rats, dogs, rabbits, and birds) and some animals do not (for example, man, apes, monkeys, and guinea pigs). The animals that do not synthesize vitamin C must get it in their diet. The chief dietary sources of vitamin C are fresh fruit, green vegetables, and vitamin C tablets.

Analytic methods for vitamin C generally take advantage either of its reducing power (ease of oxidation) or of the keto products that result when it is oxidized (dehydroascorbic acid and diketogulonic acid). Thus, vitamin C may be determined with 2,6-dichlorophenolindophenol [5] because vitamin C reduces this colored substance to a colorless derivative. Alternatively, vitamin C may be heated in acid in the presence or absence of metal salts to form keto products which, in turn, form colored derivatives upon reaction with 2,4-dinitrophenylhydrazine [9]. The method presented here is of the latter type.

Vitamin C Method

METHOD
Roe and Kuether [9].

MATERIAL
Serum, plasma, whole blood, or urine.

REAGENTS
Trichloroacetic acid, 6% (w/v). Dissolve 12 g of trichloroacetic acid in water and dilute to 200 ml.

DNP reagent. To 75 ml of water, slowly add 25 ml of concd sulfuric acid with swirling and cooling. Dissolve 2 g of 2,4-dinitrophenylhydrazine (DNP) in this solution and filter if necessary. Store in a refrigerator.

Sulfuric acid 85% (v/v). Dilute 170 ml of concd sulfuric acid to 200 ml with water. (*Caution!* The acid must be added slowly and with swirling and cooling.)

Thiourea. Mix 10 ml of absolute ethanol with 10 ml of water. Dissolve 2 g of thiourea in this solution.

PROCEDURE

To 12 ml of 6% trichloroacetic acid in a 16 × 150 mm tube, add 3 ml of serum, whole blood, or urine with shaking.

Mix (5 seconds on a vortex mixer) until a fine white suspension is obtained.

Centrifuge for 5 minutes, and decant the supernatant into a clean tube of the same size.

Add a spatula tip (approximately 130 mg) of Norit (for example, Norit A, activated charcoal powder, acid-washed Norit) and mix thoroughly.

Filter carefully using an 11 cm, no. 40 filter paper, so that no Norit comes through.

Measure 4 ml of the filtrate into each of 2 tubes (Blank and Test).

Add 1 drop of thiourea solution to each tube.

To the Test, add 1 ml of DNP reagent and mix for 5 seconds on a vortex mixer.

Incubate all tubes for 3 h at 37 C.

At the end of the incubation, place all tubes in an ice bath.

To the Blank (in the ice bath) add 5 ml of 85% sulfuric acid slowly and with mixing. Then add 1 ml of DNP reagent and mix.

To the Test, add 5 ml of 85% sulfuric acid similarly.

Allow all tubes to stand for 30 minutes.

Read the absorbance of each Test against its corresponding Blank in 19 mm cuvets at 540 nm.

CALCULATIONS

Serum.

$$\text{absorbance of Test} \times \frac{\text{concentration of Standard (2 mg/dl)}}{\text{absorbance of Standard}}$$

$$= \text{vitamin C (mg/dl)}$$

Urine.

$$\text{absorbance of Test} \times \frac{\text{concentration of Standard (2 mg/dl)}}{\text{absorbance of Standard}}$$

$$\times \frac{\text{24-h urine volume (ml)}}{1000} = \text{Vitamin C (mg/d)}$$

STANDARDIZATION

Both Test and Blank tubes for standards should be included with each set of tests. These solutions are introduced into the procedure after the filtration step; that is, they are not treated with Norit, but simply measured in 4 ml amounts for treatment with thiourea, etc.

Vitamin C stock standard, 100 mg/dl. Dissolve 25 mg of ascorbic acid in 25 ml of 6% trichloroacetic acid. Prepare fresh each day.

Vitamin C working standard, 0.4 mg/dl. Dilute 0.1 ml of stock standard
to 25 ml with 6% trichloroacetic acid. Prepare fresh each day.

The absolute concentration of the working standard is 0.4 mg per dl.
However, since the sample is diluted 1:5 (3:15) while the standard
is not, the value of the standard in this method is 5 × 0.4 = 2.0 mg
per dl. For urine this value is multiplied by 10 to convert to milligrams
per liter, then by the 24-hour volume (ml) divided by 1000 to give
milligrams per day:

$$\text{mg/dl} \times 10 \times \frac{\text{total volume (ml)}}{1000} = \text{mg/d}$$

NORMAL VALUES

Normal serum, plasma, and whole blood values for ascorbic acid are
approximately 0.5 to 2.0 mg per dl. Urinary excretion of vitamin A
corresponds to intake.

DISCUSSION

A protein-free filtrate of the sample is prepared with trichloroacetic acid.
Acid-washed charcoal is added to a portion of the acidic supernatant to
clarify the solution and to facilitate the rapid oxidation of ascorbic acid
to dehydroascorbic acid. This product then reacts with 2,4-dinitrophenyl-
hydrazine to form 2,4-dinitrophenylosazone of dehydroascorbic acid,
which, in turn, forms a red chromophore when treated with sulfuric acid.

The purpose of the thiourea, a mild reducing agent, is to protect
against oxidizing interferences from substances like iron.

Since untreated charcoal is a source of oxidizing contaminants, acid-
washed charcoal (commercially available) is used, because it is free of
many of these contaminants. The addition of sulfuric acid to the tests
must be carried out without any appreciable rise in temperature (hence,
the ice bath) to avoid analogous oxidative side reactions which may
contribute an interfering color.

If there is to be a delay before the analysis, samples should have a
preservative added to maintain the ascorbic acid concentration. Oxalic
acid crystals (20 mg/ml) may be used for this purpose, but sometimes
a gel forms when they are added to serum, making accurate pipetting
difficult. A few crystals of potassium or sodium cyanide is an effective
preservative, but the poisonous nature of the cyanides warrants extra
caution.

PHYSIOLOGICAL SIGNIFICANCE OF VITAMIN C CONCENTRATIONS

Vitamin C deficiency leads to scurvy, a disease characterized by general
weakness, loosening of the teeth, many small hemorrhages from capillary defects, and brittle bones. Although no particular tissue or organ
seems to store vitamin C, deficiency symptoms take about 4 months to

appear after ascorbic acid is removed from the diet. Vitamin C levels decrease during pregnancy.

Vitamin B₁₂

The vitamin B_{12} molecule is distinguished by the fact that it contains a cobalt atom. Absorption of vitamin B_{12} from the intestine requires that it be bound to a special protein called *intrinsic factor,* which is secreted by the gastric parietal cells of the stomach. A deficiency of intrinsic factor results in a deficiency of vitamin B_{12} and subsequently causes pernicious anemia. General malabsorption diseases also can cause vitamin B_{12} deficiency.

One type of method for quantitation of vitamin B_{12} in serum depends on the influence of the vitamin on the growth of the microorganism, *Euglena gracilis* [1]. In the proper medium, these bacteria will multiply in proportion to the concentration of vitamin B_{12} present. After a suitable incubation period, the concentration of bacteria is estimated by measuring the turbidity of the solution. Calculations are made with reference to appropriate standards.

The competitive binding technic, as explained in Chapter 8, is employed also for the measurement of vitamin B_{12}. In this technic [7], intrinsic factor is used as the binding protein, ^{57}Co-vitamin B_{12} is employed as the tracer, and hemoglobin-coated charcoal is used to separate the free and bound forms.

If the microbiological assay is used, care must be taken to prevent bacterial contamination of specimens since this will result in spuriously high results. In addition, falsely low values may result if antibiotics are present in the patient's serum, which will inhibit the growth of *E. gracilis.*

Normal values for vitamin B_{12} are 150–1000 pg per ml. In pernicious anemia and some neurological diseases values less than 100 pg per ml usually are found. Increased levels may occur in hepatitis, hepatic necrosis, chronic granulocytic leukemia, erythroleukemia, and polycythemia vera.

Folic Acid

The term *folic acid,* or *folate,* refers collectively to several folate compounds present in the blood. One such compound, pteroylglutamic acid, frequently referred to as folic acid, is composed of pterin, *p*-aminobenzoic acid, and L-glutamic acid groups.

Folic acid is a vitamin necessary for the production of red cells. In this regard, its function is similar to that of vitamin B_{12}.

The concentration of folic acid in serum may be measured by a microbiological assay similar to that outlined in the preceding section for vitamin B_{12}. In the folic acid assay the microorganism *Lactobacillis casei* [6] is used.

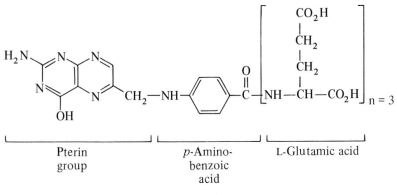

| Pterin group | p-Amino-benzoic acid | L-Glutamic acid |

Folic Acid

Competitive binding technics, as explained in Chapter 8, also are employed for measurement of folic acid. In these technics [11], a milk protein can be used as the binder, tritiated folic acid as the tracer, and hemoglobin-coated charcoal is used to separate the free and bound forms.

If the microbiological assay is used, care must be taken to prevent bacterial contamination of specimens since this will result in falsely high results. In addition, spuriously low values may result if antibiotics which inhibit the growth of *L. casei* are present in the patient's serum.

Normal values for folic acid in serum are 5–16 ng per ml. Folic acid levels may be decreased in pernicious anemia, intestinal malabsorption, and chronic alcoholism. Folate deficiency is the most common vitamin deficiency and measurement of the folate level is the most sensitive parameter for detecting this disorder.

Since folic acid is a common constituent of many vitamin preparations, high folate values may be due to ingestion of vitamins by the patient.

References

1. Adams, J. F., and McEwan, F.   Activities of various cobalamins for *Euglena gracilis* with reference to vitamin B_{12} assay with Euglena. *J. Clin. Pathol.* 24:15, 1971.
2. Baker, H., and Frank, O.   *Clinical Vitaminology: Methods and Interpretation.* New York: Interscience, 1968.
3. Bessey, O. A., Lowry, O. H., Brock, M. J., and Lopez, J. A.   The determination of vitamin A and carotene in small quantities of blood serum. *J. Biol. Chem.* 166:177, 1946.
4. Carr, F. H., and Price, E. A.   Color reaction attributed to vitamin A. *Biochem. J.* 20:498, 1926.
5. Farmer, C. J., and Abt, A. F.   Determination of reduced ascorbic acid in small amounts of blood. *Proc. Soc. Exp. Biol. Med.* 34:146, 1936.
6. Herbert, V.   Aseptic addition method for *Lactobacillus casei* assay of folate activity in human serum. *J. Clin. Pathol.* 19:12, 1966.

7. Lau, K. S., Gottlieb, C., Wasserman, L. R., and Herbert, V.   Measurement of serum vitamin B_{12} level using radioisotope dilution and coated charcoal. *Blood* 26:202, 1965.
8. McCormick, D. B., and Wright, L. D. (Eds.).   Vitamins and Coenzymes. In *Methods in Enzymology*. New York: Academic, 1971. Vol. 18, Part A.
9. Roe, J. H., and Kuether, C. A.   The determination of ascorbic acid in whole blood and urine through the 2,4-dinitrophenylhydrazine derivative of dehydroascorbic acid. *J. Biol. Chem.* 147:399, 1943.
10. Sobel, A. E., and Snow, S. D.   Estimation of serum vitamin A with activated glycerol dichlorohydrin. *J. Biol. Chem.* 171:617, 1947.
11. Waxman, S., Schreiber, C., and Herbert, V.   Radioisotopic assay for measurement of serum folate levels. *Blood* 38:219, 1971.

25. Iron

Serum Iron

Most of the practical methods for determination of iron in serum involve liberation of iron from proteins, removal of the proteins, and subsequent coupling of the reduced iron with substances such as orthophenanthroline [1], α,α'-dipyridyl [3], bathophenanthroline [5], or 2,4,6-tripyridyl-s-triazine [2] to form a colored complex that can be quantitated colorimetrically. The method presented below is based on these principles but is simpler and more sensitive than many of the earlier methods.

SERUM IRON METHOD

METHOD

Peters et al. [4]; Yee and Zin [6].

MATERIAL

Serum.

REAGENTS

Iron-free water should be used in preparing all reagents. If the purified water gives a high blank, it should be repurified by distillation or ion exchange for use in this method.

Trichloroacetic acid, 30% (w/v). Solution prepared from reagent-grade trichloroacetic acid crystals gives a high blank; therefore the acid crystals should be distilled before use. The following technic may be used:

Weigh a 500 ml Erlenmeyer flask.

To a 500 ml side-arm boiling flask (with ground-glass stopper) add the entire contents of a $\frac{1}{4}$ lb bottle of trichloroacetic acid crystals (CCl_3COOH).

Set the Erlenmeyer flask in an ice bath and the boiling flask over a burner as shown in Figure 25-1. Distillation should be carried out in a fume hood if possible. No stopper is needed on the Erlenmeyer flask as the fumes condense rapidly. Distill almost all of the acid, leaving approximately 20 ml behind to avoid distilling over contaminants.

Weigh the Erlenmeyer flask and its contents and subtract from this the weight of the flask to find how much trichloroacetic acid was distilled.

Add 2.8 ml of water for each gram of acid to make a 30% (w/v) solution. (This takes into account the volume occupied by the acid.)

Reducing reagent. Add 17 ml of concentrated hydrochloric acid to 983 ml of water. Dissolve 5 g of ascorbic acid in this solution.

Sodium acetate, saturated. Add reagent grade sodium acetate ($NaC_2H_3O_2 \cdot 3H_2O$) to approximately 200 ml of water with stirring until no more will dissolve. Crystals should always be visible to ensure saturation.

TCA

TCA

crushed ice

Figure 25-1
Distillation of Trichloroacetic Acid

Ferrozine, 20 mg/dl. Dissolve 200 mg of ferrozine [3-(2-pyridyl)-5,6-bis(4-phenylsulfonic acid)-1,2,4-triazine, disodium][1] in 100 ml of water.

PROCEDURE

Note: All glassware and pipets should be specially cleaned with dichromate cleaning solution or $6N$ nitric acid and rinsed well with iron-free water. This equipment should be kept separate from the rest of the laboratory glassware.

Into each of three 16×125 mm test tubes measure 3 ml of reducing reagent.

To one tube add 2 ml of water (Blank).

To another tube add 2 ml of working standard (Standard).

To the third tube add 2 ml of serum (Test).

Mix well using a vortex mixer, and allow to stand for 30 minutes.

Add 1 ml of the 30% trichloroacetic acid solution, and mix well. Allow to stand for at least 15 minutes. Then centrifuge at 2500 rpm for 15 minutes. The Blank and Standard need not be centrifuged since they contain no precipitate.

After centrifugation, pipet 4 ml of each supernatant fluid into 19 mm cuvets. This transfer requires careful technic since the pipet must be placed close to the precipitate, but must not disturb it. If desired, this may be done by using a pipetting bulb or by removing the supernatant

[1] Available from Hach Chemical Co., Ames, Iowa.

fluid with a Pasteur pipet and rubber bulb into a clean tube before measuring the 4 ml.

Measure 4 ml of the Blank and Standard into cuvets and treat these in the same manner as the Test.

Add 0.5 ml of sodium acetate solution to each and mix well, using a vortex mixer.

Add 2 ml of the ferrozine solution, mix well, and allow to stand for at least 10 minutes at room temperature.

Read the absorbances of the Standard and Test in a spectrophotometer at 560 nm, setting the Blank at zero.

CALCULATIONS

$$\text{absorbance of Test} \times \frac{\text{concentration of Standard } (100 \ \mu g/dl)}{\text{absorbance of Standard}}$$

$$\times \frac{1}{1.015} = \text{serum iron } (\mu g/dl)$$

If a result greater than 400 μg per dl is obtained, the procedure should be repeated using 0.5 ml of serum and 1.5 ml of water. The final result must then be multiplied by 4.

The 1.015 is a volume correction factor. After precipitation of the proteins, the total volume remains the same but the liquid volume is smaller due to the volume occupied by the precipitate. Since the iron is distributed only in the liquid phase, protein precipitation results in a slight increase in the concentration of iron in the supernatant fluid. The authors of this method found that this error amounted to approximately 1.5%, thus the correction factor 1.015.

STANDARDIZATION

Stock standard (100 mg of iron per 100 ml). Weigh exactly 100 mg of pure iron wire and allow it to dissolve in 100 ml of 1N hydrochloric acid. This solution takes several days to prepare, but it is stable indefinitely.

Working standard (100 μg of iron per 100 ml). Dilute 0.5 ml of the stock standard to 500 ml with water. This standard is stable indefinitely at room temperature.

NORMAL VALUES

Serum from healthy adults contains approximately 50 to 180 μg of iron per 100 ml.

DISCUSSION

Iron is a ubiquitous substance. Since the quantities being measured are quite small, it is essential that certain precautions be observed. All purified water used in preparation of the reagents, in performance of the procedure, and in rinsing the glassware must be iron free. Some sources of purified water meet this requirement, but in some cases it may be necessary to both distill and deionize the water in order to render it suitable for use in the serum iron method.

Glassware and pipets for iron determinations should be kept separate from other laboratory glassware. The best solution to use for cleaning the glassware is dilute ($6N$) nitric acid, although dichromate cleaning solution, kept only for this purpose, is suitable.

Ideally the Blank, when read against water, should show very little absorbance. A high reading in the Blank indicates iron contamination in one or more reagents. Although this may not invalidate the determinations, the cause should be located and corrected.

Iron is liberated from its combination with globulin by hydrochloric acid, and it is reduced by the ascorbic acid. Proteins are removed with trichloroacetic acid, and the excess acid is buffered with sodium acetate. The free iron then combines with ferrozine to give a purple-colored complex. With the procedure presented here, the reaction is linear up to approximately 400 μg per dl, and the color is stable for many hours.

Iron is an integral part of the hemoglobin molecule. Although traces of hemoglobin in the serum will not have a significant effect on the validity of the iron determination, serum with visible hemolysis should not be used. With hemolysis, positive errors of 10 μg per dl or more may be expected, the magnitude of the error being in proportion to the degree of hemolysis.

Iron-Binding Capacity

The iron in serum is bound to a globulin called *transferrin*. However, only approximately one-third of the transferrin in serum normally is in combination with iron. Occasionally, knowledge of the quantity of transferrin available is helpful in making a differential diagnosis involving a problem in iron metabolism. A method for determination of total iron-binding capacity (TIBC) utilizing the iron procedure described above is presented here.

IRON-BINDING CAPACITY METHOD

METHOD
Young et al. [7].

MATERIAL
Serum.

REAGENTS
The same reagents as described above for serum iron in the first part of the chapter, plus the following:

Magnesium carbonate $(MgCO_3)_4 \cdot Mg(OH)_2 \cdot 4H_2O$.

Iron solution. Dilute 3.5 ml of iron stock standard and 5 ml of $1N$ HCl to 1 liter with water.

PROCEDURE

Add 3 ml of iron solution to 1 ml of serum. Mix and allow to stand for 5 minutes.

Add approximately 50 mg of magnesium carbonate and allow to stand for 30 minutes, mixing vigorously at 5-minute intervals.

Centrifuge for 10 minutes.

Carefully measure 2 ml of the supernate into a 16×125 mm tube.

Add 3 ml of the reducing agent and proceed as for serum iron (see page 330), starting with the addition of 30% trichloroacetic acid solution.

CALCULATIONS

Calculate as serum iron (page 331). Then multiply by 4 to get TIBC in mg per dl.

In the serum iron method, 2 ml of serum are used. For TIBC, 2 ml of 1:4 dilution (or 0.5 ml) of the serum are used. Hence, the correction factor of 4.

NORMAL VALUES

Total iron-binding capacity values in the serum of healthy adults usually range between 250 and 420 mg per dl.

DISCUSSION

When the iron solution is added to serum, ferric iron is taken up by any transferrin that was not originally bound to iron. Since the ferric solution is added in excess, all the transferrin in the serum ultimately is bound to iron.

The function of the magnesium carbonate is to adsorb the free iron left after the binding process is complete.

The iron determination performed subsequently is actually an indirect measure of the protein-binding globulin transferrin. If a serum iron determination is performed at the same time and the result subtracted from the TIBC value, the difference represents the transferrin in the serum that was not bound to iron. This quantity is known as the *unsaturated iron-binding capacity (UIBC)*.

Physiological Significance of Serum Iron and TIBC Values

Most of the iron in the body is in the red cells (hemoglobin) and in tissue. However, quantitation of the relatively small amount of iron in plasma is helpful in making certain diagnoses, and the TIBC is useful in establishing the cause of abnormal serum iron concentrations.

The concentration of iron in serum is low with iron-deficiency anemia and with malignancies. However, the TIBC value usually is high with iron-deficiency anemia and low with malignancies. Abnormally high serum iron values are found with hemochromatosis and liver disease.

With the former disorder the TIBC usually is low. The serum iron is not a very specific test since low serum-iron levels accompany many types of infections and debilitating disease.

References

1. Barkan, G., and Walker, B. S.   Determination of serum iron and pseudo-hemoglobin iron with ortho-phenanthroline. *J. Biol. Chem.* 135:37, 1940.
2. Caraway, W. T.   Macro and micro methods for the determination of serum iron and iron-binding capacity. *Clin. Chem.* 9:188, 1963.
3. Kitzes, G., Elvehjem, C. A., and Schuette, H. A.   The determination of blood plasma iron. *J. Biol. Chem.* 155:653, 1944.
4. Peters, T., Giovanniello, T. J., Apt, L., and Ross, J. F.   A new method for the determination of serum iron-binding capacity: I. *J. Lab. Clin. Med.* 48:274, 1956.
5. Peters, T., Giovanniello, T. J., Apt, L., and Ross, J. F.   A simple improved method for the determination of serum iron: II. *J. Lab. Clin. Med.* 48:280, 1956.
6. Yee, H. Y., and Zin, A.   An autoanalyzer procedure for serum iron and total iron-binding capacity, with use of ferrozine. *Clin. Chem.* 17:950, 1971.
7. Young, D. S., and Hicks, J. M.   Method for automatic determination of serum iron. *J. Clin. Pathol.* 18:98, 1965.

26. Drugs and Poisons[1]

Drugs are compounds, not normally present in the body, which are administered or taken for their therapeutic (beneficial) value.

Poisons are substances which have a toxic (harmful) effect on the body. Almost any substance can be toxic under certain conditions and in certain concentrations. Usually poisons are absorbed or ingested into the body accidentally.

Drugs themselves constitute a common class of poisons when they are taken, either deliberately or accidentally, in toxic quantities.

Biological samples, such as serum or urine, are analyzed for drugs either to guide therapy or to determine the cause of illness or death. In the case of poisons, only the latter purpose is involved.

The three branches of analytical science currently involved in drug and poison analysis are toxicology, forensic science, and clinical chemistry. Traditionally, the toxicologist has been concerned only with poisons and poisonings. Although this has always included analyses of drugs either abused (like narcotics) or taken in overdose, some toxicologists have become involved in therapeutic monitoring of drugs as well and consider that toxicology (or clinical toxicology) covers all aspects of drug and poison analysis.

Forensic or police scientists are concerned with the physical and chemical analysis of evidence in criminal investigations. Although drug and poison analysis is only one of the types of testing in which they are involved, it is one of their major areas of investigation.

Drug and poison analysis is only one of the types of testing carried out by clinical chemists. Currently this area of analysis is expanding rapidly in regard to both the compounds being analyzed and the types of methodology employed. For this reason, it is appropriate here to develop at least an introductory understanding of some of the concepts and principles involved in the classification and analysis of drugs. This chapter covers those aspects of drug and poison analysis which are more pertinent to the clinical chemistry laboratory. Although a separate section later in this chapter deals with poisons, some discussion of these substances also is included, when appropriate, in the drug section.

For more detailed information about drugs and poisons, the reader is referred to appropriate sources [3, 8, 9, 12, 13, 19–21, 24, 26, 28, 30, 31, 33, 36, 40, 41].

Classes of Drugs

Drugs may be classified in a number of different ways. They can be grouped according to their molecular structures, chemical properties, methods of analysis, pharmacological effects, or medicinal usage. Because drugs grouped together by one property may not be in the same

[1] The authors are grateful to Louis A. Williams for his assistance with the preparation of this chapter.

group when classified by a different property, no classification scheme is ideal. The drugs commonly encountered in the clinical chemistry laboratory usually are grouped with reference to their clinical or pharmacological properties rather than to properties related to the methods of quantitative analysis. An outline of some such groupings is presented in Table 26-1.

Drugs sometimes are referred to by their generic (chemical) names and sometimes by trade (brand) names. A given drug may be produced by several pharmaceutical houses and, therefore, may be available under several brand names. Although clinicians usually refer to drugs by brand names, analysts should use generic names wherever practical. Brand names are always capitalized whereas generic names are not. Examples of some common drugs and their brand names are given in Table 26-2. Several reference books are of value to the chemist in identifying or classifying drugs [15, 16].

Table 26-1. Classification of Drugs

Group	Clinical Use	Examples
Amphetamines	Stimulant	Amphetamine Dextroamphetamine
Analgesics	Pain reliever	Aspirin Morphine Codeine
Antibiotics	Bacterial inhibitor	Penicillin Streptomycin
Anticoagulants	Clotting retardant	Heparin Coumarin
Anticonvulsants	Prevent convulsions	Diphenylhydantoin Barbiturates
Cardiac glycosides	Heart stimulant	Digitoxin Digoxin
Diuretics	Increase urine flow	Thiazides Mercurials
Hallucinogens	Produce hallucinations	LSD Amphetamines
Narcotics	Anesthetic	Morphine Codeine Heroin
Psychotherapeutics	Tranquilizer	Phenothiazines Meprobamate
Sedative-hypnotics	Sedation; inducing sleep	Barbiturates Ethanol Chloral hydrate

Table 26-2. Examples of Drug Nomenclature

Generic Name	Brand Name
amobarbital	Amytal
phenobarbital	Luminal
pentobarbital	Nembutal
secobarbital	Seconal
chlordiazepoxide	Librium
diazepam	Valium
diphenylhydantoin	Dilantin
glutethimide	Doriden
meprobamate	Equanil; Miltown
methyprylon	Noludar
propoxyphene	Darvon

Drug Metabolism

Almost all drugs are chemically altered in the body before they are excreted. These alterations are enhanced by enzymes. Some of these enzymes catalyze chemical changes of drugs because the drugs resemble the normal body substrates of the enzymes. This accounts for a wide variety of physiological drug transformations. There also is a special class of enzymes, particularly in the liver, whose function it is to alter all foreign substances (including drugs) which enter the body. The types of reactions carried out generally serve to enhance the water solubility of the substances involved. This promotes their excretion into the urine because they tend to dissolve more freely in the blood and, therefore, are rapidly cleared from the blood by the kidneys. This process constitutes a protective mechanism against foreign substances. As a consequence of this metabolic process, some of the tests for drugs are directed toward metabolites (altered forms) of the drugs rather than toward the drugs themselves. This is more often the case for drug analyses with urine than with blood samples because drug metabolites tend to be excreted rapidly into the urine once they are formed. Although a wide variety of chemical modifications of drugs occur in the body, conjugation (bond formation) to glucuronate (a glucose derivative) and sulfate groups is particularly important.

This process enhances the water solubility of the drug, of course, because the sulfate and glucuronate groups are charged and, therefore, are highly water soluble. Because the presence of either of these groups on a drug frequently interferes with its analysis in urine, an acid hydrolysis or glucuronidase treatment often is employed in the analytic method. Acid hydrolysis removes both the sulfate and glucuronate groups from drugs and can be used as long as the drug is stable in hot

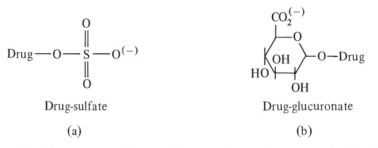

Drug-sulfate   Drug-glucuronate

(a)   (b)

acid. The enzyme glucuronidase catalyzes the removal of only the glucuronate group but is a very mild procedure.

Therapeutic and Toxic Drug Levels

Unless drugs are used in proper amounts, they can produce discomforts (side effects) and dangers (toxicity). Drug dosages required for the best therapeutic effect vary greatly among individuals. Although the "usual" dose of most potent drugs will be optimal for some people, the same dose may be ineffective or toxic in others. Fortunately, the therapeutic and toxic effects of some drugs correlate reasonably well with their serum concentrations. Thus, the dosage schedule can be guided by measurements of the level of the drug in the serum.

Therapeutically effective serum concentrations have been established for some drugs. Levels in excess of these concentrations are considered potentially toxic but not necessarily fatal. Stated therapeutic and toxic levels usually cannot be interpreted too closely; in many cases there is some overlap of the highest therapeutic and minimum toxic levels.

The exact effect of a given drug level may vary from person to person due to differences in the rates and extents of absorption, organ distribution, metabolism, and excretion of the drug. In addition, the effect of a certain drug on one individual can depend on such factors as diet, state of activity, presence of disease, and concomitant administration of other drugs. The last two factors are particularly important since many drugs will enhance or diminish the effects of other drugs (drug interaction). Also many drugs undergo altered rates of metabolism and excretion during disease.

Interferences with Clinical Laboratory Tests

A large number of tests in clinical chemistry are subject to interferences from drugs [5, 6, 7, 18, 25, 29, 35, 39, 42]. An extensive computer list has been published as a special issue of the journal *Clinical Chemistry* [48]. These sources include information about the effects of drugs on specific tests and about the modifying effects of other routine factors such as certain foods, the menstrual cycle and menopause in women, state of physical activity, and change in posture before blood collection.

Two types of drug interference with clinical laboratory tests occur: physiological (pharmacological) and methodological (chemical or physical). In the case of a physiological interference, the drug produces a change within the body of the concentration of the substance being measured. For instance, the drug morphine restricts the passage of pancreatic amylase into the small intestine, resulting in a rise in the serum level of this enzyme. The increased amylase activity would falsely suggest pancreatitis unless the drug effect was suspected.

A methodological interference occurs when the drug (or a drug metabolite) has a direct effect on the chemical test. For instance, the drug may react similarly to the substance being measured, may quench some of the color development, or it may give rise to an interfering color or turbidity. This means that a drug may affect one method of analyzing a certain substance but have no effect on another. For this reason, the specific procedure affected is always specified for a methodological interference.

Drug Abuse

The term *abuse* implies a deleterious effect, either for the individual, society, or both. Drug abuse is a complex concept since it involves not only a medical evaluation, but also social and cultural judgments. The concept generally applies to drugs which are used by people to modify their mental state, for example, to produce a sense of well-being or to cause hallucinations. The classes of drugs which tend to be abused are sedative-hypnotics, narcotics, psychotherapeutics, hallucinogens, and amphetamines. Prominent examples are the alkaloids (morphine, codeine, heroin), barbiturates, amphetamines, and ethanol. These drugs usually are addictive.

Poisons

As has been mentioned, almost anything can be a poison. However, the term usually is applied to substances which are particularly harmful in their physiological effects. Most poisonings are accidental, resulting from occupational or environmental exposure. The need for expediency in the analysis of samples for poisons is obvious.

Although there are numerous poisons, the three categories which include most poisons encountered in clinical chemistry are drugs, heavy metals, and environmental poisons.

DRUGS

Most poisonings are caused by drugs which are taken in excessive amounts. In fact, relatively few drugs—the barbiturates, the benzodiazepams, glutethimide, methaqualone, salicylates, amphetamines, opiates, synthetic analgesics, phenothiazines—are responsible for a large percentage of all poisonings.

HEAVY METALS

Although the heavy metals could be classified with the environmental poisons, they are sufficiently distinctive in their properties and important as a type of poison to be considered as a separate category. The most common ones are lead, arsenic, mercury, antimony, beryllium, bismuth, cadmium, chromium, gold, manganese, nickel, and thallium. Generally, industrial exposure is most often the cause of heavy-metal poisoning, although concern has also been expressed about nonindustrial ingestion of lead and mercury.

ENVIRONMENTAL POISONS

This category includes all of the potential poisons which contaminate the air, water, and natural foodstuffs in the environment. For the clinical chemist, the most important poisons in this group are lead and carbon monoxide. A method is given for the latter in this chapter.

The major types of substances falling into this category are insecticides, food poisonings, organic (industrial and household) solvents, strong acids and bases, gases (for example, hydrogen cyanide, carbon monoxide), and industrial dusts (for example, silica, which causes silicosis).

Methodology of Drugs and Poisons

Considerable overlap occurs in the methods which are used for analysis of drugs and of poisons. In the following sections, some of the analytic approaches used in the clinical chemistry laboratory are discussed.

QUALITATIVE VS. QUANTITATIVE ANALYSIS

When drug levels are monitored for therapeutic purposes, exact quantitative methods are required. However, in cases of poisoning or drug abuse usually qualitative analysis (identification) is adequate. If the patient is alive, rapid identification of the poison is essential for prompt, appropriate therapy. In the case of a death from poisoning, identification of the causal agent is important and possibly a semiquantitative estimation of its concentration may be required.

SEPARATION TECHNICS

I. EXTRACTION

Extraction of drugs or poisons from biological samples (serum, urine, pulverized tissue) with organic solvents is the preliminary step in many types of drug and poison methods. The objectives are to not only separate the drug or poison from interfering and contaminating substances but also to provide a means of concentrating the substance under analysis so as to increase the sensitivity of the assay. This latter objective is achieved by evaporating (usually with the aid of heat and a stream of air) the organic solution of the drug or poison to a very small volume.

For instance, if a drug is extracted from 10 ml of urine with ether and the ether layer then is concentrated to a volume of 0.10 ml before being applied to a thin-layer chromatography plate or injected into a gas chromatograph, the drug has been concentrated 100 times (10/0.1) prior to the detection or quantitation step.

An important consideration in regard to organic extractions is whether the substance to be extracted is acidic, basic, neutral, or amphoteric, and whether this property can be changed by altering the pH. The drug must be devoid of any charge before it can be extracted efficiently into an organic solvent. Organic substances with no ionizing groups (for example, no $-CO_2H$ or $-NH_2$ groups) will tend to extract into an organic solvent at any pH. Those with a $-CO_2H$ group (or other type of acidic group, that is, a group which loses a proton easily) will extract better at a low (acidic) pH because only then will the $-CO_2H$ group exist in this unionized form. At a higher pH (above pH 6 for $-CO_2H$), the $-CO_2H$ group exists largely as $-CO_2^-$. So the organic substance will show much less tendency to extract into the organic phase.

In contrast, the organic substance with an $-NH_2$ type of group will extract into the organic phase more efficiently at a higher pH (above pH 9) because such a group tends to exist largely as $-NH_3^+$ below pH 9, thereby placing a charge on the organic substance at a lower pH.

The neutral drugs are best extracted at a neutral pH (approximately 7).

Some drugs, like morphine, are amphoteric; that is, they may behave as either acids or bases. To extract such a drug, the pH first must be adjusted close to the appropriate isoelectric point (net charge equals zero) of the particular drug.

For these reasons, extraction technics usually require the addition of acid or base to the aqueous phase before an extraction step. A succession of pH adjustment and organic extraction steps sometimes is used. For instance, an initial acidification and extraction may be carried out to remove certain impurities (for example, impurities with $-CO_2H$ groups) prior to addition of base. A second organic extraction then can be performed to remove the substance of interest (for example, a substance with a $-NH_2$ group). This concept is illustrated in Figure 26-1.

2. ION EXCHANGE RESINS

Drugs can also be removed from biological samples with ion exchange resins. Drugs bind to ion exchange resins not only because of ionic interactions (for example, a positively charged drug is bound by a negatively charged ion exchange resin), but also because of hydrophobic binding (a binding interaction between the aromatic or hydrocarbon backbone of the resin and the corresponding types of groups in the drug). After the drug is bound by the resin, the interfering substances are washed away. Then the drug is eluted with a strong solvent such as aqueous acid or base, aqueous acid or base containing organic solvent, or an aqueous solution containing a high concentration of salt.

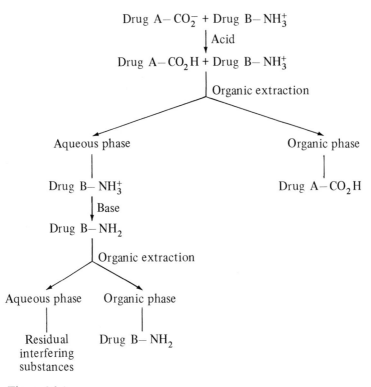

Figure 26-1
Separation of Drugs by Extraction

3. XAD-2

Amberlite XAD-2 is structurally and functionally analogous to ion exchange resins: it is fabricated as insoluble, plastic (polystyrene) beads, and it adsorbs drugs. However, one important difference exists. Unlike ion exchange resins, XAD-2 lacks ionic groups; it is purely a porous, hydrophobic polymer. It therefore binds drugs solely by a hydrophobic attraction rather than by the combination of ionic and hydrophobic binding which occurs with ion exchange resins. XAD-2 is used to simultaneously bind acidic, neutral, and basic drugs by hydrophobic binding. After an aqueous washing step to remove non-binding, interfering substances, the drugs are eluted with organic solvents.

4. THIN LAYER CHROMATOGRAPHY

Thin layer chromatography is essentially a qualitative technic and not really a very rapid one, since 1 to 2 hours usually are required before an answer is available. It is, however, convenient, sensitive, and capable of separating a number of substances at the same time. For these reasons, it is used frequently to screen for the possible presence of certain drugs, particularly drugs of abuse. Sometimes two or three detection sprays can be used successively to reveal more than one class of drugs. Although a negative result is a good indication that a certain drug is absent at the detection limit for the method, a positive result only *suggests* that a drug is present. Positive findings with thin layer chromatography should be confirmed by another technic such as ultraviolet spectrophotometry, fluorometry, or gas chromatography.

The general principles of thin layer chromatography are given in Chapter 7.

5. GAS CHROMATOGRAPHY

When applicable to a drug or poison, gas chromatography often is the method of choice since it offers sensitivity, speed, quantitation, and the ability to separate a number of substances at the same time. Only volatile substances (or substances which can be rendered volatile by a chemical modification) can be analyzed. The disadvantages of gas chromatography as compared with thin layer chromatography are greater cost, maintenance problems, and lack of suitability for handling large numbers of samples. However, a greatly expanded capability is provided.

Further discussion of gas chromatography may be found in Chapter 7.

SPECTROPHOTOMETRIC TECHNICS

Spectrophotometric technics include spectrofluorometry, ultraviolet- and visible-absorption spectrophotometry, and flame emission and atomic absorption spectrometry. Although these are quantitative technics, the sensitivity depends on the particular substance being measured or on the sensitivity of the chemical reaction being used to generate a colored or fluorescent product from the substance. Most methods require removal of interfering substances, usually by extraction steps or chemical reactions, or both, before the actual spectrophotometric readings are obtained. However, for many drugs and poisons, convenient, specific methods culminating in a spectrophotometric analysis are available. The advantage of absorption spectrophotometry is that a complete spectral absorption curve can be obtained (with a double-beam instrument) to provide both a qualitative and quantitative analysis when interfering substances are absent.

Spectrophotometric methods are useful for identifying as well as quantitating drugs or poisons. The principles of these analytic technics are discussed in Chapters 4 and 5.

IMMUNOCHEMICAL TECHNICS

Whenever antibodies can be generated against a drug by injecting an animal with a large molecule, like a protein, to which the drug is attached (see page 102), an immunochemical assay can be devised. Both semiquantitative and quantitative methods are available. The sensitivity of these methods is extremely high, but lack of specificity sometimes is a problem, especially if drugs of closely related molecular structure are present. Preparation of specific antibodies is a difficult process. The use of a radioactive label (radioimmunoassay) or an enzyme provides quantitative results, whereas a hemagglutination-inhibition type of assay provides semiquantitative results. These assays employ the competitive binding principle which is discussed in Chapter 8.

CHEMICAL TECHNICS

A number of other chemical technics (such as steam distillation, ashing, and wet digestion) also are used to prepare biological samples for quantitative analysis of drugs.

Blood Alcohol

Ethyl alcohol (ethanol) usually is not considered to be a drug. However, because it is intoxicating, and sometimes toxic, it is grouped with drugs having similar effects. Usually it is measured either by reaction with dichromate or alcohol dehydrogenase, or by a gas chromatographic method.

In the dichromate method [44], the alcohol is oxidized by a yellow solution of chromic acid, forming colorless acetic acid and green chromous ion. The reaction can be quantitated by measuring the intensity of the green color or by a second reaction of the remaining chromic ions with brucine to form an orange-red color.

The alcohol dehydrogenase enzymatic method [27] involves the oxidation of alcohol by the coenzyme NAD to form acetaldehyde (CH_3CHO) and NADH. The reaction is measured by the increase in absorbance at 340 nm from NADH production. This reaction is explained further on page 242.

The gas chromatographic method [2] is the method of choice, being rapid, specific, and sensitive. Either an organic extract or a direct injection of blood may be made onto the column.

The method described here, though not specific, is sensitive, relatively simple to perform, and meets the usual needs for estimation of blood alcohol.

ALCOHOL METHOD

METHOD
Williams et al. [44].

MATERIAL
Blood, serum.

REAGENTS
Sodium carbonate, 20% (w/v). Dissolve 20 g of sodium carbonate (Na_2CO_3) in water and dilute to 100 ml.

Dichromate reagent. Measure 300 ml of water into a 1 liter Erlenmeyer flask. Cool the flask in ice or cold water and slowly add 300 ml of concentrated sulfuric acid (H_2SO_4) with mixing and intermittent cooling. When cooled to room temperature, pour into a 1 liter graduated cylinder and dilute to 600 ml with water.

Pour 500 ml of this solution into a beaker, add 0.355 g of potassium dichromate ($K_2Cr_2O_7$), and stir until the dichromate is dissolved. Store in a glass-stoppered bottle.

Brucine reagent. To 45 ml of water add 5 ml of concentrated sulfuric acid and mix. Add 0.5 g of brucine ($C_{23}H_{26}N_2O_4$), and stir until solution is complete. This solution is stable for approximately 1 week in the refrigerator. Ideally, this reagent should be prepared fresh each time the method is used.

White Vaseline.

APPARATUS
Conway microdiffusion cells (see Figure 12-1).
Oven set at 50 C to 60 C.

PROCEDURE
Note: It is advisable to perform all tests in duplicate.
Lubricate the covers of 5 Conway microdiffusion cells with white Vaseline.
Prop up one edge of each cell approximately 5 mm.
To the lower side of the outer chamber add 1 ml of sodium carbonate solution.
To the inner chamber add exactly 2.0 ml of dichromate solution.
Seal the cover on one cell and set it aside (Blank).
Set the covers in place on the other cells, leaving a small opening to the outer chamber at the lower side.
Using to-deliver pipets, add 50, 100, and 200 μl of working standard to 3 cells, sealing the cover on each as soon as the standard is added (Standards).
Using the same technic, add 100 μl of blood to the remaining cell.
Mix the contents of each cell by *gently* rotating it while it is in a level position, taking care not to spill the dichromate from the center well into the outer well.
Place all the cells in an oven preheated to 50 C to 60 C, and allow them to incubate for 1 hour.
Carefully slide the cells out of the oven, holding both the cells and their covers.[2]
For each cell proceed as follows:
Remove the cover and, using a Pasteur pipet, transfer the contents of

[2] At this point the Vaseline is in a liquid state, but it gels rapidly when removed from the oven.

the center chamber to a 25 ml glass-stoppered volumetric flask or
graduated cylinder. Tilt the cell slightly to get all the solution.
Rinse the chamber 3 times with 2 ml amounts of water, adding these
rinsings to the flask.
To all flasks add 2 ml of brucine reagent, swirl to mix, and allow to stand
for 5 minutes.
Dilute to 25 ml with water, stopper, and mix.
Pour into 19 mm cuvets and set the Blank at 0.800 absorbance in a
spectrophotometer at 475 nm.
Read the absorbances of the Standards and Test, rechecking the 0.800
setting of the Blank between readings.

CALCULATIONS

Using linear graph paper, plot the Standard concentrations[3] against their
respective absorbances. Zero concentration (the Blank) is plotted at
0.800 absorbance.

Locate the concentration corresponding to the Test absorbance on the
standard curve.

If a test reads lower[4] than the 200 mg per dl standard, dilute a portion
of the blood 1:4 and repeat the test. Multiply the calculated result by 4.

STANDARDIZATION

The stock standard for this method is absolute ethyl alcohol, which is
commercially available. The specific gravity of absolute ethanol is 0.798,
which means that each 1 ml weighs approximately 0.8 g or 800 mg.

Alcohol Working Standard, 100 mg/dl. Dilute 1.0 ml of absolute ethanol
to 20 ml with water. Redilute 0.5 ml of this solution to 20 ml with
water. Prepare this dilution fresh each time.
When 100 μl is used as a sample, this standard is equivalent to a blood-
alcohol concentration of 100 mg per dl. Thus, 50 and 200 μl equal
alcohol concentrations of 50 and 200 mg/dl respectively.

DISCUSSION

The sensitivity of this method permits detection of blood alcohol concentrations as low as 10 mg per dl.

White Vaseline is recommended as the sealing agent for the cells since
this substance does not interfere with the method whereas some types of
grease may be oxidized by dichromate. Furthermore, Vaseline is easily
removed with hot water and detergent.

During incubation, alcohol is liberated from the carbonate solution in
the outer chamber and absorbed by the acid dichromate solution in the

---

[3] See Standardization, immediately following.
[4] With this method, the amount of color is inversely proportional to the concentration of alcohol.

center well. The reaction which takes place is best represented in two steps as follows:

$$K_2Cr_2O_7 + 3CH_3CH_2OH + 4H_2SO_4 \rightarrow$$
$$3CH_3CHO + Cr_2(SO_4)_3 + K_2SO_4 + 7H_2O$$

$$K_2Cr_2O_7 + 3CH_3CHO + 4H_2SO_4 \rightarrow$$
$$3CH_3COOH + Cr_2(SO_4)_3 + K_2SO_4 + 4H_2O$$

The alcohol first is converted to the aldehyde, then to the corresponding acid. Dichromate is present in excess and most methods measure either the remaining dichromate or the formed chromic ion. However, much greater sensitivity is achieved when brucine is used. This reagent is oxidized by the excess dichromate to form an orange-colored chromophore. Both the dichromate and brucine are added in excess. Quantitation is based only on the *change* in dichromate concentration from the beginning to the completion of the reaction. When brucine is added, a purple color forms, but this rapidly changes to a stable orange color. The final chromophore has its peak absorbance approximately at 425 nm. However, the color of the blank is so intense that, if it is read against water, its absorbance is too great to give reliable readings with most spectrophotometers. Furthermore, if the solution is diluted to give a lighter Blank, the sensitivity of the method is increased, and the range decreased to impractical levels. Although a wavelength of 475 nm[5] is on the descending part of the spectral absorption curve, it does permit setting the Blank as described in the method and obtaining readings up to a concentration of 200 mg per dl. Accuracy will not be sacrificed provided standards accompany each test or group of tests. The curve obtained is not a straight line so it must be plotted, and a calculation factor (slope constant) cannot be derived for the method.

This method is not specific for ethanol; it also measures methanol, isopropanol, acetaldehyde, and formaldehyde. However, the latter three substances do not undergo the first reaction as is shown above and, therefore, react with much less dichromate than does ethanol. Furthermore, the occurrence of these compounds in the blood is quite rare as compared with that of ethanol.

It is preferable to report alcohol concentrations in units of milligrams per 100 ml to conform with the reporting of other blood constituents. Because of long-standing policies set by forensic chemists, the unit "grams per 100 ml" or simply "percent," sometimes is used, resulting in some confusion. A level stated as 0.1% would be 100 mg per dl.

---

[5] It is best to use a wavelength as close to 425 nm as the conditions will permit. This will vary from one instrument to another.

PHYSIOLOGICAL SIGNIFICANCE OF
BLOOD ALCOHOL CONCENTRATIONS

Measurable amounts of alcohol are not present in blood unless alcohol has been ingested. The tolerance to alcohol varies markedly among individuals, and it is impossible to draw a strict correlation between blood alcohol levels and specified clinical conditions. However, some broad generalities may be stated. Levels over 400 mg per dl usually are associated with coma and death. Between 100 and 200 mg per dl, intoxication is apparent in almost all individuals, and the legal definitions of intoxication usually are based on blood alcohol levels in this range. However, many individuals show definite clinical signs of intoxication at levels well below 100 mg per dl.

If alcohol is ingested along with other drugs (such as barbiturates), a synergistic effect may occur, that is, the pharmacological effect of the drug will be enhanced. In these cases, knowledge of the blood alcohol level and the nature and concentration of other drugs present often helps the physician to evaluate the condition of the patient.

Barbiturates

Barbiturates are important as sedative-hypnotics in clinical medicine. Since large doses are toxic, it is important that the clinical laboratory have available a means of determining the barbiturate concentration in blood. Thin layer chromatography [23] often is used for the detection of barbiturates. Confirmation and quantitation of positive results then is usually carried out by ultraviolet spectrophotometry or gas chromatography [32]. The method presented here involves chloroform extraction and differential ultraviolet spectrophotometry.

BARBITURATE METHOD

METHOD
Goldbaum [14]; Williams and Zak [46].

MATERIAL
Blood, serum.

REAGENTS
Sodium hydroxide, 0.45N.[6] Dilute 90 ml of 1N sodium hydroxide (NaOH) to 200 ml with water.

Phosphate solution.[6] Dissolve 62 g of monobasic sodium phosphate, monohydrate ($NaH_2PO_4 \cdot H_2O$) in water and dilute to 200 ml.

Chloroform (CHCl₃). Fill a 2 liter volumetric flask, containing a magnetic stirring bar, to the bottom of the neck with chloroform. Add 10% (w/v) NaOH to the mark, and stir on a mixer for 5 minutes. Aspirate the top layer. Add water to the mark. Mix 5 minutes. Then

6 These solutions are so formulated that 3 ml of the NaOH plus 0.5 ml of phosphate solution should have a pH of 10.5. If necessary, adjust the concentration of the NaOH to produce this condition.

aspirate. Add 1N HCl to the mark. Mix 5 minutes. Then aspirate. Wash 3 times with water in the same fashion. Filter into a dark bottle.

Phosphate buffer, pH 7.4. Dissolve 5.96 g of anhydrous dibasic sodium phosphate (Na_2HPO_4) and 1.09 g of anhydrous monobasic potassium phosphate (KH_2PO_4) in water and dilute to 500 ml.

APPARATUS

Screw-cap tubes, 25 × 150 mm.
Ultraviolet spectrophotometer, double-beam, recording.

PROCEDURE

Into each of two 25 × 150 mm screw-cap tubes measure 50 ml of chloroform.

To one add 3 ml of sample (Test).

To the other add 3 ml of standard and 0.5 ml of phosphate buffer, pH 7.4 (Standard).

Cap and shake vigorously for 1 minute.

Allow to stand for at least 1 minute, remove the caps, then aspirate off the aqueous (top) layer and discard it. Pour the bottom layer (chloroform) through Whatman no. 42 (11 cm) filter paper into clean tubes.

Add 8 ml of 0.45N NaOH, cap, and shake vigorously for 3 minutes.

Allow the solutions to separate. Then aspirate and discard the bottom (chloroform) layer.

Pour the sodium hydroxide layer into 12 ml conical centrifuge tubes, and centrifuge at 2000 rpm for 5 minutes.

Carefully pipet 3 ml of each supernatant fluid into each of two 10 mm silica cuvets, taking care not to transfer any of the chloroform from the bottom of the tubes.

To one of each pair add 0.5 ml of phosphate solution (sample cuvet) and to the other 0.5 ml of 0.45N sodium hydroxide (reference cuvet).

Mix the contents of both cuvets well.

Using a double-beam, ultraviolet recording spectrophotometer, set zero absorbance (nothing in either cuvet well) at approximately mid scale. Then scan each sample cuvet against its corresponding reference cuvet from 400 nm to 220 nm. A positive peak at 260 nm and a negative one at 240 nm confirms the presence of a barbiturate (see Figure 26-2).

The absorbance difference between the two peaks is directly proportional to the barbiturate concentration.

CALCULATIONS

If a barbiturate is present, calculate its concentration as follows:

$$(A_T^{260} - A_T^{240}) \times \frac{\text{concentration of Standard (20 mg/l)}}{A_S^{260} - A_S^{240}}$$

$$= \text{barbiturate[7] (mg/liter)}$$

[7] Since identification of the type of barbiturate present is not made, it should be reported only as "barbiturate."

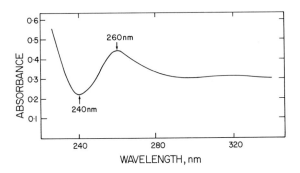

Figure 26-2
Barbiturate Difference Spectrum (pH 13 vs. pH 10.5)

where A_T is the absorbance of the Test, A_S is the absorbance of the Standard, and 260 and 240 are the wavelengths at which the readings are made.

STANDARDIZATION

Ideally, standardization should be performed using the specific barbiturate being analyzed. However, in most instances this is impractical; therefore, the method usually is standardized using phenobarbital. When other barbiturates are quantitated using phenobarbital as a standard, the error usually is less than $\pm 10\%$.

Barbiturate standard 20 mg/liter. Weigh 20.0 mg of pure phenobarbital and dissolve it in 1 liter of water. Store in a refrigerator. Since phenobarbital decomposes in solution, this standard should not be kept more than 1 week. A constant may be derived from the following relationship:

$$\frac{\text{concentration of Standard}}{\text{net absorbance of Standard}} = \text{constant}$$

This constant should be checked periodically with freshly prepared standard.

DISCUSSION

Barbiturates are not present in the blood except when ingested. Occasionally, barbiturate-free blood may give small artifactual absorbance readings that might suggest that barbiturates are present. For this reason, measurements below 2 mg per liter should not be taken as positive evidence that barbiturates are present.

Earlier versions of this method employed ammonium chloride rather than the phosphate solution. However, the use of this reagent sometimes causes deposits to form on the mirrors of the spectrophotometer with a significant loss of sensitivity.

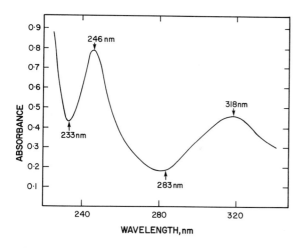

Figure 26-3
Salicylate Difference Spectrum (pH 13 vs. pH 10.5)

At 260 nm barbiturates absorb light more strongly at pH 13 (sample cuvet) than at pH 10.5 (reference cuvet). Thus, the net absorbance results in a positive peak. At 240 nm barbiturates absorb light more strongly at pH 10.5 than at pH 13. Therefore, a negative absorbance peak is observed. The intensity of both peaks increases with greater barbiturate concentration. Background absorbance interferences tend to cancel out with this differential technic.

The method, as described, calls for the use of a double-beam record-ing spectrophotometer to automatically plot the differential curve be-tween the sample (pH 13) and the reference portion (pH 10.5). The method may also be used with a single-beam instrument with individual readings being made at selected wavelengths. However, this is not as simple or as informative as using a double-beam instrument.

Since this technic involves both positive and negative peaks, the reference point (nothing in either cuvet) must be set at some absorbance greater than zero. A setting of 0.5 absorbance is satisfactory if it can be attained, but this capability varies from one instrument to another.

With some barbiturate technics, salicylates and other acid drugs are extracted along with the barbiturates and may give erroneous results. However, at the pH of blood (7.4) these drugs are not extracted.[8] Buffer is added to the standard so that its pH during the extraction will approximate that of the blood. If other materials, such as urine or gastric washings, are used in the method, they should be buffered to pH 7.4 before extraction. For purposes of comparison, a salicylate curve is shown in Figure 26-3.

[8] At extremely high concentrations some salicylate may be extracted but, since its absorption curve differs from that of barbiturate, usually it can be recognized.

Some drugs, such as sulfonamides, bis-hydroxycoumarin (Dicumerol), and diphenylhydantoin (Dilantin), may be extracted along with the barbiturates. However, since their absorption properties are the same at pH 13 and pH 10.5, they do not interfere with the method.

It has been suggested that the various barbiturates may be distinguished from each other by the differences in their spectral absorption curves [14]. However, these differences are so subtle as to raise serious doubts about the reliability of this approach.

In Figure 26-2 it will be noted that a large positive absorbance difference exists at 230 nm; this is because sodium hydroxide absorbs light rather strongly at this wavelength.

Thiobarbiturates show absorption properties closer to those of salicylate than to those described above [43]. However, since these compounds are used primarily as anesthetics, they are rarely encountered in clinical chemistry.

PHYSIOLOGICAL SIGNIFICANCE OF BARBITURATE CONCENTRATIONS

Usually barbiturate levels are requested to (1) determine whether a person has ingested barbiturate, (2) determine whether a serious clinical condition is due to barbiturate toxicity, or (3) regulate therapeutic dose. In the first instance, the sensitivity and specificity of the method are important factors. If the concentration of barbiturates in the sample being analyzed is below 2 mg per liter, it is difficult to be certain that any drug is present at all. However, even though the differences are slight, if the differential curve has the appropriate characteristics, as shown in Figure 26-2, barbiturates probably are present. With the method given here, interference from other drugs is easily recognized, although it may not be possible to identify or quantitate barbiturates in the presence of the interfering compound.

Evaluating barbiturate toxicity is both a qualitative and a quantitative problem. Some of the barbiturates, such as phenobarbital, are "long-acting" and may be tolerated in rather high concentrations in the blood. Other barbiturates, such as secobarbital, are "short-acting," and toxicity occurs at relatively low blood levels. A third group of barbiturates, such as amobarbital, are intermediate-acting drugs. Information regarding the correlation of blood levels of various barbiturates with the clinical conditions of patients has been published. The therapeutic range for serum phenobarbital levels is approximately 15–40 mg per liter. Generally, a comatose state may occur when the concentrations of barbiturate are 60 mg per liter long-acting; 15 mg per liter intermediate-acting; or 10 mg per liter short-acting. Various barbiturates may be distinguished from each other by thin layer or gas chromatographic technics.

Of course, the possibility always exists that other drugs may be ingested along with barbiturates, which may complicate the clinical evaluation of a barbiturate level. Ethyl alcohol has a synergistic (enhancing) effect that can produce a clinical condition more severe than the barbiturate alone would cause.

Bromide

Bromide is an important component of many patent medicines (particularly sedatives) and is toxic in high concentrations. Therefore, it is important to the physician that a method be available for the determination of a serum bromide level.

Most methods for the quantitative determination of bromide in blood depend upon the reaction of bromide with gold chloride to produce an orange-brown color.

BROMIDE METHOD

METHOD
Wuth [47].

MATERIAL
Serum.

REAGENTS
Gold chloride. Weigh 0.5 g of gold chloride ($AuCl_3 \cdot HCl \cdot 3H_2O$), dissolve, and dilute to 100 ml with water. Store in a refrigerator.

Trichloroacetic acid, 10% (w/v). Weigh 10 g of trichloroacetic acid (CCl_3COOH), dissolve, and dilute to 100 ml with water.

PROCEDURE
Blow 8 ml of 10% trichloroacetic acid into a 16 × 125 mm tube containing 2 ml of serum (Test).
Shake vigorously. Centrifuge for 5 minutes. Then filter through small (7 cm) filter paper.
Add 5 ml of filtrate to a 19 mm cuvet.
To another cuvet add 5 ml of trichloroacetic acid (Blank).
To a third cuvet add 5 ml of working standard (Standard).
Add 1 ml of gold chloride solution to each tube, mix, and allow to stand for 5 minutes.
Read the absorbances of the Test and the Standard in a spectrophotometer at 440 nm, setting the Blank at zero.

CALCULATIONS

$$\text{absorbance of Test} \times \frac{\text{concentration of Standard (50 mg/dl)}}{\text{absorbance of Standard}}$$
$$= Br\ (mg/dl)$$

With bromide levels higher than 100 mg per dl, the test should be repeated using 2 ml of filtrate plus 3 ml of 10% trichloroacetic acid. The calculated result should then be multiplied by $\frac{5}{2}$ or 2.5.

STANDARDIZATION

Stock standard (*2000 mg/dl as Bromide*). Weigh 2.575 g of sodium bromide (NaBr), dissolve, and dilute to 100 ml with water.

The molecular weight of sodium bromide is 103 but the atomic weight of bromine is 80. Therefore, 2.575 g of sodium bromide contain

$$\frac{80}{103} \times 2.575 = 2.00 \text{ g of bromide}$$

Working standard. Dilute 0.5 ml of the stock standard to 100 ml with 10% trichloroacetic acid.

The bromide concentration in this solution actually is 10 mg per dl. However, since the filtrate is a 1:5 dilution of serum, 5 ml of this standard, treated as filtrate, represent a bromide concentration of 50 mg per dl in the serum.

DISCUSSION

The minute quantity of bromide normally present in the blood (approximately 1 μg per dl) is not significant and cannot be measured with this method. Bromide levels are of value to the physician only during bromide therapy or in cases of suspected bromide poisoning (bromism). The sensitivity of the method is adequate for these purposes. The lower limit of detectability for this method is approximately 3 mg per dl. Values lower than this may be artifactual and do not necessarily indicate the presence of bromide.

The concentration of bromide capable of producing toxic symptoms varies with individual circumstances, but concentrations higher than 100 mg per dl of bromide in the blood should be regarded as being potentially toxic.

Bromide imparts a brownish color to gold chloride in solution. The intensity of this color is directly proportional to the concentration of bromide. The sensitivity and accuracy of the method leave something to be desired, but it is adequate for the usual clinical purposes.

Large quantities of bromide present in blood will introduce positive errors in the chloride methods. If the true chloride concentration is desired with a serum containing bromide, the bromide concentration should be converted to milliequivalents per liter and subtracted from the apparent chloride concentration:

$$\text{apparent Cl (meq/liter)} - \left(\frac{\text{Br (mg/dl)} \times 10}{80} \right)$$
$$= \text{true Cl value (meq/liter)}$$

Sometimes bromide results are expressed in terms of sodium bromide (NaBr) rather than bromide ion. The conversion factors are as follows:

Br (mg/dl) $\times$ 1.29 = NaBr (mg/dl)
NaBr (mg/dl) $\times$ 0.78 = Br (mg/dl)

Salicylate

Accidental salicylate poisoning from an overdose of aspirin or some other salicylate-containing medication is common, particularly in children.

Salicylate may be identified and quantitated by the same differential spectrophotometric method as presented for barbiturates. The characteristic spectrum is shown in Figure 26-3. However, salicylate usually is measured by a colorimetric method in which added ferric ion forms a purple complex with salicylate [10].

In this colorimetric method, the salicylate can be either measured directly or after extraction with an organic solvent. Although the extraction method is more accurate, the direct method yields acceptable results and is the method which is presented here.

SALICYLATE METHOD

METHOD
Keller [22].

MATERIAL
Serum.

REAGENTS
Stock ferric nitrate
1. Prepare approximately 0.07N nitric acid by adding 1.8 ml of concentrated nitric acid to 400 ml of water.
2. Dissolve 2 g of ferric nitrate, $Fe(NO_3)_3 \cdot 9H_2O$, in 200 ml of the 0.07N nitric acid.

Dilute ferric nitrate. Just before use, mix 10 ml of stock ferric nitrate solution with 8 ml of water. Prepare more solution in the same proportions if more than one determination is to be made.

Nitric acid 0.04N (approximately). Mix 100 ml of the 0.07N nitric acid with 80 ml of water.

PROCEDURE

To each of two 19 mm cuvets add 5 ml of dilute ferric nitrate reagent.

To one cuvet add 0.5 ml of working standard (Standard), and to the other add 0.5 ml of serum (Test).

To a third cuvet add 5 ml of 0.04N nitric acid and 0.5 ml of serum (Blank).

Mix the contents of each tube and read the absorbances against a water blank at 535 nm in a spectrophotometer.

CALCULATIONS

(absorbance of Test − absorbance of Blank)

$$\times \frac{\text{concentration of Standard (200 mg/liter)}}{\text{absorbance of Standard}}$$

$$= \text{salicylate (mg/liter)}$$

If a reading greater than 0.600 occurs, add 5.5 ml of 0.04N nitric acid to the Test and Blank, mix, and read again. Multiply the result by 2.

STANDARDIZATION

Stock standard (20,000 mg/liter as salicylate). Weigh 2.34 g of reagent grade sodium salicylate ($C_7H_5NaO_3$). Dissolve and dilute to 100 ml with water. Store in a refrigerator.

The molecular weight of *sodium* salicylate is 160, while that of salicylate is 137. Therefore, 2.34 g of sodium salicylate contain

$$\frac{137}{160} \times 2.34 = 2.00 \text{ g of salicylate}$$

Working Standard (200 mg/liter). Dilute the stock standard 1:100 with water. Prepare this standard fresh each time and use as serum.

DISCUSSION

Normally, there is no salicylate present in the blood, but since salicylate (or salicylic acid) is a common and widely used drug, it is sometimes necessary that the blood salicylate level be determined to guide therapy or to diagnose salicylate poisoning. The optimal therapeutic range of salicylate concentration in serum is 150 to 300 mg per liter. Salicylates are relatively harmless, but may produce toxic effects when a concentration higher than 300 mg per liter is attained in the plasma.

Aspirin is the acetylated form of salicylic acid, and is converted to the free form in the body. Since this method determines only *free* salicylic acid, aspirin itself would not participate in the reaction.

According to the directions given above, the results are expressed as *salicylate* (or salicylic acid). If the same results are to be expressed as the sodium salt, mathematical conversions may be made based on the molecular weights:

$$\text{sodium salicylate} \times \frac{137}{160} \text{ (or 0.86)} = \text{salicylate}$$

$$\text{salicylate} \times \frac{160}{137} \text{ (or 1.17)} = \text{sodium salicylate}$$

Digoxin

Digoxin is one of the more common cardiac glycosides. It is used primarily to treat heart failure since it increases the force of contraction of the failing heart and corrects certain arrhythmias. It should not be confused with digitoxin, a closely related compound, or digitalis, which is a powdered leaf preparation containing digoxin, digitoxin, and other cardiac glycosides.

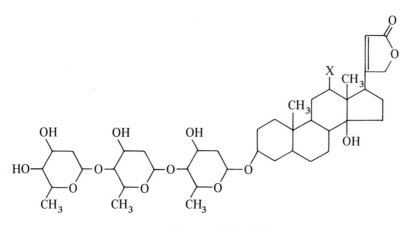

Digoxin (X = OH)
Digitoxin (X = H)

The method of analysis for digoxin presented here is a modification of the reference radioimmunoassay method [37].

DIGOXIN METHOD

METHOD
Smith et al. [37]; Zettner and Duly [49].

MATERIAL
Serum.

REAGENTS
Ethanol, 95%.

Ethanol, 50% (v/v). Add 500 ml of 95% ethanol to 500 ml of water.

Sodium hydroxide, approximately 5N. Dissolve 20 g of NaOH in water and dilute to 100 ml.

Bovine serum. If turbid, filter before use.

Buffer, pH 7.4. Dissolve 8.77 g of sodium chloride, 1.74 g of monobasic potassium phosphate (KH_2PO_4), and 1 g of sodium azide (NaN_3) in approximately 950 ml of water. Adjust the pH to 7.4 ± 0.05 using 5N NaOH. Dilute to 1 liter with water.

Antibody solution. Portion the antiserum[9] into 1 ml amounts and freeze. Each week, thaw 1 ml of the antiserum and dilute to 160 ml with buffer. This solution is stable for 1 week when stored in a refrigerator.

Charcoal suspension. Suspend 5 g of charcoal (Norit A) in 500 ml of buffer and stir for 10 minutes. Store in a refrigerator. Stir vigorously for 5 minutes before use and constantly during use. The suspension can be used either cold or at room temperature but should not be stirred continuously on a stirrer which generates heat and therefore warms the suspension to a temperature above room temperature.

Stannous chloride, approximately 4%. Dissolve 8 g of stannous chloride ($SnCl_2$) in 200 ml of water containing 1.7 ml of concd HCl.

Tritiated digoxin. Dilute the tritiated digoxin[10] with 50% ethanol to a final concentration of 16 ng per ml.

For example, if the stock material contains 0.080 mg of digoxin per 1.0 ml, this would be 80,000 ng per ml. Since a concentration of 16 ng per ml is sought, the following calculation is made:

$$80,000 \cdot 1 = 16 \cdot x$$

$$x = 5,000$$

This means that 1 ml of the stock must be diluted to 5,000 ml to give the desired working solution. In practice, 10 μl would be diluted to 50 ml and this solution stored in a refrigerator. A 50 μl portion of this final solution contains 0.8 ng of tritiated digoxin.

Counting solution. Remove 800 ml of toluene from a full 1 gallon bottle of scintillation-grade toluene. To the toluene remaining in the bottle, add 125 ml of ScintiPrep 2,[11] cap, and mix well. Then add 600 ml of Scintisol-GP,[12] cap, and mix well again.

---

[9] From Antibodies, Inc., Davis, Calif., or other suitable source.
[10] From New England Nuclear Corp., Boston, Mass., or other suitable source.
[11] Fisher Scientific Co., Pittsburgh, Pa.
[12] New England Nuclear Corp., Boston, Mass.

PROCEDURE

Note: All Standards, Tests, etc., should be set up in duplicate.

Set up two 12 × 75 mm polystyrene tubes for each nonicteric or non-hemolyzed sample and four of these tubes for each icteric or hemo-lyzed sample.

Set up eighteen additional polystyrene tubes.

Add 0.2 ml of nonicteric or nonhemolyzed serum to each of two tubes (Test).

Add 0.2 ml of icteric or hemolyzed sample to each of four tubes (Test and Corrected Test).

Add 0.2 ml of bovine serum to each of six tubes (Total Count, Blank, and Zero Standard).

Add 1.1 ml of buffer to the Total Count tubes.

Add 0.6 ml of buffer to the Blank tubes.

Add duplicate 0.2 ml amounts of each working standard (S1 through S6) to appropriately labeled tubes.

Add 600 µl of antibody solution to each tube *except* the Blank, Total Count, and Corrected Test, and mix on a vortex mixer.

Allow to stand for 5 minutes at room temperature.

Add 50 µl of tritiated digoxin to every tube and mix on a vortex mixer.

Allow to stand for 5 minutes at room temperature.

Add 500 µl of rapidly stirring (magnetic stirrer) charcoal suspension to all tubes *except* the Total Count and Corrected Test.

Allow to stand for 15 minutes at room temperature.

Centrifuge all tubes *except* the Total Count and Corrected Test for 5 minutes at 2,000 rpm.

Decant the contents of all tubes into appropriate counting vials containing 0.2 ml of stannous chloride solution. Take care to pour each tube to the same extent and to avoid transfer of the charcoal precipitate when decanting the tubes which contain it.

Add 12 ml of counting solution to each vial, cap, mix well, and let stand for 10 minutes at room temperature in the dark.

Obtain the counts per minute (cpm) of the vials (count the vials) using a beta scintillation counter. Count for 3% error (precision) or 5 minutes maximum counting time.

CALCULATIONS

Average the counts from each pair of duplicates.

Plot the average counts per minute (cpm) of the Standards against their respective concentrations on linear graph paper. A typical plot is shown in Figure 26-4.

If any of the samples were icteric or hemolyzed, correct their cpm values as follows:

$$\frac{cpm\ of\ Test}{cpm\ of\ Corrected\ Test} \times cpm\ of\ Total\ Count = cpm\ for\ this\ Test$$

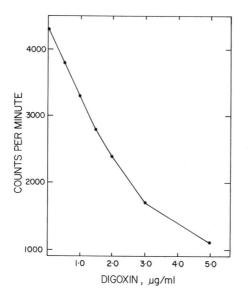

Figure 26-4
Digoxin Standard Curve

Use the standard curve to determine the digoxin concentrations of the Tests in μg per liter from their cpm values.

STANDARDIZATION

Stock standard, 1 mg/ml. Dissolve 25 mg of digoxin in 95% ethanol and dilute to 25 ml, warming to effect solution if necessary. Store in the freezer.

Intermediate standard, 1 μg/ml. Dilute 0.1 ml of the stock standard to 100 ml with 95% ethanol. Store in the freezer.

Digoxin working standards (store refrigerated).

S6 (5 μg/liter): Dilute 250 μl of the intermediate standard to 50 ml
with bovine serum (BS)
S5 (3 μg/liter): 9 ml of S6 + 6 ml of BS
S4 (2 μg/liter): 6 ml of S6 + 9 ml of BS
S3 (1.5 μg/liter): 6 ml of S6 + 14 ml of BS
S2 (1.0 μg/liter): 3 ml of S6 + 12 ml of BS
S1 (0.5 μg/liter): 2 ml of S6 + 18 ml of BS

A series of standards must accompany each group of tests because the standard curve is not linear, and it varies from time to time. A number of suggestions have been made, including the use of log functions and various log graph papers, in an attempt to produce a linear calibration curve. However, since the standards must be analyzed and plotted each time, the use of linear graph paper is as practical a method as any. A typical standard curve is shown in Figure 26-4.

A commonly accepted therapeutic range for digoxin in serum is 0.9 to 2.0 μg per liter. Values below this range may be therapeutically ineffective, and values above this range may be toxic. These criteria do not apply in all situations.

The therapeutic serum range for *digitoxin,* a substance closely related to digoxin in both structure and type of effect, is 14–30 μg per liter, which is about ten times that of digoxin.

DISCUSSION

The general principles of competitive binding assays are discussed in Chapter 8. In the present case, essentially all of the digoxin in the serum and standards first is bound by antidigoxin antibody. The amount of un-reacted antibody then is determined by adding tritiated digoxin (^{3}H-digoxin) and allowing it to bind to the free antibody. The more digoxin present in the serum sample, the more antibody which reacts with this digoxin, leaving less antibody to react with the tritiated, or labeled, digoxin. Since it is the bound labeled material which is finally measured, the counts are inversely related to the digoxin concentration in the serum.

When charcoal is added to the mixture only the free (not bound to antibody) digoxin is adsorbed by the charcoal and removed from the solution. The antibody-bound tritiated digoxin left in the supernatant then is counted.

Since the *Total Counts* tube receives neither antibody nor charcoal, the labeled digoxin remains free but is not removed from the solution. Therefore this tube should have the maximum number of counts.

The *Blank* tube receives no antibody. So all of the labeled digoxin remains free and, therefore, is removed by the charcoal. Theoretically there should be no counts in this tube, but in practice the cpm in the Blank usually are approximately 5% of the Total Counts. A greater percentage than this may indicate a reagent problem.

The difference in counts between the Zero Standard and the Total Counts reflects the degree of binding of the antibody to the labeled digoxin with no unlabeled digoxin present. Ideally the percent binding

$$\left(\frac{\text{cpm Zero Standard}}{\text{cpm Total Counts}} \times 100 \right)$$

should be 50–60%. Marked deviations from this range may indicate a reagent problem.

Icteric (yellow from elevated bilirubin) or hemolyzed (somewhat red from hemoglobin) serum requires correction tubes because this color will reduce the efficiency of the scintillation counting. As discussed in Chapter 8, beta particles (radiation) from radioisotopes like tritium are counted by using special organic molecules called *scintillators*. Such a substance is dissolved in the counting solution and releases a small flash of light when it is struck by a beta particle from an ^{3}H atom. What is actually counted by the scintillation counter then are the light pulses produced by the radiation pulses. When extraneous color is present, the number of light flashes which escape from the vial is reduced because some of these flashes are absorbed or quenched. Thus the counting efficiency is less. The overall effect, if an experimental correction is not carried out, is a falsely high digoxin value for the patient sample.

Although most antibody preparations are fairly specific for the respective glycosides, they do cross react (anti-digoxin antibody with digitoxin or anti-digitoxin antibody with digoxin) to some extent. *Since the usual therapeutic levels of digitoxin are about ten times those of digoxin, a digoxin determination performed with serum from a patient treated with digitoxin (or digitalis) might yield an erroneously high result. Conversely, the result of a digitoxin determination performed with serum from a patient receiving digoxin would be very low (or nondetectable).* For these reasons, it is important that the proper drug (digoxin, digitoxin, or digitalis) be specified when a test is ordered from the laboratory.

PHYSIOLOGICAL SIGNIFICANCE OF DIGOXIN CONCENTRATIONS

Digoxin and the other cardiac glycosides are widely used in the treatment of cardiac disorders. However, the margin of safety with these drugs generally is small. Serum concentrations only slightly above the levels which are therapeutically effective can cause toxic effects such as loss of appetite, nausea, vomiting, visual disturbances, and irregularity of the heart beat. Higher concentrations sometimes can cause these effects to be fatal. Yet there is considerable variation from one person to another in the dose of digoxin which will produce a therapeutic effect, largely because the same dose in two different persons may result in widely different serum levels. For instance, the status of renal function is an important consideration. With severe renal disease, excretion of digoxin is decreased, resulting in its accumulation in the blood, that is, relatively high serum levels from relatively low doses.

The primary value of an analytic method for digoxin, then, is to provide therapeutic monitoring for patients on digoxin therapy.

Carbon Monoxide

Carbon monoxide gas (CO) is a very common and dangerous poison. It is both colorless and odorless. Even low concentrations in the air can be fatal. It is toxic because carbon monoxide combines much more strongly with hemoglobin than does oxygen. This results in the formation of carboxyhemoglobin, rather than oxyhemoglobin, thereby causing deprivation of oxygen in the victim.

Generally, two types of methods are used to measure carbon monoxide levels: (1) spectrophotometric measurement of carboxyhemoglobin, and (2) chemical release of the carbon monoxide as a gas from the carboxy-hemoglobin, followed by quantitation.

Carboxyhemoglobin, which is cherry pink, can be determined spectro-photometrically [1] in the presence of hemoglobin, which is red, because distinctive differences exist in the positions of the spectral absorption peaks of these two substances. To simplify the determination, a reducing agent is added to convert any oxyhemoglobin (O_2-hemoglobin complex) to hemoglobin.

The quantitation technics employed for liberated carbon monoxide gas include gas chromatography [34] and microdiffusion [45]. In micro-diffusion, the gas is absorbed by a second solution in a Conway diffusion cell and participates in a chemical reaction which can be quantitated. In the method presented below, released carbon monoxide is absorbed by a solution of palladium chloride, and quantitation is based on the decrease in absorbance which results from the reduction of palladium chloride by the carbon monoxide.

CARBON MONOXIDE METHOD

METHOD
Williams et al. [45].

MATERIAL
Blood.

REAGENTS
Hydrochloric acid, 0.1N. Dilute 8.3 ml of concentrated HCl to 1 liter.

Sulfuric acid, 10% (v/v). Dilute 100 ml of concentrated H_2SO_4 to 1 liter.

Palladium chloride, 0.005N. Dilute 0.44 g of high-purity palladium chloride ($PdCl_2$) in 500 ml of 0.1N HCl in a 1 liter flask. Allow to stand overnight. Dilute to 1 liter with 0.1N HCl.
One milliliter of this solution contains sufficient palladium to react with all of the carbon monoxide in 0.5 ml of blood which has a carbon monoxide value of 11.2 ml per dl.

PROCEDURE
Note: A hemoglobin determination must be performed with each blood sample.

Pipet 3 ml of the palladium chloride solution into the center well of a Conway diffusion cell (see page 162) that has a greased outer rim.

Add 1 ml of 10% H_2SO_4 to the outer well, and place the cover on the cell, leaving just enough of an opening to add the blood.

Add 0.5 ml of blood through the opening and then slide the cover over to cover the cell completely.

Mix the contents of the outer well by means of gentle rotation, and let stand for 2 h at room temperature.

Transfer the contents of the center well quantitatively to a 50 ml volumetric flask with 3 rinses of 3 ml each of 0.1N HCl.

Dilute to 50 ml with 0.1N HCl and mix well.

Read the absorbance in a spectrophotometer, using a 10 mm cuvet, at 278 nm, setting zero absorbance with 0.1N HCl.

Dilute 3 ml of the 0.005N palladium chloride solution to 50 ml in a volumetric flask with 0.1N HCl, and read the absorbance of this solution in the same way. Any sample which has a zero concentration of CO will give an absorbance value the same as that obtained here.

CALCULATIONS

The carbon monoxide value is determined from a standard curve (see Standardization).

If the absorbance of the palladium chloride solution (3 ml diluted to 50 ml) differs from the value given on the standard curve when this solution was prepared and measured previously, the standard curve needs to be corrected. It is satisfactory to start with the new zero concentration point and draw a new standard curve parallel to the old one.

Perform a hemoglobin determination with the blood sample. Then complete the calculations as follows:

$$\frac{\text{carbon monoxide (ml/dl)}}{\text{hemoglobin (g/dl)}} \times \frac{100}{1.35} = \text{percent of hemoglobin saturated with carbon monoxide}$$

NORMAL VALUES

Normally carbon monoxide may be present in the blood of nonsmokers up to a concentration of 2%. In moderate smokers concentrations up to 5% may be found, and even higher concentrations may occur in heavy smokers.

STANDARDIZATION

Dilute amounts of 0.5, 1.0, 1.5, 2.0, 2.5, and 3.0 ml of 0.005N palladium chloride to 50 ml with 0.1N HCl.

Read the absorbances against 0.1N HCl at 278 nm in a 10 mm cuvet.

Each palladium chloride solution gives an absorbance value which corresponds to a certain level of carbon monoxide in 0.5 ml of blood. The carbon monoxide (CO) values assigned to these solutions are as follows:

ml of 0.005N PdCl₂ diluted to 50 ml	Assigned CO value (ml/dl)
0.5	28.0
1.0	22.4
1.5	16.8
2.0	11.2
2.5	5.6
3.0	0

Draw a standard curve by plotting (on linear graph paper) the absorbances along the ordinate against their respective CO values along the abscissa. This curve should be linear and, although its position, as monitored by the zero standard, may change, the slope remains the same.

DISCUSSION

Carbon monoxide is released from hemoglobin by the dilute sulfuric acid in the outer well. It diffuses into the center well and reacts with the palladium ions, reducing them to palladium metal, which collects as a silvery black film (sheen) on the surface of the palladium chloride solution.

$$PdCl_2 + CO + H_2O \rightarrow Pd + CO_2 + 2HCl$$

The reaction is quantitated based on the decrease in absorbance at 278 nm which occurs when $PdCl_2$ reacts with carbon monoxide. ($PdCl_2$ has an absorption peak at 278 nm in this solution.)

Since a palladium chloride solution can be reduced in its concentration either by dilution or by exposure to carbon monoxide, with the same spectral consequences at 278 nm, a series of palladium chloride solutions of increasing dilution can be used as standards and assigned carbon monoxide values. These values are equivalent to the amount of CO which would be released from 0.5 ml samples of blood with increasing CO concentrations.

PHYSIOLOGICAL SIGNIFICANCE OF CARBON MONOXIDE

Carbon monoxide binds to hemoglobin nearly 200 times stronger than does oxygen. The product is called carboxyhemoglobin. Once a significant quantity of carboxyhemoglobin is formed in the blood, it is no longer possible for the blood to transfer oxygen to the tissues. The result is anoxia and possibly death. Even a concentration of 0.02% of carbon monoxide in the air can cause toxicity in humans. A concentration of 0.5% in the air may cause death in 1 to 2 minutes.

Symptoms (headache, dizziness, muscular weakness, and nausea) may occur when the saturation in the blood (% hemoglobin saturated with CO) reaches 20%. The symptoms may be severe with levels that are over 30%, and death can result with levels as low as 40%. Death usually results with levels over 60%.

The main sources of carbon monoxide are automobile exhausts and defective stoves, furnaces, gas heaters, and motors.

Only whole blood samples are suitable for analysis because the other common fluids (serum, urine) do not contain enough hemoglobin to retain a significant quantity of carbon monoxide.

Lead

All lead compounds are potential poisons, particularly the more soluble ones. Lead poisoning in children usually results from children's chewing on objects painted with lead-containing paints. Industrial exposure is the usual cause of lead poisoning in adults. Clay (ceramic) dishes and cups contain lead which may be leached into foods or fluids over a period of time and result in poisoning. Since lead accumulates in body tissues, long-term ingestion of small amounts of lead results in the same type of symptoms of toxicity as a sudden, large exposure. These symptoms may include cramps, weakness, vomiting, and headache. Lead poisoning in children, when not fatal, may lead to permanent brain damage.

The two most common methods for lead analysis of biological specimens are atomic absorption spectrophotometry and dithizone extraction. The principles of atomic absorption are discussed in Chapter 5. Both flameless and semiflameless technics are satisfactory for lead analysis. One flameless method involves extraction of the lead from a whole-blood sample with the APDC-MIBK reagent. (APDC is ammonium pyrrolidine dithiocarbamate, a chelating agent; MIBK is methyl isobutyl ketone, an organic solvent). The APDC, which is soluble only in the MIBK organic phase, complexes the lead and carries it into the organic phase. A portion of the organic lead solution then is added to the ignition container (for example, carbon rod, graphite furnace or cup, tantalum ribbon or boat) for flameless atomic absorption analysis. Details of the instrumental procedures are available from the manufacturers of each of the various models of these instruments.

The semiflameless atomic absorption technic frequently used for lead is the Delves cup method [11]. A microsample of whole blood first is heated briefly in a nickel cup in the presence of hydrogen peroxide to oxidatively volatilize the organic matter so that it does not interfere (as smoke) in the next step. The cup then is thrust into a flame to volatilize the lead out of the cup and into the light beam of the instrument.

Dithizone is a chelating agent which reacts with lead to form an intensely red-colored complex which can be quantitated spectrophotometrically [38]. The blood or urine sample is first degraded either in a muffle furnace or with hot acid prior to an organic extraction of the lead into an organic solvent solution of dithizone. Interferences by various other metal ions can be eliminated by additional extraction steps. The sensitivity and specificity are much less than with the atomic absorption technics and, therefore, much more sample is required (for example, 10 ml instead of 10 μl of blood).

Samples for lead determination should always be collected in acid-washed containers. Normal limits for lead concentrations in whole blood usually are stated as 60 μg per dl for adults and as 40 μg per dl for children. Up to 80 μg of lead may be excreted in the urine per 24 h in the absence of toxicity. Since virtually all of the lead in blood normally is in the red cells, serum or plasma is not suitable for analysis.

Lithium

Lithium is used to treat manic-depressive psychoses. It is administered as lithium carbonate and produces tranquility in the manic stage of these illnesses. Serum levels of lithium frequently are determined during treatment in order to maintain a therapeutic level and avoid toxic effects. The stated therapeutic range for lithium concentrations in serum is 0.5 to 1.3 meq per liter. Toxic effects may occur when serum lithium levels exceed 1.5 meq per liter and may include nausea, anorexia (loss of appetite), vomiting, diarrhea, mental confusion, convulsions, or a combination of these symptoms. Death from coma may occur at values above 4 meq per liter.

Both flame emission [4] and atomic absorption [17] spectrophotometry are used for lithium measurement. The principles of these methods are discussed in Chapter 5. Flame emission spectrophotometry requires standards which approximate serum in their composition of sodium, potassium, and calcium ions because these ions tend to enhance the lithium signal. The preferred technic for lithium is atomic absorption. In a typical atomic absorption method, 1 ml of serum or 0.1 ml of urine is diluted with water, or sometimes with an EDTA solution, and measured directly against aqueous lithium standards treated similarly. The operating details depend on the particular instrument which is used.

References

1. Amenta, J. S.   The Spectrophotometric Determination of Carbon Monoxide in Blood. In Seligson, D. (Ed.), *Standard Methods of Clinical Chemistry*. New York: Academic, 1963. Vol. 4.
2. Baker, R. N., Alenty, A. L., and Zaek, J. F., Jr.   Simultaneous determination of lower alcohols, acetone, and acetaldehyde in blood by gas chromatography. *J. Chromatogr. Sci.* 7:312, 1969.
3. Baselt, R. C., Wright, J. A., and Cravey, R. H.   Therapeutic and toxic concentrations of more than 100 toxicologically significant drugs in blood, plasma, or serum: A tabulation. *Clin. Chem.* 21:44, 1975.
4. Bender, G. T.   *Chemical Instrumentation: A Laboratory Manual Based on Clinical Chemistry*. Philadelphia: Saunders, 1972. P. 85.

5. Caraway, W. T., and Kammeyer, C. W.   Chemical interference by drugs and other substances with clinical laboratory test procedures. *Clin. Chim. Acta* 41:395, 1972.

6. Christian, D. G.   Drug interferences with laboratory blood chemistry determinations. *Am. J. Clin. Pathol.* 54:118, 1970.

7. Constantino, N. V., and Kabat, H. F.   Drug-induced modifications of laboratory test values—revised 1973. *Am. J. Hosp. Pharm.* 30:24, 1973.

8. Curry, A.   *Poison Detection in Human Organs* (2nd ed.). Springfield, Ill.: Thomas, 1969.

9. Davidsohn, I., and Henry, J. B.   *Todd-Sanford Clinical Diagnosis* (15th ed.). Philadelphia: Saunders, 1974. P. 665.

10. Davidsohn, I., and Henry, J. B.   *Todd-Sanford Clinical Diagnosis* (15th ed.). Philadelphia: Saunders, 1974. P. 693.

11. Delves, H. T.   A micro-sampling method for the rapid determination of lead in blood by atomic absorption spectrophotometry. *Analyst* 95:431, 1970.

12. Dreisbach, R. H.   *Handbook of Poisoning: Diagnosis and Treatment* (7th ed.). Los Altos, Calif.: Lange, 1971.

13. Frings, C. S.   Drug screening. *CRC Crit. Rev. Clin. Lab. Sci.* 4:357, 1973.

14. Goldbaum, L. R.   Determination of barbiturates. *Anal. Chem.* 24:1604, 1952.

15. Goodman, C. S., and Gilman, A. (Eds.).   *The Pharmacological Basis of Therapeutics* (4th ed.). London: Macmillan, 1970.

16. Goth, A.   *Medical Pharmacology* (6th ed.). St. Louis: Mosby, 1972.

17. Hansen, J. L.   The measurement of serum and urine lithium by atomic absorption spectrophotometry. *Am. J. Med. Technol.* 34:625, 1968.

18. Hansten, P. D.   *Drug Interactions.* Philadelphia: Lea & Febiger, 1973. P. 257.

19. Jacobs, M. B.   *The Analytical Toxicology of Industrial Inorganic Poisons.* New York: Interscience, 1967.

20. Kaye, S.   *Handbook of Emergency Toxicology* (3rd ed.). Springfield, Ill.: Thomas, 1970.

21. Kaye, S.   Rapid, simple, reliable tests for poisons. *Lab. Med.* 3:28, 1972.

22. Keller, W. J.   A rapid method for the determination of salicylates in serum or plasma. *Am. J. Clin. Pathol.* 17:415, 1947.

23. Lehmann, J., and Karamustafaoglu, V.   Rapid differentiation of barbiturates in blood serum by thin-layer chromatography. *Scand. J. Clin. Lab. Invest.* 14:554, 1962.

24. Loomis, T. A.   *Essentials of Toxicology.* Philadelphia: Lea & Febiger, 1968.

25. Lubran, M.   The effects of drugs on laboratory values. *Med. Clin. North Am.* 53:211, 1969.

26. Maickel, R. P.   Techniques for the microassay of drugs in biological materials. *CRC Crit. Rev. Clin. Lab. Sci.* 4:383, 1973.

27. Malmstadt, H. W., and Hadjiioannou, T. P.   Specific enzymatic determination of alcohol in blood by an automatic spectrophotometric reaction rate method. *Anal. Chem.* 34:455, 1962.

28. Marks, V., Lindup, W. E., and Baylis, E. M.   Measurement of Therapeutic Agents in Blood. In Sobotka, H., and Stewart, C. P. (Eds.). *Advances in Clinical Chemistry.* New York: Academic, 1973. Vol. 16.

29. Mayers, F. H., Jawetz, E., and Goldfien, A.   *Medical Pharmacology.* Los Altos, Calif.: Lange, 1972. P. 643.

30. McBay, A. J.   Toxicological findings in fatal poisonings. *Clin. Chem.* 19:361, 1973.
31. Mule, S. J.   Routine identification of drugs of abuse in urine: I. Application of fluorometry, thin-layer and gas-liquid chromatography. *J. Chromatogr.* 55:255, 1971.
32. Parker, K. D., and Kirk, P. L.   Separation and identification of barbiturates by gas chromatography. *Anal. Chem.* 33:1378, 1961.
33. Polson, C. J., and Tattersall, R. N.   *Clinical Toxicology* (2nd ed.). Philadelphia: Lippincott, 1969.
34. Rodkey, F. L.   The Measurement of Carbon Monoxide in Biological Fluids. In Sundermon, F. W., and Sunderman, F. W., Jr. (Eds.), *Laboratory Diagnosis of Disease Caused by Toxic Agents.* St. Louis: Green, 1971.
35. Schwartz, M.   Interferences in Diagnostic Biochemical Procedures. In Sobotka, H., and Stewart, C. P. (Eds.), *Advances in Clinical Chemistry.* New York: Academic, 1973. Vol. 16.
36. Scott, R. M.   *Clinical Analysis by Thin-layer Chromatographic Techniques.* Ann Arbor, Mich.: Ann Arbor-Humphrey Science, 1969.
37. Smith, T. W., Butler, V. P., Jr., and Haber, E.   Determination of therapeutic and toxic serum digoxin concentrations by radioimmunoassay. *N. Engl. J. Med.* 281:1212, 1969.
38. Stolman, A., and Stewart, C. P.   Metallic Poisons. In Stewart, C. P., and Stolman, A. (Eds.), *Toxicology, Mechanisms and Analytical Methods.* New York: Academic, 1960. Vol. 1.
39. Sunderman, F. W.   Drug interference in clinical biochemistry. *CRC Crit. Rev. Clin. Lab. Sci.* 1:427, 1970.
40. Sunshine, I. (Ed.).   *Handbook of Analytical Toxicology.* Cleveland: Chemical Rubber, 1969.
41. Thienes, C. H., and Haley, T. J.   *Clinical Toxicology* (5th ed.). Philadelphia: Lea & Febiger, 1972.
42. Van Peenen, H. J., and Files, J. B.   The effect of medication on laboratory test results. *Am. J. Clin. Pathol.* 52:666, 1969.
43. Williams, L. A., Hardy, A. T., Cohen, J. S., and Zak, B.   Ultraviolet differential spectrophotometry of thiobarbiturates. *J. Med. Pharm. Chem.* 2:609, 1960.
44. Williams, L. A., Linn, R. A., and Zak, B.   Determination of ethanol in fingertip quantities of blood. *Clin. Chim. Acta* 3:169, 1958.
45. Williams, L. A., Linn, R. A., and Zak, B.   Ultraviolet absorptiometry of palladium for determination of carbon monoxide hemoglobin. *Am. J. Clin. Pathol.* 34:334, 1960.
46. Williams, L. A., and Zak, B.   Determination of barbiturates by automatic differential spectrophotometry. *Clin. Chim. Acta* 4:170, 1959.
47. Wuth, O.   Rational bromide treatment. *J.A.M.A.* 88:2013, 1927.
48. Young, D. S., Pestaner, L. C., and Gibberman, V.   Drug interferences with clinical laboratory tests. *Clin. Chem.* (Special Issue). April, 1975. (Entire issue).
49. Zettner, A., and Duly, P. E.   Principles of competitive binding assays (saturation analyses). II. Sequential saturation. *Clin. Chem.* 20:5, 1974.

27. Cerebrospinal Fluid (CSF) and Gastric Analysis

Of the various biological fluids submitted for chemical examination the commonest, with the obvious exceptions of blood and urine, are cerebrospinal fluid (CSF) and gastric secretions. Although these fluids contain numerous chemical compounds, only a few are of clinical interest. The tests described in this chapter are those requested most frequently.

Cerebrospinal Fluid (CSF) Analysis

Most of the chemical constituents of blood are also present in cerebrospinal fluid (CSF), but the concentrations in this fluid of only a few such substances are of interest to the clinician. The concentrations of some of the more commonly analyzed constituents are listed in Table 27-1.

Cerebrospinal fluid is obtained by a procedure called *lumbar puncture,* in which the fluid is sampled with a special needle. The technician should always be aware of the fact that the repetition of a lumbar puncture, if at all possible, is no simple matter for the physician or the patient. For this reason, the technician should make every effort to perform *all* analyses requested even though the amount of fluid available may be quite small. Moreover, he should perform the tests with extra care, in the realization that repetition may be impossible.

Another point to be considered when working with cerebrospinal fluid is that pathogenic microorganisms may be present. Therefore caution should be exercised in pipetting the fluid. Pipets used for this purpose should be soaked in a bactericidal agent before cleaning and reuse.

GLUCOSE

The concentration of glucose in CSF is determined in the same manner as blood glucose (see Chapter 11).

If bacteria are present in CSF, they utilize glucose rapidly. Therefore it is important to prepare a filtrate, or add fluoride, as soon as possible after the fluid is obtained.

The glucose concentration in the CSF of healthy persons is 50 to 70 mg per dl. The presence of bacteria results in a low glucose value. In cases of severe bacterial meningitis this value may drop to an indeterminably low concentration (0 to 5 mg per dl).

Since the glucose concentration in CSF usually parallels the blood glucose level, diabetic patients with high blood-glucose concentrations may have correspondingly high levels of glucose in the CSF.

CHLORIDE

The chloride method described in Chapter 10 may be used for determining the chloride concentration in CSF. Normally, the concentration of chloride in CSF is 120 to 130 meq per liter, but lower values often are

370

Table 27-1. Partial Composition of Cerebrospinal Fluid

Substance	Normal Range
Bicarbonate	25–28 meq/liter
Calcium	4.6–5.6 mg/dl
Chloride	120–130 meq/liter
Gamma globulin	Up to 15% of total protein
Glucose	50–70 mg/dl
Potassium	2.2–3.4 meq/liter
Sodium	142–150 meq/liter
Total protein	20–45 mg/dl
Urea nitrogen	5–15 mg/dl

found in meningitis and other conditions in which the CSF protein concentration is increased. Changes in the blood-chloride concentration usually are reflected directly in the CSF.

PROTEINS

The most common tests for proteins in cerebrospinal fluid are total protein, total gamma globulin, gamma G globulin (IgG), electrophoresis, and A/G (albumin-globulin) ratio. The latter four tests measure one or more of the fractions included in the total protein measurement.

TOTAL PROTEIN

Total protein usually is determined in CSF by a turbidimetric [1, 6, 8] or Lowry-Folin [5] technic.

The principles of turbidimetry are presented in Chapter 4. An agent like trichloroacetic acid or sulfosalicylic acid is used to precipitate the proteins as a fine suspension. The extent of the precipitate reflects the amount of protein present. Although a high degree of reliability is maintained only with difficulty, the method is sensitive and rapid. There are many theoretical objections to turbidimetric methods; when properly set up and standardized, however, these technics are adequate for measuring CSF total protein.

The Lowry-Folin method is based on the ability of proteins to reduce a phosphotungstic-phosphomolybdic acid reagent to a blue color, especially when Cu^{2+} ions also are present. The actual moieties on the protein which carry out this reaction are the amino acids tyrosine and tryptophan, plus the complexed Cu^{2+} ions. Although this method offers good sensitivity, it is subject to interferences from various drugs and certain nonprotein material. An arbitrary value (correction factor) sometimes is subtracted from all of the readings to correct for interference from nonprotein material, assuming an average amount of these nonprotein interfering materials are present.

The micro Kjeldahl or biuret technics (see Chapter 15) may be used to determine total protein in CSF but, due to the relatively low protein concentration in the fluid, 2 or 3 ml are necessary for an accurate determination. This requirement plus interference problems render these methods impractical for routine use.

The method presented below is a turbidimetric method employing sulfosalicylic acid.

TOTAL PROTEIN METHOD

METHOD
Denis-Ayer [1] modified.

MATERIAL
Cerebrospinal fluid.

REAGENTS
Sulfosalicylic acid, 1.5%. Dissolve 3.0 g of sulfosalicylic acid hydrate $(HO_3SC_6H_3OHCO_2H \cdot 2H_2O)$ in water and dilute to 200 ml. Store in a dark bottle.

Sodium chloride, 0.85%. Dissolve 8.5 g of NaCl in water and dilute to 1 liter.

PROCEDURE
Into each of three 19 mm cuvets measure 5.5 ml of sulfosalicylic acid.
To one add 0.5 ml of water (Blank), to another 0.5 ml of CSF (Test), and to the third 0.5 ml of standard (Standard).
Mix well and allow to stand for 10 minutes.
Read the absorbance of the Test and Standard in a spectrophotometer at 420 nm using the Blank to set zero absorbance.

CALCULATIONS

Total protein value of Standard in mg/dl $\times$ $\dfrac{\text{absorbance of Test}}{\text{absorbance of Standard}}$

$$= \text{CSF total protein (mg/dl)}$$

STANDARDIZATION
Standard. Dilute 0.1 ml of serum of known total protein concentration to 20 ml with 0.85% sodium chloride. Prepare fresh each day.

$$\frac{\text{Total protein value of serum (g/dl)}}{0.2}$$

$$= \text{Total protein value of Standard in mg/dl}$$

NORMAL VALUES
The total protein concentration in CSF normally is 20 to 45 mg per dl.

DISCUSSION
The proteins in CSF are precipitated by sulfosalicylic acid. The resulting turbidity is determined with a spectrophotometer. At low concentrations,

turbidities may be determined with reasonable accuracy with a spectro-photometer, but if the turbidity is dense a nephelometer should be used. With the method described here, reproducible curves, which follow Beer's law closely up to 100 mg per dl, are obtained using a spectro-photometer.

The first few drops of CSF should be added to the sulfosalicylic acid solution slowly; if a dense cloud forms immediately, the protein concentration probably is very high. In this event the rest of the 0.5 ml sample should not be added to the acid but be returned to its original tube, and the determination should be set up using 0.2 ml or 0.1 ml of CSF. This precautionary measure sometimes makes possible the accurate determination of total protein without the need for an additional sample of CSF. Occasionally a sample will read too high even with 0.1 ml of fluid. In this case, the determination may be made using the *serum* total protein method described on page 184.

Sulfosalicylic acid produces different degrees of turbidity with equal concentrations of different types of protein [4, 6]. A solution of pure albumin, in this method, yields about twice as much turbidity as a solution of the same concentration of globulin. Because the amount of turbidity produced is dependent, to some extent, upon the albumin-globulin (A/G) ratio, standard solutions for this method should be prepared, as nearly as possible, with the same percentage of albumin and globulin as is found in CSF. Since the A/G ratio in CSF is of very nearly the same order as that in serum, the best standard is normal human serum.

If a CSF protein standard is purchased from a commercial source, the analyst should seek assurance that the standard has *not* been prepared with pure albumin since this introduces an error of approximately 85% in the CSF protein determination. Some of the commercially available control sera contain only albumin as a protein and are, therefore, unsuitable for use in standardizing the turbidimetric CSF protein method.

Trichloroacetic acid or a solution of an acid and a salt sometimes are used in place of sulfosalicylic acid methods. Although these reagents are unaffected by the A/G ratio, they lack the sensitivity and precision of sulfosalicylic acid. Under proper conditions, the latter reagent gives results that are reasonably accurate and reliable.

If red cells are present in a sample of CSF, the fluid should be centrifuged and decanted before protein analysis, to avoid the possibility of obtaining falsely high values due to hemoglobin.

PHYSIOLOGICAL SIGNIFICANCE OF
CSF TOTAL PROTEIN CONCENTRATIONS

The total protein concentration in CSF is elevated in most types of meningeal inflammations and is, therefore, of little value in differential diagnosis. Elevations in CSF protein concentrations may also result from brain tumors, subarachnoid hemorrhage, or occasionally from contamination of the CSF with blood as the result of trauma during withdrawal of the fluid. High protein concentrations due to the latter cause are erroneous and may be misleading; therefore, the technician should record such contamination.

ELECTROPHORESIS

Electrophoresis is the method of choice for fractionation of the major groups of proteins in biological fluids. Essentially, the same five groups of proteins are encountered in CSF as in serum, in about the same relative amounts (see Chapter 15). The only major complication with CSF electrophoresis is the necessity to concentrate the sample prior to the analysis because of the low concentration of protein which is present. Two commercial devices[1] are particularly useful. One is an ultrafiltration cone which fits into the top of a test tube. The cone is inserted into the tube, the sample is placed in the cone, and the entire unit is centrifuged to force most of the liquid and small solutes through the cone, leaving essentially all of the protein in a drop or two of residual fluid at the bottom of the cone. Because the pores in the cone must be very small in order to retain the proteins, the rate of fluid flow through the cone is negligible without the driving force provided by centrifugation.

A second unit is a vertical cell which tapers to a point on the bottom like a centrifuge tube, except it is flat and one side is an ultrafiltration membrane backed by an absorbent pad. In this case, fluid and small solutes are pulled through the membrane by the absorbent pad rather than by centrifugal force.

Even when these concentrating technics are used, sometimes the final protein concentration still is too low for reliable electrophoretic analysis. However, more rigorous concentrating methods may result in denaturation of some of the proteins.

TOTAL GAMMA GLOBULINS

Total globulins may be measured as a percentage of total protein by electrophoresis or by precipitation with zinc sulfate [9]. As discussed above, a concentration step is necessary before an electrophoresis pattern can be obtained. Even then quantitation may not be accurate. In the case of zinc sulfate precipitation, quantitation can be based either on the turbidity of the precipitate or on a colorimetric test for protein after the precipitate is washed free of interfering substances. However, this is neither a particularly convenient nor sensitive method. Because the immunoglobulin G (IgG) component of total gamma globulins can be measured with much greater sensitivity and convenience than total gamma globulins, and also provides basically the same diagnostic in-

[1] Amicon Corp., Lexington, Mass.

formation, the IgG method has replaced methods for total globulins or gamma globulins in many laboratories.

IMMUNOGLOBULIN G (IgG)

Of the five types of immunoglobulins (IgG, IgA, IgM, IgD, IgE), immunoglobulin G (gamma G, IgG) is the one which occurs in highest concentration in most of the fluids of the body, including the blood and cerebrospinal fluid. It can be measured conveniently in CSF using a low-level IgG radial immunodiffusion plate.

IgG METHOD

METHOD
Berner et al. [2].

MATERIAL
Cerebrospinal fluid.

APPARATUS
Commercial, low-level IgG immunodiffusion plates (Hyland or other suitable brand).

PROCEDURE
Open the immunodiffusion plate. If the plate contains droplets or an excessive film of moisture on the surface of the agar, allow the plate to remain uncovered at room temperature until these droplets or excess film of water disappear (15–20 minutes). (A large excess of water also may be absorbed carefully with a tissue after tipping the plate to collect the water on one side.)
Fill the wells with the Test and Standards, and replace the cover.
Incubate for the recommended time and temperature, for example, 48 h at 37 C, in a moisture chamber.
Remove the cover and, with a calibrated viewer, measure to the nearest 0.1 mm the diameter of each precipitin ring.
Prepare a standard curve by plotting the diameters or computed areas of the Standards against their IgG concentrations.
Determine the concentrations of the Tests from the standard curve.
If the IgG level of a Test approaches or exceeds that of the highest Standard, dilute the CSF with 0.85% NaCl, and repeat the analysis. Multiply the final result by the dilution factor.

STANDARDIZATION
Use commercially available reference solutions of IgG.

NORMAL VALUES
The concentrations of IgG in cerebrospinal fluid are best stated in terms of percentages of total protein. Normal values for IgG are up to 15% of the CSF total protein concentration.

DISCUSSION

The basic principles of the immunodiffusion technic are discussed in Chapter 15. In the procedure presented here, commercially available agar plates are used. These are flat plastic dishes which contain a layer of agar in which a small amount of anti-IgG antibody has been dissolved. This is accomplished by adding anti-IgG antiserum to a warm solution of agar, then allowing the agar to cool to form a gel. The sample and standards are placed in precut wells in the agar and allowed to diffuse outward in all directions. The antibody recognizes only the IgG in these solutions and reacts with it to form a visible precipitate once the IgG from the sample has been diluted sufficiently by the diffusion process to reach a precipitating concentration. The result is a ring of precipitate around the sample well. Samples with higher IgG concentrations will give rise to larger precipitin (precipitate) rings. Either the diameters or areas of the rings are then measured to determine the IgG concentration. Because of the sensitivity of the results to both the conditions and reagents used, it is important to always include several standards and construct a standard curve.

All rings should be symmetric, and no extra bands or spots of precipitate should occur. An asymmetric ring indicates improper sample application and requires that the test or standard involved be repeated. Extra precipitate also indicates improper sample application, usually spillage of the sample onto the surface of the agar directly from the applicator or indirectly from an overfilled well.

PHYSIOLOGICAL SIGNIFICANCE OF
CSF GAMMA GLOBULIN CONCENTRATIONS

This test is used primarily for the diagnosis of multiple sclerosis. With this disease, gamma globulin (or IgG) concentrations usually are greater than 15% of the CSF total protein concentration.

A/G RATIO

Although the A/G, or albumin-globulin ratio, can be calculated after an electrophoretic analysis, a simple screening test is available in which the results depend on the relative amounts of albumin and globulin which are present. This is the colloidal gold test by Lange.

COLLOIDAL GOLD TEST

Since its publication, there have been many modifications proposed which have been directed mainly toward adjusting the sensitivity of the method and the stability of the gold sol. The method described here has been found suitable in several laboratories.

METHOD

Lange [7], modified by Borowskaja [3].

MATERIAL
Cerebrospinal fluid.

REAGENTS

Sodium chloride, 0.4 percent (*w/v*). Dissolve 4 g of sodium chloride (NaCl) in water and dilute to 1 liter.

Gold sol.[2] All water used in the preparation of this solution should be highly purified. All glassware should also be rinsed with this water.
Dissolve 0.15 g of gold chloride ($AuCl_3 \cdot HCl \cdot 3H_2O$) in 15 ml of water (use a plastic spatula).
Dissolve 0.5 g of sodium citrate ($Na_3C_6H_5O_7 \cdot 2H_2O$) in 50 ml of water.
To a 2 liter Erlenmeyer flask add 940 ml of water and 10 ml of the gold chloride solution. Heat to 90 C. Then add the 50 ml of citrate solution. Allow the solution to boil gently for 2 to 3 minutes during which time a clear red color should develop. Remove from the heat and cool to room temperature. Store in a clean, tightly stoppered bottle and do not expose to strong sunlight. The pH should be approximately 6.4. This solution is stable at room temperature.

PROCEDURE

Set up eleven 13 × 100 mm test tubes (disposable tubes are recommended).
Add 0.9 ml of 0.4% NaCl to the first tube and 0.5 ml to each of the other 10 tubes.
Add 0.1 ml of CSF to the first tube, mix, and serially dilute 0.5 ml through the next 9 tubes discarding the last 0.5 ml. Tube 11 receives no CSF.
Add 2.5 ml of colloidal gold solution to each of the 11 tubes, mix well, and allow to stand overnight (away from light).
Since tube 11 receives no CSF, it serves as a Blank. Any color changes in the other tubes are recorded according to the following scale:

0 = No color change (as Blank)
1 = Pink
2 = Lilac or purple
3 = Blue
4 = Pale blue with precipitin
5 = Dense precipitate, colorless supernatant

DISCUSSION AND PHYSIOLOGICAL SIGNIFICANCE
OF THE COLLOIDAL TEST

Normally a colloidal gold test should result in no discoloration or precipitation in any of the tubes (0000000000). If a change does occur,

[2] If difficulty is experienced in the preparation of this solution, it might be advisable to purchase it from a reliable commercial source.

Table 27-2. Abnormal Colloidal Gold Curves

Type of Curve	Example
First zone	5555431000
Mid zone	0003331100
End zone	0000234444

usually it indicates some type of central nervous system disease. Differential diagnosis sometimes may be made by noting the segment of the curve in which the precipitation occurs. The curve is arbitrarily divided into three zones beginning with the first tube; typical examples of positive curves are illustrated in Table 27-2.

Due to the sensitivity of colloidal gold solution, falsely positive tests sometimes may occur. A false positive usually is indicated when the curve does not follow the curve patterns described here but rather appears as a high reading flanked by low readings. For example, readings of 0002000500 suggest that tubes number 4 and 8 are false positives.

Colloidal gold curves sometimes are helpful in the differential diagnosis of various diseases of the central nervous system. However, since there are no absolutely characteristic changes in the curves in any such disease, the clinical value of the test is somewhat limited.

Colloidal gold is influenced by albumin and globulin, the albumin acting to preserve the colloidal state while the globulin tends to precipitate the gold. When the two protein fractions are present in normal concentrations and ratio in CSF, the colloidal gold is unaffected. With protein disturbances, however, the colloidal gold is precipitated at certain dilutions of CSF. The range of dilutions affected is dependent upon the nature of the protein alteration, which is in turn related to the type of disease.

Successful preparaion of the gold sol depends upon the use of very pure water and scrupulously clean glassware. Sometimes ordinary distilled water will suffice, but sometimes it must be redistilled from an all-glass apparatus. The glassware used for preparation and storage of the reagent should be cleaned with acid cleaning solution and rinsed well with purified water.

Any disturbance of the colloidal state results in the formation of a purple precipitate. Thus, the slightest amount of contaminant present in a tube may cause precipitation. For this reason, the use of clean, disposable tubes is recommended.

When a colloidal gold test is left overnight, the tubes should be covered with a clean towel to prevent contamination by dust.

The concentration of the sodium chloride solution should be verified by sodium or chloride analysis after it is prepared, since gross deviations from 0.4% may result in precipitation of the gold.

Gastric Analysis

Gastric analysis is concerned with the function of the stomach. Gastric juice is a solution of hydrochloric acid, electrolytes (sodium, potassium, calcium, chloride, magnesium, phosphate), digestive enzymes (mainly pepsin), intrinsic factor (the protein which facilitates vitamin B_{12} absorption), and other miscellaneous substances that all are secreted by the stomach. Usually only the concentration of acid being secreted is of clinical interest. This determination may be made either directly with a sample of gastric juice obtained through a tube which the patient has swallowed or indirectly with a special technic using urine samples.

In the direct approach, the acidity of the sample can be measured with a pH meter or by an acid-base titration. Either phenolphthalein or a pH meter can be used to monitor the titration procedure. The value obtained is called *total acidity*. The acidity can be determined in two parts known as *free* and *combined acid*. Free acid is calculated from the amount of base necessary to obtain an endpoint with Topfer's reagent, an indicator which turns from red to orange at pH 3.5. Combined acid is calculated by the amount of base necessary to titrate the rest of the acid to a phenolphthalein endpoint (pH 8.5). The names *free* and *combined* derive from the idea that the HCl which is titrated to pH 3.5 is largely free, while that which is titrated between pH 3.5 and 8.5 is largely combined with (loosely bound to) protein. Obviously, free plus combined acidity equals total acidity. Usually only the total acidity is measured when a titration procedure is performed.

In the tubeless (indirect) technic, a substance is taken orally whose degree of intestinal absorption into the blood depends on the amount of acid in the stomach. The substance then is rapidly cleared from the blood by the kidneys into the urine. The urinary level of the substance reflects the acidity level in the stomach. One such compound is Diagnex Blue. This is a fine suspension of ion exchange plastic beads to which the blue dye, azure A, is bound ionically. When this material reaches the stomach, the azure A is released from the resin (ion exchange beads) by the action of the H^+ in the stomach. The degree of this release depends on the acid level in the gastric juice. Only the azure A that is released can make its way into the blood and then into the urine. The urinary level is quantitated spectrophotometrically. Although not all of the dye ends in its blue form in the urine, a brief heating step after addition of acid and a copper salt converts all of it into this form.

Certain features of these two approaches (direct and tubeless technics) are analogous. For instance, both a control and one or more samples either of gastric juice, or, in the tubeless procedure, urine, may be obtained for analysis. In these cases, the patient fasts overnight, a control gastric or urine sample is collected, and a stimulant of gastric secretion such as Histalog (a histamine analog also known as Betazole) or gastrin (a hormone) is administered. Samples of the gastric juice are collected at timed intervals thereafter, for the direct approach. The tubeless reagent is given and a urine sample is collected 2 hours later in the tubeless method.

The advantages of direct analysis of gastric juice is that the results are more directly quantitative, and analysis can be carried out as a function of time after administration of a stimulant. The distadvantages are the discomfort to the patient from the stomach tube and the difficulty in obtaining a pure sample of gastric juice without contamination from saliva and duodenal contents. Although the tubeless method offers less patient discomfort, the disadvantages are many: (1) the results are only qualitative, (2) the results are dependent upon normal intestinal and renal function, (3) the procedure actually is rather time consuming, and (4) procedural errors such as failure of the patient to completely swallow the reagent and failure to collect all of the urine can produce large errors.

TOTAL ACID (DIRECT) METHOD

SAMPLE
Gastric juice.

REAGENTS
Phenolphthalein solution. Dissolve 1 g of phenolphthalein in 100 ml of 95% ethanol.

Sodium hydroxide, 0.1N. A standardized solution should be used. Obtain this commercially or prepare as described in Chapter 1.

PROCEDURE
If the specimen contains solid food particles, filter it through several layers of cotton or glass wool.

Measure 10 ml of sample and approximately 10 ml of water into a white porcelain evaporating dish.

Add 3 drops of the phenolphthalein solution and titrate with 0.1N sodium hydroxide to a pink color. (If a pH meter is used, titrate to pH 8.5.)

CALCULATIONS

ml of 0.1N NaOH × 10 = Total Acid (meq/liter)

NORMAL VALUES AND PHYSIOLOGICAL SIGNIFICANCE OF GASTRIC ACID CONCENTRATIONS
The total acid concentration usually falls in the range 10 to 50 meq per liter. The value is substantially higher after stimulation with Histalog, gastrin, or some other stimulant.

The normal pH of gastric juice is 1.5 to 4.0. The use of a stimulant will drop the pH to a value of 2.0 or lower in normal individuals.

If acid is absent (achlorhydria) or present in very low concentration (hypochlorhydria), this may be diagnostic of pernicious anemia or gastric carcinoma. The gastric juice from patients with gastric ulcer usually contains abnormally large amounts of hydrochloric acid (hyperchlorhydria).

References

1. Ayer, J. B., Dailey, M. E., and Fremont-Smith, F.   Denis-Ayer method for the quantitative estimation of protein in the cerebrospinal fluid. *Arch. Neurol. Psychiatry* 26:1038, 1931.
2. Berner, J. J., Ciemins, V. A., and Schroeder, E. F.   Radial immunodiffusion of spinal fluid. *Am. J. Clin. Pathol.* 58:145, 1972.
3. Borowskaja, D. D.   Zur methodik der goldsolbereitung. *Z. Immunitaetsforsch.* 82:178, 1934.
4. Bossak, H. N., Rosenberg, A. A., and Harris, A.   A quantitative turbidimetric method for the determination of spinal fluid protein. *J. Vener. Dis. Inf.* 30:100, 1949.
5. Daughaday, W. H., Lowry, O. H., Rosebrough, N. J., and Fields, W. S.   Determination of cerebrospinal fluid protein with the folin reagent. *J. Lab. Clin. Med.* 39:663, 1952.
6. Henry, R. J., Sobel, C., and Segalove, M.   Turbidimetric determination of proteins with sulfosalicylic and trichloroacetic acids. *Proc. Soc. Exp. Biol. Med.* 92:748, 1956.
7. Lange, C.   Die Ausflockung kolloidalen goldes durch Zerebrospinalflüssigkeit bei luetischen Affekitonen des Zentralnervensystems. *Z. Chemother.* 1:44, 1912.
8. Meulemans, O.   Determination of total protein in spinal fluid with sulphosalicylic acid and trichloroacetic acid. *Clin. Chim. Acta* 5:757, 1960.
9. Papadopoulos, N. M., Hess, W. C., O'Doherty, D., and Wakeman, L.   Spinal-fluid protein determination for the differentiation of neurologic disorders. *Clin. Chem.* 9:97, 1963.

28. Kidney Function Tests and Analysis of Urine

The kidneys excrete waste products from the body and regulate the chemical composition of body fluids. There are several constituents of the blood whose concentrations may be used to assess kidney function, but much information may also be gained by examining the urine produced by the excretory process of the kidneys.

To help explain the clinical significance of the several available kidney function tests, the physiology of the kidney is considered here in a very brief and general way. Each kidney contains approximately one million working units called *nephrons*. Each nephron consists of a ball of capillaries, known as a *glomerulus,* connected to a smaller tubular channel, which is called a *tubule*. Three distinct processes take place in each nephron (Figure 28-1): (1) The blood flowing through the capillaries of the glomerulus is filtered, the filtrate passing into the tubule through the glomerular membrane. This process is called *glomerular filtration*. The glomerular filtrate contains most of the constituents of the blood with the exception of protein, which does not pass through the normal glomerular membrane in any appreciable quantity. (2) As the glomerular filtrate passes along the tubule, more solutes are added by secretion through the walls of the tubule from adjoining blood vessels. This process is called *tubular secretion*. (3) Solutes and water also pass back through the walls of the tubule into the blood by the process of *tubular reabsorption.*

The end product of these three processes passes through larger collecting tubules into the ureters, which carry it to the bladder, from which it is excreted through the urethra as urine. The constituents of urine are the products of glomerular filtration *plus* the products of tubular secretion *minus* constituents returned to the blood through tubular reabsorption.

Urine is a complex aqueous solution of inorganic salts and organic waste products of body metabolism. Some 60 g of solids in solution normally are excreted every 24 hours, including 35 g of organic matter such as urea, creatinine, and uric acid, and 25 g of mineral salts, which are mainly chlorides and phosphates. General analysis of some of the physical and chemical properties of urine can yield useful information to the physician. There is a small group of simple tests and observations which is widely used and adopted as a good indication of general kidney function. These tests constitute what is called a *routine urinalysis.* Since this subject is covered in detail in several books [1, 2, 6], only some of the more important aspects are considered here.

Routine Urinalysis

The urine sample most suitable for routine examination is an early morning specimen collected before breakfast because this is usually a

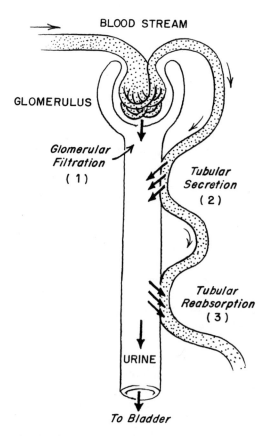

BLOOD STREAM

GLOMERULUS

Glomerular
Filtration
(1)

Tubular
Secretion
(2)

Tubular
Reabsorption
(3)

URINE

To Bladder

Figure 28-1
Basic Functions of the Nephron

concentrated specimen with an acid pH. The urine should be examined while it is fresh; on standing, bacteria multiply, glucose decomposes, the urine becomes alkaline due to the liberation of ammonia from urea by bacteria, and many of the microscopic elements disintegrate. If a urine sample cannot be analyzed while fresh, it may be preserved to some extent by refrigeration.

VOLUME

The volume of a random urine sample, such as that usually received for routine analysis, is of no significance and consequently need not be recorded. The average amount of urine excreted by a healthy adult is approximately 1200 to 1500 ml per 24 hours, but even in health this volume is variable, depending upon such factors as water intake, diet, environmental temperature, and activity. In diseases such as diabetes mellitus and diabetes insipidus, gross increases in 24-hour urine volumes may occur (polyuria), while diminished urine output (oliguria) occurs with certain types of kidney damage and with dehydration.

CLARITY AND COLOR

Normally urine is clear, although turbidity may appear occasionally due to mucus or precipitated salts (particularly urates or phosphates). Abnormal turbidity may be due to large quantities of cells (red, white, or epithelial) or to bacteria. A report of the clarity of urine in terms of "clear," "cloudy," or "turbid" is adequate.

The color of urine excreted by healthy individuals is due to urochrome pigments and ranges from almost colorless to dark yellow. The color depends mainly upon the concentration of the sample, the more concentrated specimens usually being a darker yellow. Certain foods may also impart color to the urine. The most dramatic common example of this is the red color that may appear in urine after ingestion of beets; frequently this is mistaken for hematuria (blood in the urine).

In disease there are many compounds that may appear in the urine and impart an abnormal color. The commonest of these are bilirubin (amber color) and blood (red or rust color). The color of the urine should be described as accurately and simply as possible.

pH

The pH of urine may be determined with adequate accuracy with pH paper which, when immersed in an aqueous solution, takes on a characteristic color dependent upon the pH of the solution. The pH may be estimated to the nearest 0.5 pH unit by comparing the test paper with a chart.

The pH of urine normally varies between 4.5 and 8.0, with an average pH of approximately 5.5. The early morning specimen usually is acid, but later in the morning the urine frequently becomes alkaline.

Test Strips and Tablets

A number of test strips and tablets are available[1] for testing urine. In each case, a mixture of chemicals has been either mounted as a thin pad onto the end of a plastic strip or fabricated into a tablet. A positive test is revealed by a characteristic color change. It is important to follow the directions carefully, since erroneous results can occur with poor technic. The user should also be aware of the interferences associated with each procedure; this information is available from the manufacturer. A few of the more commonly used products are discussed here briefly.

GLUCOSE

Clinitest tablets[1] and Clinistix strips[1] are both used to test for glucose in urine. The tablets contain all of the reagents involved in Benedict's

[1] Ames Co., Elkhart, Indiana.

reaction for reducing sugars. The addition of a tablet to a urine sample produces a hot alkaline solution because the tablet contains sodium hydroxide. The cupric sulfate in the tablet dissolves and reacts with any reducing sugars like glucose to afford a mixture of cuprous salts including yellow cuprous hydroxide (CuOH) and red cuprous oxide (Cu_2O). With increasing concentrations of reducing sugars, the colors range from blue (no reaction) to green to brown to orange. Specific directions and a color chart are provided by the manufacturer.

It is important to observe the reaction carefully since a very high glucose level can cause the reaction to first turn green, then orange, then back to green again. Under these circumstances, the urine should be diluted with water and the test repeated. A positive test merely indicates the presence of reducing substances in the urine. A number of other sugars besides glucose as well as certain other compounds (for example, vitamin C) can give false positive results.

The Clinistix strips utilize the glucose oxidase reaction for glucose. The reagents and steps involved in this reaction are discussed on page 145. The enzymes glucose oxidase and peroxidase are impregnated into the test-pad portion of the strip along with a chromogen. The strip is dipped briefly in urine. The strip turns blue (oxidized chromogen) if glucose is present. Because of their specificity, the strips can be helpful in distinguishing glucose from other reducing substances, for example, lactose, which commonly is present in the urine of pregnant women.

PROTEIN

A number of products are commercially available for the semiquantitative detection of protein in urine. The most common of these products are impregnated test strips[2] which rely on the tendency of the protein in a urine sample to bind the blue form of the indicator, bromphenol blue, rather than the yellow (protonated) form. Thus, the intensity of the blue color of the test portion of the strip increases with the concentration of protein in the urine.

Other tests, available as tablets or powdered mixtures of reagents, precipitate the proteins and, therefore, produce a turbidity when increased protein concentrations are present. These tests are analogous to the turbidimetric method for urine proteins given on page 388.

KETONE BODIES

The ketone bodies sometimes found in urine are acetoacetic acid, β-hydroxybutyric acid, and acetone. Their presence indicates excessive metabolism of body fats, which may occur in diabetic acidosis or malnutrition.

Reagent tablets (Acetest[3]) and strips (Ketostix[3]) are available for

---

[2] Albustix, Ames Co., Elkhart, Indiana.
[3] Ames Co., Elkhart, Indiana.

detection of ketone bodies in urine. Both employ the nitroprusside reaction. Ketones in the presence of an amine like glycine form a lavender-purple colored complex with nitroprusside.

BLOOD

Occult blood (blood not visible to the eye) can be revealed in urine with a Hematest (Ames Co.) strip. The test portion of the strip is impregnated with an organic peroxide and orthotolidine. The hemoglobin of the blood catalyzes the oxidation of the orthotolidine by the organic peroxide to green-blue-black compounds which subsequently turn red.

Total Solutes

The major solutes in urine are sodium, potassium, calcium, chloride, sulfate, phosphate, urea, uric acid, and creatinine. The urine that is formed initially in the interior of the kidney is much less concentrated in these substances than that which leaves this organ. This is because the kidney recovers much of the water that is present in the initially formed urine before it is discarded, thereby conserving body water. This concentrating ability of the kidneys is one of its most important functions, and, in cases of renal tubular damage, is one of the first functions to be lost. Thus a measure of the solute content of urine is an indication of kidney function.

The three conveniently measured properties of urine that depend on its solute content are specific gravity, refractive index, and osmolality. Although an increased solute content will increase the values of each of these properties, the observed increases are not always proportional to one another because some solutes affect each of these properties differently. For instance, protein (an abnormal urine solute) contributes more to specific gravity and refractive index readings than to osmolality values.

Tests for total solutes sometimes are performed under special dehydration or hydration conditions. These procedures are called *concentration tests* and *dilution tests,* respectively. In the former, the patient may be asked to avoid all fluids with and between meals during a 24-hour period before a series of urine samples is collected for analysis. In the dilution test, after an overnight fasting period, the patient may be asked to drink, for example, 1.4 liters of water, during a 45-minute period prior to collection and analysis of a series of urine samples. The specimens should attain a specific gravity of 1.025 or more in a concentration test and reach a lower level of 1.002 or less in a dilution test if the kidneys possess their normal functional capacity to cope with water challenges (excess or deficiency) to the body.

Specific Gravity

Specific gravity depends on the number, density, and weight of dissolved particles. Measurements usually are made with a urinometer, that is, a hydrometer that has been adapted to measure specific gravity in urine. The urinometer is a float calibrated from 1.000 (the specific gravity of purified water) to 1.040, which is as high a specific gravity as one would expect with human urine. The test consists merely in dropping the float into a container of urine and reading the level at which it equilibrates. The minimum volume of urine required usually is approximately 15 ml. The float must not be touching the sides or bottom of the vessel when the reading is taken, and the urine should be at the temperature for which the float was calibrated, usually room temperature. If the urine is colder than the specified temperature, positive errors will be introduced, and the reverse is true at higher temperatures. When necessary, corrections may be made by noting the temperature of the urine and adding to the observed specific gravity 0.001 for every 3 C over the calibration temperature, or subtracting 0.001 for every 3 C under the specified temperature.

If desired, a highly accurate specific gravity may be determined by carefully weighing a small amount of urine and comparing its weight with the weight of the same volume of water. For example, suppose 1 ml of a given urine sample, at 25 C, weighs 1.016 g. The weight of 1 ml of water at that temperature is 0.996 g. Therefore, the specific gravity of the urine is:

$$\frac{1.016}{0.996} = 1.020$$

The specific gravity of urine is variable over a 24-hour period. Normally, the range is between 1.008 and 1.025, but variations are induced by such factors as fluid intake, activity, environmental temperature, and the presence of glucose or protein. The results of a specific gravity determination are of most value when obtained under the controlled conditions of a concentration or dilution test as discussed above.

Occasionally a specific gravity of 1.000 will be observed. The kidney may produce urine of a very low specific gravity, but it cannot produce pure water (specific gravity 1.000). The urinometer float is a relatively crude measure of specific gravity, so readings of 1.000 should be reported as "less than 1.001." To minimize the incidence of such results, the floats should be checked frequently to ensure that they read 1.000 in purified water at the prescribed temperature.

Refractive Index

Refractive index is influenced by the weight of dissolved particles. It is measured with a refractometer. The principle involved is that the degree to which a beam of light is bent when it enters the urine depends on the total solute content of the urine. The readings actually are used to estimate the specific gravity of the urine because specific gravity values provide a more familiar expression of urine solute content. The correlation between refractive index and specific gravity values is good except when abnormal constituents like proteins or glucose are present. Some urinary refractometers are calibrated in terms of specific gravity values, which is convenient, but can be misleading. The advantage of a refractometer over a urinometer is that the former requires only a drop or two of sample.

Osmolality

Osmolality depends on the effective number of dissolved particles. Measurements usually are made with an osmometer, which is an instrument that measures freezing point. Only very small amounts of sample (for example, 0.2 ml) are required, and the measurements are very reliable. The principle involved is that the solutes in urine lower its freezing point. More exactly, the freezing point is lowered in proportion to the effective number of solute particles which are present, that is, in proportion to the osmolality of the solution. The instrument can be calibrated in terms of osmolality because the freezing points of solutions are determined by their osmolality values. Although the correlation between osmolality and specific gravity values is poor for urine, probably osmolality values are a better indication of kidney function because the kidney likely concentrates the urine based on the osmolality rather than on the specific gravity of the urine.

One advantage of an osmolality over a specific gravity measurement is that a correction for an abnormal protein level is necessary only in the latter case. This is because the specific gravity measurement depends largely on the total weight of solute which is present, whereas the osmolality value depends only on the effective number of solute particles. Since proteins are such big molecules, a relatively small number of them can contribute a sizable solute weight, thereby making a significant contribution to the specific gravity but not the osmolality.

The normal range for urine osmolality in adults is 300 to 1000 milliosmols per kilogram. During periods of dehydration values of 800 to 1400 milliosmols per kilogram may be found, and during diuresis values of 40 to 80 milliosmols per kilogram may be found.

The unit for urine osmolality is milliosmoles (mosmol) of solute per kilogram of urine water. The values are approximately equal to the number of millimoles of dissolved particles per kilogram of urine water for the smaller and simpler particles like Na^+, Cl^-, and urea molecules.

Total Protein

Usually turbidimetric methods are used for quantitating protein in urine because these methods are simple and sensitive. Only a moderate degree of accuracy is provided, but this is adequate for the usual clinical purposes. Sulfosalicylic acid and trichloroacetic acid are two of the agents used to precipitate the protein in turbidimetric methods for urine protein. Although the use of sulfosalicylic acid results in a greater total turbidity and, therefore, a greater sensitivity, the precipitate produced with trichloroacetic acid is not influenced by the relative amounts of albumin and globulin present.

TOTAL PROTEIN METHOD

METHOD
Henry et al. [3].

MATERIAL
Urine.

REAGENTS

Trichloroacetic acid, 12.5%. Dissolve 12.5 g of trichloroacetic acid in water and dilute to 100 ml.

Hydrochloric acid, 1N. Dilute 8.3 ml of concd hydrochloric acid to 100 ml with water.

Sodium chloride, 0.85%. Dissolve 8.5 g of sodium chloride in water and dilute to 1 liter.

Hydrochloric acid, 0.5N. Add 50 ml of 1N HCl to 50 ml of water.

PROCEDURE
Record the total volume of the 24-h urine collection.
Filter or centrifuge 10 ml of the urine if it is cloudy.
Check the pH of the urine with pH paper or a pH meter. If it is alkaline, adjust the pH into the range of 5 to 7 using 1N hydrochloric acid.
Measure 6 ml of clear urine (possibly supernatant or filtrate) into each of two 19 mm cuvets (Test and Blank).
Into another cuvet measure 6 ml of standard (Standard).
Add 1.5 ml of trichloroacetic acid to the Test and the Standard.
Add 1.0 ml of 0.5N hydrochloric acid to the Blank.
Mix all tubes well and let stand for 7 minutes.
Add 6 ml of water to another cuvet (Water Blank).
Mix the Standard and read its absorbance in a spectrophotometer at 520 nm, using the Water Blank to set zero absorbance.
Mix, and read the absorbance of the Test at 520 nm using its Blank (not Water Blank) to set zero absorbance.

CALCULATIONS

$$\text{total protein value of standard (mg/liter)} \times \frac{\text{absorbance of Test}}{\text{absorbance of Standard}}$$

$$= \text{total protein (mg/liter)}$$

If a value over 500 mg per liter occurs, repeat the Test using a 1:10 dilution of the urine and multiply the final result by 10.

Report as milligrams per day:

$$\text{total protein (mg/liter)} \times \frac{\text{24-h urine volume (ml)}}{1000}$$
$$= \text{total protein (mg/d)}$$

STANDARDIZATION

Standard. Add 0.1 ml of serum with a known total protein concentration to 39.9 ml of 0.85% sodium chloride. Prepare fresh each day. The concentration of this standard is calculated as follows:

$$\frac{\text{total protein value of serum in g/dl}}{0.04}$$
$$= \text{total protein value of Standard in mg/liter}$$

NORMAL VALUES

The normal range of excretion of total protein in urine is approximately 0–150 mg per 24 hours.

DISCUSSION AND PHYSIOLOGICAL SIGNIFICANCE OF
URINE PROTEIN CONCENTRATIONS

The albumin and globulin proteins are large, delicate molecules. They are precipitated by agents like trichloroacetic acid because this results in their denaturation (unfolding). This precipitation, of course, produces the turbidity. Turbid solutions pass less light than clear solutions. Therefore turbidity can be quantitated in terms of absorbance units in a spectrophotometer even though light is being blocked rather than absorbed. The wavelength which is used is not critical. A wavelength of 520 nm is recommended because: (1) the absorbance value of a turbid solution is relatively constant in this spectral region, and (2) this visible wavelength is far removed from the absorption peaks of the most common colored (yellow) substances in urine.

Generally, increased protein excretion in the urine (proteinuria) indicates renal disease, for example, nephrotic syndrome, because the filtration mechanism of the kidney is impaired, and plasma proteins are leaking into the urine. Multiple myeloma (a bone cancer) frequently causes proteinuria because of the production and urinary excretion of Bence Jones proteins. The normal kidney can prevent large blood proteins like albumin and the globulins from appearing in the urine in significant quantities because its filtration mechanism (glomeruli) can hold most of them back. Bence Jones proteins, on the other hand, appear in the urine whenever they occur in multiple myeloma because they are small enough to pass through the glomeruli.

Glucose

Quantitation of the total excretion of glucose per 24 hours sometimes is of clinical interest. The *o*-toluidine method (page 148) may be used for quantitative analysis of glucose in urine, with the following modifications:

Measure the volume of the 24-hour urine collection.
Perform a Clinitest[4] analysis with the urine and designate the resulting color according to the color chart provided by the manufacturer.
Dilute the urine with water as shown below:

Designation (%)	Dilution
0 to ½	1:3
¾ to 1	1:5
2	1:10

Perform the *o*-toluidine procedure for glucose exactly the same as directed for serum on page 148, treating the diluted urine the same as serum.

CALCULATIONS
Calculate the glucose concentration as given on page 149. Then correct for the dilution (either 1:3, 1:5, 1:10), and multiply by [0.01 × total volume of 24-h urine (ml)] to obtain glucose in milligrams per day.

NORMAL VALUES
The normal range of excretion of glucose in urine is 0 to 250 mg per 24 hours.

DISCUSSION AND PHYSIOLOGICAL SIGNIFICANCE OF
URINE GLUCOSE CONCENTRATIONS
Although the *o*-toluidine method is much more specific for glucose than the Clinitest method, nevertheless there are a few sugars which sometimes occur in the urine which will be measured by both technics. These other sugars primarily are lactose and galactose.

Lactose often appears in the urine during lactation. It gives about one-third as much color with *o*-toluidine as does glucose. Physicians are cognizant of the fact that positive urine sugar results that are obtained with specimens from nursing mothers usually indicate lactosuria, and little attention is paid to this normal occurrence. Lactose can be distinguished from glucose by the use of a glucose oxidase method like Clinistix, which is discussed on page 385.

Galactose occurs in the urine only in infants with galactosuria, a very

4 Ames Co., Elkhart, Indiana.

rare genetic disease. The color yield of galactose with *o*-toluidine is the same as with glucose. Galactose can be distinguished from glucose by thin-layer chromatography.

The occurrence of an elevated glucose level in the urine generally is indicative of a disease process, particularly diabetes mellitus. Glucose passes through the glomerular membrane but is reabsorbed from the tubules. The kidney has a definite limit for this process. In diabetes mellitus, glucose may appear in the urine because often there is too much of this substance present in the blood to permit complete reabsorption. A few individuals have low renal thresholds for glucose and consequently in such patients glycosuria may occur with relatively low blood glucose levels. This may also occur in healthy women during the latter part of pregnancy.

Phenolsulfonphthalein (PSP) Test

Another important function of the kidneys, besides water reabsorption, is the secretion of waste products into the urine. This function can be evaluated with the phenolsulfonphthalein (PSP) test. Phenolsulfonphthalein, also called phenol red, is a dye that is readily removed from the blood by the secretory mechanism of the kidneys. After a control urine sample is obtained, and the patient is properly hydrated, a solution of PSP (for example, 6 mg of PSP in a 1 ml volume) is injected intravenously. The time of injection is noted and the patient is asked to empty his bladder completely at periods of exactly 15 minutes, 30 minutes, 1 hour, and 2 hours after injection of the dye. Each of the four specimens is analyzed spectrophotometrically.

Normal kidneys excrete at least 25% of the dye in 15 minutes and a total of at least 65% of the dye in 2 hours. The percentage of excretion diminishes with each successive collection, but occasionally a higher percentage of dye may appear in a later sample. Since tubular function will not change measurably during the test, the only explanation for this finding is incomplete emptying of the bladder on the part of the patient. This may be accidental, but it may also indicate obstruction of the urinary tract.

A small amount of the dye normally is excreted by the liver, but in the presence of liver damage *all* the dye is excreted by the kidneys. Therefore an exceptionally high PSP excretion may occur in the urine of patients suffering from disease of the liver.

Renal Clearance Tests

Clearance tests are designed to measure the efficiency with which the kidneys remove certain substances from the blood. There are many substances, both endogenous and exogenous, which are determined in clearance tests, but only urea and creatinine are commonly measured for clinical purposes. The tests for these two substances are used for the same purpose, they have the same significance, and they are performed in a similar manner.

The creatinine clearance test is less influenced by the rate of urine flow than is the urea clearance test; to be valid, the latter requires urine flows of more than 2 ml per minute. Although there is no such physiological restriction on the creatinine clearance test, adequate urine flows are desirable to minimize errors in collection. The creatinine clearance test is described in detail here. The urea clearance test may be conducted in the same manner, and the results should be of similar value.

CREATININE CLEARANCE TEST

Ideally the creatinine clearance test is performed with a 24-hour specimen, although samples collected over any period of time may be used. Periods of less than 30 minutes are not recommended. If a relatively short period is to be used (for example, 30 minutes or 1 hour), it is advisable to collect two successive samples so that the test may be performed and calculated in duplicate. At the outset of the collection period the bladder should be emptied completely, the specimen discarded and the time noted. If short periods are to be used, the patient should be hydrated in advance, and each voiding should represent a *complete* emptying of the bladder. In these cases the exact time lapse between voidings is very important and should be carefully calculated. At some time during the performance of the test a blood sample should be obtained for creatinine analysis.

The complete urine sample (or samples) and the blood specimen should be delivered to the laboratory for analysis. It is essential that the times marking the beginning and end of each collection period be carefully noted.

The urine and blood samples should then be analyzed for creatinine concentration as described in Chapter 13.

CALCULATIONS

The creatinine concentration of each sample is calculated and the results are used in the following formula:

$$C_{cr} = \frac{UV}{P}$$

where C_{cr} is the creatinine clearance in milliliters per minute, U is the creatinine concentration in the urine in milligrams per 100 ml, P is the creatinine concentration in the serum in milligrams per 100 ml, and V is the urine flow in milliliters per minute.

The rate of urine flow (in milliliters per minute) and the creatinine concentration are calculated for the urine sample. These data are used to calculate the clearance. If two urine samples are collected, the clearances should be calculated separately and then averaged.

SAMPLE CALCULATIONS

The creatinine clearance calculations may be best understood from examples. The data given here represent a normal clearance test and an abnormal one.

Test Number 1	*Time*
Bladder emptied (beginning of test)	9:10 (sample discarded)
Blood drawn	9:30
Sample voided	9:45

The volume of the urine sample was 120 ml. The creatinine concentration in the urine was found to be 27.8 mg per dl, while the serum creatinine level was 0.8 mg per dl.

Clearance Calculations for Test Number 1
The time interval is from 9:10 to 9:45, or 35 minutes.
The volume is 120 ml. The urine flow is therefore 120 ml per 35 minutes
    or 3.4 ml per minute (V).
Substitute these data in the clearance formula:

$$C_{cr} = \frac{27.8 \times 3.4}{0.8}$$
$$C_{cr} = 118 \text{ ml per minute}$$

Test Number 2 (24-hour)	*Time*
Bladder emptied (beginning of test)	8:15 AM (sample discarded)
Blood drawn	8:00 AM (2nd day)
Last sample collected	8:15 AM (2nd day)

The volume of the 24-hour urine sample was 1500 ml. The creatinine concentration in the urine was found to be 94 mg per dl, while the serum creatinine level was 1.4 mg per dl.

Clearance Calculations for Test Number 2
The time interval is 24 hours or 1440 minutes.
The volume is 1500 ml. The urine flow is therefore 1500 ml per 1440
    minutes or 1.04 ml per minute (V).
Substitute these data in the clearance formula:

$$C_{cr} = \frac{94 \times 1.04}{1.4}$$
$$C_{cr} = 70 \text{ ml per minute}$$

Some physicians prefer to express the 24-hour creatinine excretion in *liters per day*. To obtain this quantity, divide the urine concentration by the serum concentration and multiply the resulting figure by the number of *liters* of urine excreted. In the above example (2) the value would be:

$$C_{cr} = \frac{94}{1.4} \times 1.5$$
$$C_{cr} = 101 \text{ liters per day}$$

NORMAL VALUES

For healthy adults the range for creatinine clearance is approximately 100 to 180 ml per minute, while normal urea clearances are 50 to 100 ml per minute. These values multiplied by 1.4 give liters per day. Ideally, clearance should be corrected for body size. The patient's surface area may be estimated if the height and weight are known. This value, divided into the standard area, 1.73 square meters, gives the correction factor to be applied to the calculated clearance.

DISCUSSION

The error for clearance tests should not be much greater than the error inherent in the determination of urea or creatinine. However, there are perhaps no other clinical tests which are performed improperly so often. It is for this reason that the collection of two urine samples and individual calculations for each are recommended if relatively short intervals are to be used. If a wide discrepancy occurs between the two calculated clearances, they are of no value and should not be averaged, unless both are unequivocally in either the normal or abnormal range.

Whenever possible, it is recommended that a period of 24 hours, 12 hours, or at least 6 hours be used.

The most common and troublesome errors encountered in conducting clearance tests are

1. incomplete voiding by the patient;
2. inaccurate timing of the periods;
3. failure to deliver complete specimens to the laboratory, with consequent errors of volume measurement;
4. inadequate urine flow due to lack of prehydration, resulting in small volumes and proportionately large errors if the bladder is not emptied completely.

The clearance tests are simple but informative procedures. If careful attention is paid to the details of performance, the results should be reliable and reproducible.

PHYSIOLOGICAL SIGNIFICANCE OF
RENAL CLEARANCE TESTS

The significance of a clearance test depends upon the way in which physiological processes taking place in the kidneys affect the material being measured. Creatinine is filtered through the glomerulus and, to a lesser degree, secreted in the tubules. Urea is also filtered through the glomerulus but is subject to some tubular reabsorption. For these reasons, the creatinine clearance is somewhat higher than the urea clearance, but both tests are used as guides to the glomerular filtration rate. A normal creatinine clearance of 120 ml per minute means that in 1 minute the kidneys are capable of completely clearing the creatinine from 120 ml of blood by glomerular filtration. This, of course, does not actually happen, but it is a convenient method of expressing the filtration rate.

The creatinine clearance tests and urea clearance tests are considered sensitive guides to renal function, the former being slightly more reliable. In early cases of kidney disease, a fall in the clearance rate may occur before the serum creatinine level or urea nitrogen level rises out of the normal range.

Urinary Calculi

Urine contains many organic and inorganic salts. Occasionally some of these salts precipitate to form stones (calculi) in the urinary tract. These stones vary widely in size, from minute grains to huge concretions that almost fill the bladder. Tiny stones sometimes are passed in the urine, but larger ones must be dissolved or removed surgically. Stones may be found in the kidney (pelvis), ureters, or bladder.

Numerous types of stones have been reported [4, 7], but over 90% of urinary calculi are composed of inorganic calcium or magnesium salts of phosphate, oxalate, and carbonate or the organic compounds uric acid and cystine. Frequently a stone is composed of more than one of these substances.

URINARY CALCULI METHOD

The procedure described here is adequate for qualitative identification of most stones. More elaborate methods [4, 5, 8] employing infrared spectroscopy or x-ray diffraction sometimes are used.

REAGENTS

Hydrochloric acid, 5% (v/v). Dilute 5 ml of concentrated hydrochloric acid (HCl) to 100 ml with water.

Sodium hydroxide, 10% (w/v). Dissolve 10 g of sodium hydroxide (NaOH) in water and dilute to 100 ml.

Molybdate reagent. See page 208.

Sulfonic acid reagent. See page 208.

Acetic acid, 5% (v/v). Dilute 5 ml of glacial acetic acid (CH_3COOH) to 100 ml with water.

Sodium carbonate, 20% (w/v). Dissolve 20 g of sodium carbonate (Na_2CO_3) in water and dilute to 100 ml.

Phosphotungstic acid reagent. See page 178.

Ammonium hydroxide, concd.

Sodium cyanide, 5% (w/v). Dissolve 5 g of sodium cyanide (NaCN) in 100 ml of water.

Sodium nitroprusside, 5% (w/v). Dissolve 5 g of sodium nitroferricyanide, $Na_2Fe(CN)_5NO \cdot 2H_2O$, in 100 ml of water.

PROCEDURE

If the stone has blood on it, it should be washed with water. If possible, some of the stone should be saved for a repeat determination or a more exhaustive study, if necessary.

Crush the stone with a mortar and pestle or with a glass rod in a test tube. Note its color, hardness, and texture.

Transfer about half of the crushed stone (or all of it if it is very small) to a 19 × 150 mm Pyrex test tube and add 20 ml of 5% HCl.

Heat gently over a flame for about 2 minutes and note whether any effervescence (not just gas bubbles from boiling) occurs.

Filter 5 ml portions of the solution[5] into four 16 × 125 mm test tubes and treat as follows:

Tube 1. Neutralize by adding 10% NaOH, testing intermittently with pH paper. This should require approximately 1 ml of NaOH solution. Add 1 ml of molybdate reagent and 0.2 ml of sulfonic acid reagent, mix, and look for a definite blue color.

Tube 2. Neutralize as described for Tube 1, and look for turbidity just before neutralization. If turbidity appears, add 5% acetic acid dropwise and note whether the turbidity persists or disappears.

Tube 3. Neutralize as described for Tube 1. Then add 1 ml of 20% sodium carbonate and 1 ml of phosphotungstic acid reagent. Look for a blue color.

Tube 4. Aspirate directly into an atomic absorption spectrophotometer to check for calcium and magnesium.

Add a small portion of the stone itself to a 13 × 100 mm test tube.

Add 1 drop of ammonium hydroxide and 1 drop of 5% sodium cyanide, mix, and let stand for 5 minutes. Add 3 drops of 5% sodium nitroprusside and look for a red color.

---

[5] Not all of the stone may go into solution, but enough will dissolve to make chemical tests possible.

INTERPRETATION

1. Uric acid stones are usually hard and are yellow or orange in color. Phosphate or carbonate stones are usually soft and powdery. Calcium oxalate stones are very hard, usually have a rough surface, and often are dark in color.
2. Marked effervescence during heating with acid indicates the presence of carbonate.
3. A deep blue color in Tube 1 (not a pale blue) indicates the presence of phosphate.
4. In Tube 2, if the turbidity persists after adding acetic acid, oxalate is present. If the turbidity disappears, it was due to phosphate.
5. A blue color in Tube 3 indicates the presence of uric acid.
6. A strong signal on the atomic absorption instrument, under the appropriate conditions, identifies the presence of calcium or magnesium or both.
7. A beet-red color in the 13 × 100 mm tube is positive for cystine.

DISCUSSION AND PHYSIOLOGICAL SIGNIFICANCE
OF URINARY CALCULI

The analyst who is uncertain about a test observation should repeat the procedure with a pure chemical known to give a positive test. Color discrepancies should be noted since they may indicate the presence of an interfering metabolite or drug in elevated concentrations. Any equivocal results should be reported as such. With experience, the analyst can perform micro versions of the tests when very small stones are submitted for analysis. The use of small test tubes and a hand magnifying glass then is advised.

The main reason for analyzing the content of calculi is to guide efforts to prevent their recurrence in the patient. Calculi form because abnormal conditions cause certain chemicals to precipitate. These chemicals are present in excess. They precipitate because their solubility level is exceeded or because abnormal pH conditions render them less soluble. For instance, calcium-containing calculi are commonly seen in hyperthyroidism because calcium excretion in the urine is increased in this condition. Also, the elevated uric acid excretion which occurs in gout can lead to uric acid calculi. Such calculi are seen also in patients with acidic urine (constantly pH 5.5 or less); calcium oxalate calculi may be seen in this condition as well. Both uric acid and calcium oxalate are less soluble at lower pH. Calcium phosphate, on the other hand, is less soluble at higher pH. Therefore calculi involving this substance are seen in patients with alkaline urine (constantly pH 6.8 or greater). Obstructions, infections, and drug excesses also can lead to formation of calculi. The content of most calculi tends to be complex because many of their chemicals are present simply because they have been absorbed or trapped into the calculus as it was formed. This appears to be part of the explanation for the complex internal patterns that are seen in many calculi.

References

1. Bauer, J. D., Ackermann, P. G., and Toro, G. *Clinical Laboratory Methods* (8th ed.). St. Louis: Mosby, 1974.
2. Davidsohn, I., and Henry, J. B. *Todd-Sanford Clinical Diagnosis by Laboratory Methods* (15th ed.). Philadelphia: Saunders, 1974.
3. Henry, R. J., Sobel, C., and Segalove, M. Turbidimetric determination of proteins with sulfosalicylic and trichloroacetic acids. *Proc. Soc. Exp. Biol. Med.* 92:748, 1956.
4. Herring, L. C. Observations on the analysis of ten thousand urinary calculi. *J. Urol.* 88:545, 1962.
5. Leonard, R. H., and Butt, A. J. Quantitative identification of urinary calculi. *Clin. Chem.* 1:241, 1945.
6. Page, L. B., and Culver, P. J. *Syllabus of Laboratory Examination in Clinical Diagnosis.* Cambridge, Mass.: Harvard University Press, 1960.
7. Prien, E. L., and Frondel, C. Studies in urolithiasis: I. The composition of urinary calculi. *J. Urol.* 57:949, 1947.
8. Weissman, M., Klein, B., and Berkowitz, J. Clinical applications of infra-red spectroscopy. *Anal. Chem.* 31:1334, 1959.

Annotated Bibliography

This annotated bibliography provides references for further reading in clinical chemistry. It is by no means exhaustive. The emphasis is on the more general texts and journals in the field. References dealing with specific subjects in clinical chemistry are listed in the bibliographies of the appropriate chapters.

Books

Advances in Clinical Chemistry, New York, Academic Press.
This is a series of yearly volumes (for example, Volume 18 in 1975) of review articles in clinical chemistry. A number of these articles are specifically cited in the bibliographies at the ends of the chapters in this book.

Bauer, J. D., Ackermann, P. G., and Toro, G. *Bray's Clinical Laboratory Methods* (7th ed.). St. Louis: Mosby, 1968.
One of the chapters in this book is devoted to methods of analysis for the more common substances in clinical chemistry.

Curtius, H. Ch., and Roth, M. (Eds.). *Clinical Biochemistry: Principles and Methods* (2 vols.). Berlin: de Gruyter, 1974.
This extensive work (more than 1800 pages) covers much of the general and specific methodology used in the clinical chemistry laboratory. Some methods are given in detail, while others are only mentioned and referenced.

Davidsohn, I., and Henry, J. B. (Eds.). *Todd-Sanford Clinical Diagnosis by Laboratory Methods* (15th ed.). Philadelphia: Saunders, 1974.
This book covers the field of medical technology, including chemical methods. Explanations of methods and interpretations of the results are given primary emphasis and the coverage of clinical chemistry is extensive.

Faulkner, W. R., and King, J. W. (Eds.). *CRC Manual of Clinical Laboratory Procedures.* Cleveland: Chemical Rubber Company, 1970.
Detailed methodology and the principles involved are given for tests in the various areas (including chemistry) of laboratory analysis.

Faulkner, W. R., King, J. W., and Damm, H. C. *CRC Handbook of Clinical Laboratory Data* (2nd ed.). Cleveland: Chemical Rubber Company, 1968.
This book contains general discussion of some basic topics and specific substances in the area of clinical chemistry. Specific methods are not presented.

Frankel, S., Reitman, S., and Sonnenwirth, A. C. *Gradwohl's Clinical Laboratory Methods and Diagnosis* (7th ed., 2 vols.). St. Louis: Mosby, 1970.
The vast field of laboratory medicine is covered. General discussions,

detailed instructions of methods, and interpretation of results are emphasized.

Glick, D. (Ed.). *Methods of Biochemical Analysis.* New York: Wiley. This series of yearly volumes offers chapters covering both well-established and recent methodology in biochemistry. Each chapter generally gives a background of previous work, a critical evaluation of various approaches, and a presentation of the procedural details for the methodology being discussed.

Henry, R. J., Cannon, D. C., and Winkelman, J. W. (Eds.). *Clinical Chemistry Principles and Technics* (2nd ed.). New York: Harper & Row, 1974.

This voluminous book (1629 pages) provides thorough and extensively referenced coverage of the principles and technics of clinical chemistry, including detailed methods, and thus constitutes a valuable information source. Hormones, toxicology, and drug analysis are not covered.

Natelson, S. *Techniques of Clinical Chemistry* (3rd ed.). Springfield, Ill.: Thomas, 1971.

Detailed methods along with procedural notes and the principles involved are given for many of the tests available in the clinical chemistry laboratory. Instrumentation and analytical principles also are discussed.

O'Brien, D., Ibbott, F. A., and Rodgerson, D. O. *Laboratory Manual of Pediatric Micro-Biochemical Techniques* (4th ed.). New York: Harper & Row, 1968.

This book provides detailed micro methods along with clinical and technical comments for many clinical chemistry tests.

Race, G. J. (Ed.). *Laboratory Medicine* (4 vols.). Hagerstown, Md.: Harper & Row, 1973.

The first volume of this four-volume series is devoted to clinical chemistry. The series is loose-leaf bound to allow insertion of updated or new material as it becomes available from the publisher. Specific methods, principles, and interpretations are presented.

Standard Methods of Clinical Chemistry.   New York: Academic Press. This series of volumes (for example, Volume 7 in 1972, Volume 1 in 1952) presents detailed methods which have been generally accepted and specially checked by clinical chemists before inclusion in the series.

Tietz, N. W. (Ed.). *Fundamentals of Clinical Chemistry.* Philadelphia: Saunders, 1970.

This book presents most of the fundamental topics, technics, and methods used in clinical chemistry. The subjects are covered in sufficient depth so that the book may be used both as a textbook for students and as a reference book for practicing clinical chemists.

Toro, G., and Ackerman, P. G.   *Practical Clinical Chemistry*. Boston: Little, Brown, 1975.
This book provides a broad coverage of most of the specific procedures and general topics of current use or interest in clinical chemistry.

Varley, H.   *Practical Clinical Biochemistry* (4th ed.). New York: Interscience, 1967.
The value of this book lies primarily in its detailed presentation of chemical methods and the inclusion of useful comments and notes. Statements of clinical interpretations also are made.

Wolf, P. L., and Williams, D.   *Practical Clinical Enzymology*. New York: Wiley, 1973.
The first two-thirds of this book includes practical as well as clinical aspects of a large number of enzyme tests. Specific methods are fully detailed. The other one-third of the book includes typical biochemical patterns (pictures of charts from automated analyzers) for 128 disease states, along with interpretive comments.

Wolf, P. L., Williams, D., Tsudaka, T., and Acosta, L.   *Methods and Techniques in Clinical Chemistry*. New York: Wiley, 1972.
Practical (detailed procedures) and clinical aspects of the methods in use in the Stanford Clinical Laboratory at the time of publication are presented.

Periodicals

American Journal of Clinical Pathology (Lippincott, Philadelphia).
This monthly journal contains research articles and reviews emphasizing methodological and interpretive aspects of clinical laboratory analyses, including clinical chemistry. It is the official journal of the American Society of Clinical Pathologists.

American Journal of Medical Technology (American Society for Medical Technology, Bellaire, Texas).
The articles in this monthly journal cover all types of clinical laboratory methods including chemical methods. This is the official journal of the American Society for Medical Technology.

Analytical Biochemistry (Academic, New York).
This monthly journal publishes research articles covering qualitative and quantitative technics for analysis of biological substances and related materials.

Analytical Chemistry (American Chemical Society, Washington, D.C.).
Both theoretical and practical articles and reviews on all types of chemical and physical analysis are published in this monthly journal. Analytical articles of interest to the clinical chemist sometimes appear.

Analytical Chemistry: Analytical Reviews (American Chemical Society, Washington, D.C.)

This collection of review articles is published as the April issue of *Analytical Chemistry* each year. The objective is to provide an exhaustive literature coverage of each subject; most citations, therefore, receive only a brief comment. The subject of clinical chemistry is reviewed and updated every two years.

Chemical Abstracts (American Chemical Society, Columbus, Ohio).

Single paragraph abstracts of articles from a large number of journals are organized according to subject. Of the five sections that are published every two weeks, the Biochemistry Section is most relevant to the clinical chemist. Although more information about the articles is provided here than in *Current Contents* (where only titles appear), the listings are less current.

Clinica Chimica Acta (Elsevier, Amsterdam)

This is one of the two foremost journals of clinical chemistry. The emphasis is on analytical methods, similar to the journal *Clinical Chemistry*.

Clinical Chemistry (American Association of Clinical Chemists, Washington, D.C.).

This is one of the two foremost journals of clinical chemistry. The emphasis is on methods of analysis. Both new and modified procedures are presented and evaluated. The journal appears monthly with occasional issues being dedicated to special subjects such as drugs, trace metals, or radioimmunoassay.

Current Contents/Life Sciences (Institute for Scientific Information, Philadelphia).

The weekly issues of this publication contain copies of the tables of contents (titles and authors) of current issues of more than 1,000 journals in the life sciences. Scanning the tables of contents of the significant journals in the field of clinical chemistry is a good way for the clinical chemist to keep up with new developments in this field. The addresses of the senior authors are included in an index.

Journal of Chromatography (Elsevier, Amsterdam).

Articles covering all types of chromatography (gas, column, paper, thin-layer, electrophoretic) are included in this journal. The journal is published twice a month.

Journal of Clinical Endocrinology and Metabolism (Lippincott, Philadelphia).

Physiological and biochemical aspects of hormones along with new methods for their analysis are reported in this monthly journal.

Index